TAKING SIDES

Clashing Views on

African Issues

FOURTH EDITION

TAKING SIDES

Clashing Views on

African Issues

FOURTH EDITION

Selected, Edited, and with Introductions by

William G. Moseley
Macalester College

McGraw Hill

Connect
Learn
Succeed™

Connect
Learn
Succeed™

TAKING SIDES: CLASHING VIEWS ON AFRICAN ISSUES, FOURTH EDITION

Published by McGraw-Hill, a business unit of The McGraw-Hill Companies, Inc., 1221 Avenue of the Americas, New York, NY 10020. Copyright © 2012 by The McGraw-Hill Companies, Inc. All rights reserved. Previous edition(s) © 2009, 2007, and 2004. No part of this publication may be reproduced or distributed in any form or by any means, or stored in a database or retrieval system, without the prior written consent of The McGraw-Hill Companies, Inc., including, but not limited to, in any network or other electronic storage or transmission, or broadcast for distance learning.

Some ancillaries, including electronic and print components, may not be available to customers outside the United States.

Taking Sides® is a registered trademark of the McGraw-Hill Companies, Inc.
Taking Sides is published by the **Contemporary Learning Series** group within the McGraw-Hill Higher Education division.

1 2 3 4 5 6 7 8 9 0 DOC/DOC 1 0 9 8 7 6 5 4 3 2 1

MHID: 0-07-805008-1
ISBN: 978-0-07-805008-4
ISSN: 1545-5327 (print)

Managing Editor: *Larry Loeppke*
Senior Developmental Editor: *Jade Benedict*
Senior Permissions Coordinator: *Shirley Lanners*
Senior Marketing Communications Specialist: *Mary Klein*
Senior Project Manager: *Jane Mohr*
Design Coordinator: *Brenda A. Rolwes*
Cover Graphics: *Rick D. Noel*

Compositor: MPS Limited, a Macmillan Company
Cover Image: © William G. Moseley

www.mhhe.com

Editors/Academic Advisory Board

Members of the Academic Advisory Board are instrumental in the final selection of articles for each edition of TAKING SIDES. Their review of articles for content, level, and appropriateness provides critical direction to the editors and staff. We think that you will find their careful consideration well reflected in this volume.

TAKING SIDES: Clashing Views on

African Issues

Fourth Edition

EDITOR

William G. Moseley
Macalester College

ACADEMIC ADVISORY BOARD MEMBERS

Preface

$\mathbf{T}$his volume's cover photo features the male members of a farm family in southern Mali whom I met in July of 2009, and they are emblematic of small-scale farmers across Africa. Although agricultural programs had fallen out of favor with donors for several decades, these initiatives came back with a vengeance in 2008 following high food prices and social unrest that rocked a number of cities. In fact, as this book is going to press in the spring of 2011, it looks as if we are set for yet another round of high global prices for basic agricultural commodities. The aforementioned events and this photo hint at a number of agricultural and food-related issues tackled in this volume, including the degree to which climate change is driving agricultural shifts on the continent, the need for a Green Revolution in Africa, and dynamics between food production and population growth. In regards to the last issue, it is important to point out that this cover image is of a multigenerational household with a grandfather, his three sons, and four grandchildren. This type of household structure is not uncommon in a setting where successful smallholders depend on sufficient numbers of able-bodied laborers. This photo is also notable for what it does not show, that is, the women of the household who also farm in addition to rearing children, cooking, collecting firewood and water, and engaging in petty commerce.

I find studying Africa so exciting and captivating in part due to the types of issues discussed above. I have been working and conducting research in Africa for over 25 years now, originally as a development professional and currently as an academic. My applied work and research have led to extended stays in several African countries (Mali, Zimbabwe, South Africa, Malawi, Niger, and Lesotho) and to travel in many others. This first-hand experience with nuts-and-bolts development issues, as well as vital research questions, has led me to care deeply about Africa as a region and to grapple intellectually with many of the issues presented in this volume.

I have been fortunate to share my fascination for the African continent with (and learn from) some incredibly bright, insightful, and engaged students. Like other academics who have been bitten by the Africa bug, it is during my Africa course each spring that I find my own and my students' enthusiasm for the subject to be most infectious. I hope this volume will serve as a useful platform for discussions in courses offered in a variety of departments dealing with contemporary African issues, including anthropology, African studies, development studies, economics, geography, history, international studies, political science, and sociology. I find that the best way to encourage students to grapple with their preconceptions about Africa is to offer readings that both support and contradict their initial leanings. Through the process of reviewing different arguments and discussing them with their classmates, individual students continually surprise me by the degree to which their positions may change during the course of a semester (far more than if

I had tried to convince them of the validity of a certain perspective). Of course, there is no obvious right or wrong to many of these issues, and some of the most rewarding class discussions occur when there is no clear majority of students for one position or the other.

Changes to this edition   The fourth edition has been extensively revised and updated. Based on the feedback of reviews of the third edition, five new issues have been added. These include Issue 6 (Do Cell Phones and the Internet Foster "Leapfrog" Development in Africa?), Issue 8 (Does Foreign Aid Undermine Development in Africa?), Issue 9 (Is Climate Change a Major Driver of Agricultural Shifts in Africa?), Issue 12 (Is Community-Based Wildlife Management a Failed Approach?), and Issue 17 (Does Increased Female Participation Substantially Change African Politics?). One of the two selections was also changed in Issue 5 (Have Free-Market Policies Worked for Africa?) and Issue 7 (Is Increasing Chinese Investment Good for African Development?). In all, 12 of the 38 selections, or about a third of the material, have changed. Additional revisions include fine-tuning the questions, reordering the material in the book, and rewriting the introductions and postscripts.

A word to the instructor   An *Instructor's Manual With Test Questions* (multiple choice and essay) is available through the publisher for the instructor using Taking Sides in the classroom. Also available is a general guidebook, *Using Taking Sides in the Classroom,* which offers suggestions for adapting the pro-con approach in the classroom setting. An online version of *Using Taking Sides in the Classroom* and a correspondence service for Taking Sides adopters can be found at http://www.mhcls.com/usingts/.

   Taking Sides: Clashing Views on African Issues is only one title in the Taking Sides series. If you are interested in seeing the table of contents for any of the other titles, please visit the Taking Sides Web site at http://www.mhcls.com/takingsides/.

Acknowledgments   I am grateful to several students at Macalester College, especially Natalie Locke, who helped me with various aspects of this text. I wish to acknowledge members of the editorial board for this text (listed below) and several anonymous reviewers who provided me with invaluable feedback during the first, second, third, or fourth editions of this volume. I thank my spouse, Julia, and children, Ben and Sophie, for their shared enthusiasm for Africa. I finally wish to acknowledge the editorial staff at McGraw-Hill, especially Jade Benedict, for his copyediting, insight, and encouragement.

William G. Moseley
Macalester College

For Julia, Ben, and Sophie

Contents In Brief

Contents

Paul E. Lovejoy, professor of history at York University, argues that the trans-Atlantic slave trade significantly transformed African society. It led to an absolute loss of population on the continent and a large increase in the enslaved population that was retained in Africa. The economic advantages of exporting slaves did not offset the social and political costs of participation, there were disastrous demographic impacts, and Africa's relative position in world trade declined. John Thornton is a professor of history at Boston University. He notes that slavery was widespread and indigenous in African society. Europeans simply worked with this existing market and African merchants, who were not dominated by Europeans, responded by providing more slaves. African leaders who allowed the slave trade to continue were neither forced to do so against their will, nor did they make irrational decisions. As such, the preexisting institution of slavery in Africa is as much responsible as any external force for the development of the trans-Atlantic slave trade.

Duncan Clinch Heyward, a former Carolina rice planter writing in the middle of the last century, represents the mainstream view that Europeans were primarily responsible for developing South Carolina's remarkable rice plantations in the eighteenth century. In his own accounting of the rise of rice cultivation in the Carolinas, Duncan suggests that the techniques and approaches must have been derived from those observed in China. Judith Carney, a professor of geography at UCLA, explains that slaves from rice-producing areas in West Africa have only recently been recognized for their intellectual contributions to the development of rice cultivation in the New World. Carney describes how her work, and that of others, challenged the view that slaves were mere field hands, "showing that they contributed agronomic expertise as well as skilled labor to the emergent plantation economy."

Jared Diamond, professor of physiology and biogeography at UCLA, argues that Europeans were able to colonize Africa (rather than vice versa) because of the advantages of guns, widespread literacy, and political organization. These advantages stem ultimately from different historical trajectories that are linked to "differences in real estate" (or differences in physical geography). Lucy Jarosz, associate professor of geography at the University of Washington, is troubled by Diamond's narrow conception of geography and asserts that explaining differences in wealth and power between regions must also take account of social, political, and economic connections. She focuses on the specific case of Madagascar and argues that the intentions of the colonizer are as or more important than their military power for determining the nature of the colonial relationship.

Marcus Colchester, director of the Forest Peoples Programme of the World Rainforest Movement, argues that rural communities in equatorial Africa are today on the point of collapse because they have been weakened by centuries of outside intervention. In Gabon, the Congo, and the Central African Republic, an enduring colonial legacy of the French are lands and forests controlled by state institutions that operate as patron-client networks to enrich indigenous elite and outside commercial interests. Robin M. Grier, assistant professor of economics at the University of Oklahoma, contends that African colonies that were held for longer periods of time tend to have performed better, on average, after independence.

UNIT 2   DEVELOPMENT   71

Fudzai Pamacheche, of the Southern African Development Community Directorate on Trade, Industry, Finance, and Investment, and Baboucarr Koma, of the Private Sector Development, Investment and Resource Mobilization Division of the African Union Commission, argue that privatization is an example of a free-market policy that has benefited African people. They suggest that the privatization of public enterprises creates efficiency gains, stable and reduced prices, lower government subsidies and the redirection of scarce resources elsewhere, the payment of dividends to government, and increased employment. They further argue that privatization programs, if well designed, are crucial for poverty alleviation and economic integration regionally and globally. Thandika Mkandawire, director of the UN Research Institute for Social Development, counters that while African governments have reshaped domestic policies to make their economies more open, growth has faltered. Mkandawire assesses structural adjustment from a developmental perspective, judging its effects on economic development and the eradication of poverty. He suggests that structural adjustment policies designed to integrate Africa into the global economy have failed because "they have completely sidestepped the developmental needs of the continent and the strategic questions on the form of integration appropriate to addressing these needs."

Joseph O. Okpaku, president and CEO of the Telecom Africa International Corporation, argues that cell phones and the Internet have fundamentally changed the lives of people and national economies in Africa by delivering needed services more efficiently. He argues that these technologies can foster sustainable economies, build on efforts to reduce poverty, and allow individuals and institutions to prosper through increased access to information. Pádraig Carmody, a senior lecturer in geography at Trinity College Dublin, questions the transformational capacity of information and communication technology (ICT) in Africa. Although he admits that ICTs can sometimes enhance welfare, their use is embedded in existing relations of social support, resource extraction, and conflict and therefore may reinforce existing power dynamics. Since Africa is still primarily a user (rather than producer or creator) of ICTs, the use of these technologies does not fundamentally alter the continent's dependent position.

Barry Sautman, associate professor of social science at The Hong Kong University of Science and Technology, and Yan Hairong, the Department of Applied Social Science, Hong Kong Polytechnic University, argue that China's links with Africa represent a distinctive "Chinese model" of foreign investment. They further suggest that this Chinese model represents a lesser evil than assistance offered by the West. Padraig Carmody, of Trinity College Dublin, and Francis Owusu, of Iowa State University, are less sanguine about Chinese involvement in Africa. Although they also perceive potential benefits from increasing trade with China, they describe how increasing resource flows are strengthening authoritarian states and fuelling conflict.

Dambisa Moyo, who has worked for both Goldman Sachs and the World Bank, argues that aid to Africa has made the poor poorer and economic growth slower. She further states that aid has left African countries more debt ridden, prone to inflation, vulnerable to currency markets, and unattractive to high-caliber investment. Apoorva Shah, research fellow at the American Enterprise Institute, does not necessarily disagree with many of Moyo's assertions. His concern is that Moyo is not offering anything new to the debate and that she ignores how new stakeholders are moving beyond the old debates. He discusses how the Millennium Challenge Corporation only works in countries where progress is being made on the consolidation of institutions, accountable governance, and freer peoples. He further finds it difficult to accept some of Moyo's broader generalizations, such as a characterization of aid as "large systematic cash transfers from rich countries to African governments."

Pradeep Kurukulasuriya, of the United Nations Development Programme, and colleagues argue that agricultural revenues for dryland crops in Africa will fall under global warming scenarios. They suggest that irrigation is a practical adaptation to climate change in Africa. Ole Mertz and colleagues, of the Universities of Copenhagen and Dakar, suggest that farmers in Africa's Sahelian region have always faced climate variability at annual and decadal time scales. Although households at their study site in Senegal are well aware of climate change, they attribute most changes in farming practices to economic, political, and social factors, rather than environmental ones.

Michael Mortimore, a geographer, and Mary Tiffen, a historian and socio-economist, both with Drylands Research, investigate population and food production trajectories in Machakos, Kenya. They determine that increasing population density has a positive influence on environmental management and crop production. Furthermore, they found that food production kept up with population growth from 1930 to 1987. John Murton, with the Foreign and Commonwealth Office of the British government, uses household-level data to show that the changes in Machakos described by Mortimore and Tiffen "have been accompanied by a polarization of land holdings, differential trends in agricultural productivity, and a decline in food self-sufficiency." As such, he argues that the "Machakos experience" of population growth and positive environmental transformation is neither homogenous nor fully unproblematic.

Kofi Annan, former UN Secretary General, deplores the fact that sub-Saharan Africa is the only region where per capita food production has declined. Annan is now leading a new organization that answers the call of many African leaders to build on the achievements and lessons learned from the Green Revolution in Asia and Latin America that began several decades earlier. He is spearheading an African Green Revolution that aims to increase African food production. Carol Thompson, professor of political economy at Northern Arizona University, suggests that increasing yields of a few targeted crops will not solve Africa's food problems. Rather, she argues that sustaining Africa's food crop diversity and indigenous ecological knowledge is the key to reducing hunger. She further eschews food security built on global market dependence in favor of food sovereignty.

Peter Balint, of George Mason University, and Judith Mashinya, of the University of Maryland, found that the situation in Mahenye, Zimbabwe, has deteriorated significantly since an earlier time period when it was deemed to be a model community-based natural resources management (CBNRM) program. They do not blame this decline on political turmoil in Zimbabwe, but rather on a failure of leadership and the departure of outside agencies responsible for oversight and assistance. As such, they argue against full devolution of authority to the community level for wildlife management. Liz Rihoy, of the Zeitz Foundation, and Chaka Chirozva and Simon Anstey, of the University of Zimbabwe, arrive at a very different conclusion about the same community in Zimbabwe. Although they acknowledge that CBNRM could be viewed as a failure in Mahenye at certain moments in time, they see the situation as an ongoing process of development in which there are the seeds of opportunity. They claim that the CAMPFIRE program has had real impact in terms of empowering local residents, providing them with incentives, knowledge, and organizational abilities to identify and address their own problems.

UNIT 4   SOCIAL ISSUES   259

Fuambai Ahmadu, an anthropologist at the London School of Economics, finds it increasingly challenging to reconcile her own experiences with female initiation and circumcision and prevailing (largely negative) global discourses on these practices. Her main concern with most studies on female initiation is the insistence that the practice is necessarily harmful or that there is an urgent need to stop female genital mutilation in communities where it is done. She suggests that "the aversion of some writers to the practice of female circumcision has more to do with deeply imbedded western cultural assumptions regarding women's bodies and their sexuality than with disputable health effects of genital operations on African women." Liz Creel, senior policy analyst at the Population Reference Bureau, and her colleagues argue that female genital cutting (FGC), while it must be dealt with in a culturally sensitive manner, is a practice that is detrimental to the health of girls and women, as well as a violation of human rights in most instances. Creel et al. recommend that African governments pass anti-FGC laws, and that programs be expanded to educate communities about FGC and human rights.

Richard A. Schroeder, an associate professor of geography at Rutgers University, presents a case study of a group of female gardeners in The Gambia who, because of their growing economic clout, began to challenge male power structures. Women, who were the traditional gardeners in the community studied, came to have greater income-earning capacity than men as the urban market for garden produce grew. Furthermore, women could meet their needs and wants without recourse to their husbands because of this newly found economic power. Human Rights Watch, a nonprofit organization, describes how women in Kenya have property rights unequal to those of men, and how even these limited rights are frequently violated. It is further explained how women have little awareness of their rights, that those "who try to fight back are often beaten, raped, or ostracized," and how the Kenyan government has done little to address the situation.

Andrew Creese and his colleagues, who work for the World Health Organization (WHO) and European universities, suggest that cost-effectiveness is an important criterion for deciding how to allocate scarce health care funding. A case of HIV/AIDS can be prevented for $11 by selective blood safety measures and targeted condom distribution with treatment of sexually transmitted diseases. In contrast, antiretroviral treatment for adults can cost several thousand dollars. They argue that a strong economic case exists for prioritizing preventive interventions and TB treatment. Philip J. Hilts, who teaches journalism at Boston University, describes a comprehensive HIV/AIDS program in Botswana. This program offered not only preventive care, but sophisticated triple drug AIDS treatments to all people of the nation, free of charge. By 2005, the program was treating 43,000 people and the cost of treatment is one-tenth of what it is in the United States.

UNIT 5   POLITICS, GOVERNANCE, AND CONFLICT RESOLUTION   329

Michael Bratton, professor of political science at Michigan State University, and Robert Mattes, associate professor of political studies and director of the Democracy in Africa Research Unit at the University of Cape Town, find as much popular support for democracy in Zambia, South Africa, and Ghana as in other regions of the developing world, despite the fact that the citizens of these countries tend to be less satisfied with the economic performance of their elected governments. Joel D. Barkan, professor of political science at the University of Iowa and senior consultant on governance at the World Bank, takes a less sanguine view of the situation in Africa. He suggests that one can be cautiously optimistic about the situation in roughly one-third of the states on the African continent, nations he classifies as consolidated democracies and as aspiring democracies. He asserts that one must be realistic about the possibilities for the remainder of African nations, countries he classifies into three groups: stalled democracies, those that are not free, and those that are mired in civil war.

Elizabeth Powley, a specialist on gender and postconflict reconstruction, argues that Rwandan women are beginning to consolidate their dramatic gains that came with a gender-sensitive constitution in 2003 and parliamentary elections, which saw females win 48.8 percent of seats in the Chamber of Deputies. These successes were built on the specific circumstances of the Rwandan genocide, a quota system, and a sustained campaign by the women's movement in Rwanda. The women's caucus in the Rwandan parliament reviews existing laws and introduces amendments to discriminatory legislation, analyzes proposed laws with an eye to gender sensitivity, and works closely with women's organizations. Carey Leigh Hogg, program officer for Vital Voices, describes how the ruling political party in Rwanda has advocated for greater inclusion of women under the premise that this will improve the political climate, yet this same party also suppresses political dissent and ethnic identification. She argues that the Rwandan case shows how women's political identities can be dangerously frozen in a situation where the ruling party is intent on building national unity by quieting dissent.

Robert I. Rotberg, director of the Program on Intrastate Conflict and Conflict Resolution at Harvard University's John F. Kennedy School of Government, holds African leaders responsible for the plight of their continent. He laments the large number of corrupt African leaders, seeing South Africa's Mandela and Botswana's Khama as notable exceptions. According to Rotberg, the problem is that "African leaders and their followers largely believe that the people are there to serve their rulers, rather than the other way around." Arthur A. Goldsmith, professor of management at the University of Massachusetts–Boston, suggests that African leaders are not innately corrupt but are responding rationally to incentives created by their environment. He argues that high levels of risk encourage leaders to pursue short-term, economically destructive policies. In countries where leaders face less risk, there is less perceived political corruption.

David C. Gompert, an adviser to Refugees International and a senior fellow at the RAND Corporation, believes that the African Union could be effective in Sudan if adequately supported. He believes that Africans are willing to commit combat forces to stop the killing in Darfur because they are more deeply affected by such abuses than Europe or North America. In fact, the unwillingness of the great powers to create a standing UN peacekeeping force is illustrative of a weak commitment of the West to intervene in Africa. Nsonurua J. Udombana of Central European University is more critical of the African Union peacekeeping mission in Sudan. He believes this mission has failed in Darfur because it suffers from several weaknesses, including problems of command and control, logistical support, operational practice, and lack of funds.

Topics Guide

This topic guide suggests how the selections in this book relate to the subjects covered in your course. You may want to use the topics listed on these pages to search the Web more easily.

On the following pages a number of Web sites have been gathered specifically for this book. They are arranged to reflect the units of this *Taking Sides* reader. You can link to these sites by going to http://www.mhcls.com.

All the issues that relate to each topic are listed below the bold-faced term.

Agriculture

2. Have the Contributions of Africans Been Recognized for Developing New World Agriculture?
9. Is Climate Change a Major Driver of Agricultural Shifts in Africa?
10. Is Food Production in Africa Capable of Keeping Up with Population Growth?
11. Does African Agriculture Need a Green Revolution?

Communication

6. Do Cell Phones and the Internet Foster 'Leapfrog' Development in Africa?
18. Is Corruption the Result of Poor African Leadership?

Conservation

11. Does African Agriculture Need a Green Revolution?
12. Is Community-Based Wildlife Management a Failed Approach?

Cultural customs and values

3. Is European Subjugation of Africans Ultimately Explained by Differences in Land, Plant, and Animal Resources?
13. Should Female Genital Cutting Be Accepted as a Cultural Practice?
14. Are Women in a Position to Challenge Male Power Structures in Africa?
15. Is the International Community Focusing on HIV/AIDS Treatment at the Expense of Prevention in Africa?
17. Does Increased Female Participation Substantially Change African Politics?

Demographics

14. Are Women in a Position to Challenge Male Power Structures in Africa?
17. Does Increased Female Participation Substantially Change African Politics?

Dependencies, international

4. Did Colonialism Distort Contemporary African Development?
8. Does Foreign Aid Undermine Development in Africa?
15. Is the International Community Focusing on HIV/AIDS Treatment at the Expense of Prevention in Africa?

Development, economic

1. Did the Trans-Atlantic Slave Trade Underdevelop Africa?
3. Is European Subjugation of Africans Ultimately Explained by Differences in Land, Plant, and Animal Resources?
4. Did Colonialism Distort Contemporary African Development?
5. Have Free-Market Policies Worked for Africa?
8. Does Foreign Aid Undermine Development in Africa?
9. Is Climate Change a Major Driver of Agricultural Shifts in Africa?
18. Is Corruption the Result of Poor African Leadership?

Development, social

4. Did Colonialism Distort Contemporary African Development?

(Continued)

Introduction

Understanding African Issues in Context: The Global and the Local as Interpreted by the Media, Scholars, and Policymakers

William G. Moseley

> People go to Africa and confirm what they already have in their heads and so they fail to see what is there in front of them.
>
> —Chinua Achebe, Nigerian author

> When the missionaries came to Africa they had the Bible and we had the land. They said "Let us pray." We closed our eyes. When we opened them we had the Bible and they had the land.
>
> —Bishop Desmond Tutu, South African spiritual leader and novelist

As the quotes above suggest, African issues cannot be fully appreciated without a deep knowledge of the region and a broad understanding of its connections to the rest of the world. Every place, country, and region in Africa has its own environmental conditions, cultural dynamics, politics, and history. Yet, these African places do not exist in a vacuum. They almost always have a history of connections to other areas of the continent and the world. As such, serious students, scholars, and policymakers with an interest in Africa must seek to understand both the specific geographic milieu in which an issue is being debated and the extent to which a local-level problem is connected to broader scale dynamics at the national, regional, or global level. The mere act of growing food crops, for example, is a deeply local process that relates to site-specific environmental conditions, agricultural practices, and cultural preferences. But what and where certain crops are grown has often been influenced by cultural exchange, colonialism, and global markets.

Scholars and policymakers seeking to understand the synergy of global and local forces that imbue most African issues often privilege one set of causes over another. This privileging of more global or local causes is often described in terms of externalist versus internalist explanations. Scholars evoking externalist (or structuralist) explanations suggest that contemporary development patterns cannot be properly understood without an understanding of historical patterns of resource extraction and political control, as well as the position of Africa within the global economic system. In particular, these explanations often look to problematic colonial legacies to elucidate contemporary economic distortions. Scholars emphasizing internalist explanations tend to look to local factors to explain an issue or problem. These local phenomena may be negative (corruption, mismanagement, incompetence, nepotism, ethnic allegiances, regional ties, obligations to the extended family, and patron–client relationships) or positive (local cultural practices, indigenous knowledge, and family support networks).

Externalist or structuralist critiques of internalist positions range from accusations of spatial and temporal myopia to "blaming the victim." Internalist-leaning critics suggest that the structuralists are apologists who deny Africans' "agency" (i.e., the ability to influence contemporary events) and responsibility, thereby fostering a sense of victimization, helplessness, and further mismanagement. The reality, of course, is that good scholarship must equally explore both sets of factors. In some instances, both types of explanations will be equally critical for understanding an issue; in other cases, either internal or external factors may rightly been seen to be more important. The rest of this introductory chapter explores the key actors who often interpret African issues and provides an introduction to the major themes explored in the reader.

Interpreting African Issues: Journalists, Scholars, and Policymakers

Our impressions of Africa, our understanding of African issues, and the actions organizations take on the ground and in the policy sphere vis-à-vis Africa are filtered and articulated by a number of key players. Most notable among these are the popular press, the academy, and policymakers.

The Popular Press

Images and descriptions in the popular press suggest that the African continent is a troubled land where corruption, ethnic warfare, poverty, hunger, environmental destruction, and pestilence prevail. Some have even suggested that Africa is a lost cause, asserting that the continent be "written off" by international development organizations. Meanwhile, commercial tour operators also hawk the region as a place of high adventure and exoticism. Even quasi-scholarly publications such as *National Geographic Magazine* often promote a vision of a primitive or wild Africa. What these popular and commercial descriptions hold in common is level of superficiality and one-sidedness. Yes, bad things do happen in Africa and there is beautiful scenery to be seen, but this is only one side of a complex and highly varied picture. It is the apparent unwillingness (or laziness) of popular commentators to provide a more nuanced view of an enormous continent that is often frustrating to scholars of Africa (or Africanists).

Africa is, after all, a place of extraordinarily diverse, vibrant, and dynamic cultures. Since the early 1990s, no other continent has seen more dramatic improvements in human rights, political freedom, and economic development—from the overthrowing of apartheid in South Africa to the revitalization of conomies in countries such as Ghana and Uganda. Although environmental threats are real, African societies have proven their capacity, when given a chance, to use resources sustainably. Some conservation efforts in Africa have even become models for progressive community-based resource management in Western societies. The importance of human relations, family, and good neighborliness in many African societies also stands in stark contrast to the more closed and individualistic tendencies in a number of Western settings.

The Academy

For the past century, the Western academy has been organized largely along disciplinary lines. The disciplinary approach encourages focused investigations from one perspective across a range of geographies. Area studies complement disciplinary inquiry by promoting multiperspective investigations of focused regions. In other words, the region or area represents a different framework around which to organize knowledge. Recent world events have revealed that there is a dearth of regional experts within the academy, government, and the private sector. The ideal scholar–practitioner may therefore be someone who not only has a firm grasp of the methods and theories of a particular discipline, but who also has a broader understanding of the issues and challenges (cutting across several disciplines) facing a particular region of the world.

Although a variety of disciplinary departments offer courses that deal with contemporary African issues, it is notable that there are now a number of interdisciplinary programs and departments devoted to African and African American studies. In the United States, the first African studies programs were established at Northwestern University (1946) and Boston University (1953). A number of other large programs were started in the early 1960s at places like Michigan State University, University of Wisconsin, Indiana University; University of California, Los Angeles; Indiana; and Ohio University (*African Studies and the Undergraduate Curriculum*, Patrica Alden, David Lloyd, and Ahmed Samatar, eds., Lynne Rienner, 1994). By the mid-1990s, there were approximately 330 programs of African and African American studies in the United States (*Directory of African and Afro-American Studies in the United States*, African Studies Association, 1993). African studies is as or more developed at institutions in other areas of the world. Although the names of these organizations are too numerous to list, there are over 1800 academic institutions, research bodies, and international organizations involved in African studies research in all parts of the world (*International Directory of African Studies Research*, 3rd ed., Philip Baker, ed., Hans Zell Publishers, 1994). In the United States, over 1300 scholars from a broad array of disciplines meet each year at the annual meeting of the African Studies Association to present their research findings and debate key African issues. African studies associations and societies also exist in a number of other countries (e.g., Australia, Canada, France, Germany, India, Japan, the Netherlands, South Africa, Spain, Sweden, Switzerland, and the United Kingdom).

Prior to the establishment of African studies as a recognized interdisciplinary field of study, some scholars encountered resistance from quarters of the academic establishment that perceived Africa to be lacking in history or otherwise unworthy of academic investigation. African studies has grown to be a dynamic realm of inquiry that regularly contributes to the broader academic discourse. Africanists have tested the validity of widely accepted notions in the African context. They have also developed new theories that have influenced thinking in other regional contexts. The disciplines contributing to African studies may roughly be divided between the humanities (mainly history and literature), the social sciences (anthropology, archaeology, education, geography, political science, and sociology), and the physical sciences (physical geography, ecology, and botany). In the North American context, African studies meetings tend to be dominated by the humanities and the social sciences, with historians and political scientists attending in the largest numbers (perhaps due to the sheer size of their disciplines). This volume largely deals with controversial African issues in the social sciences, although some of the questions have a significant historical or environmental dimension that pulls on literatures from the humanities or physical sciences.

The burgeoning field of African studies has spawned a number of academic journals, examples of which include *Africa, Africa Today, African Affairs, African Studies Review, Cahiers d' Études Africaines, African Geographical Review, South African Geographical Journal, Canadian Journal of African Studies, Journal of African History, Journal of Modern African Studies, Journal of Southern African Studies,* and *Review of African Political Economy.* There are also a number of more policy-related or popular media magazines that focus on Africa including, for xample, *Africa Analysis, Africa Confidential, Africa Now, Afrique-Express* (French), and *Jeuneafrique* (French).

Policymakers and Development Professionals

In addition to the press and the academy, the other major sphere where African issues are framed and examined is in the offices of government bureaucrats, global policymakers, and nongovernmental organizations. These include individuals within the bilateral development agencies (e.g., U.S. Agency for International Development, British Department for International Development, Cooperation Française, or Canadian International Development Agency), international financial institutions (e.g., World Bank, International Monetary Fund, and World Trade Organization), UN agencies (e.g., Food and Agriculture Organization and World Health Organization), and nonprofit community (e.g., CARE, Oxfam, and Save the Children). Although these organizations contribute to and influence discourse, their role is slightly different than the media and the academy in that this is where rhetoric is transformed into programmatic reality.

Although I present the media, the academy, and development institutions as separate spheres of debate on Africa, there is, in fact, a considerable amount of crossover in thinking, not to mention personnel, between each realm. Policymakers and development professionals do not make decisions in a vacuum, but are influenced by public opinion (shaped by the media) and the latest academic research. Academics and policymakers attempt to influence public opinion through the media (via op-eds and interviews with the press). Governments may seek to shape academic findings by funding research projects or hiring scholars as consultants. Scholars may agree to consult on development issues because they believe this is a way for them to influence decision making. There also is a certain amount of personnel exchange between the different sectors (e.g., it is to think tanks, multilateral development agencies, and universities that a number of foreign service and aid advisors flee each time the party in power changes in Washington, Ottawa, or London). Despite these exchanges, significant differences exist in tone, timeliness, depth and emphasis in analysis, and standards of knowledge production. The distribution of power (i.e., the power to have one's ideas heard) also differs within each group.

Major Themes

If this book has one overarching theme it is African development. I interpret development in the broadest sense of the term, going beyond conventional measures of economic progress, to embrace processes occurring at a variety of spatial and temporal scales that allow people to meet their full potential. When viewed in this manner, nearly all of the issues presented in this volume pertain in one way or another to development. In order to give the volume a more accessible format, it has been organized into five thematic categories of African issues that are most often debated by Africanists in the academy, as well as by policymakers and media commentators. These thematic categories are (1) history; (2) development; (3) agriculture, food, and the environment; (4) social issues; and (5) politics, governance, and conflict resolution. Although I present these themes as distinct, the reader should understand that there are often a number of connections that exist between issues in different parts of the book.

History

The number of historical debates pertaining to Africa are so numerous that they could easily constitute a separate volume. There are, however, a few issues that stand out as they highlight the connections between Africa and the rest of the world and because these questions impinge on contemporary development patterns. The first issue is understanding the impact of the trans-Atlantic slave trade on African development patterns is critical. Was this a significant historical moment with lasting impacts or was it consistent with existing practices and effectively managed by Africans? Given that so many Africans departed to the New World, it is also crucial to understand the active role they played in shaping and developing the Americas. What led the Europeans to colonize Africa, rather than vice versa? Was it something about the continents' different natural endowments that made this possible (as some have argued), or was it more a question of intent and political economy? Finally, to what extent has colonialism shaped contemporary development patterns? Some find that this led to huge and lasting distortions in

African economies, while others argue that this was beneficial for economic development.

Development

The nature of and approach to development in the African context is highly contentious. Commentators, scholars, and policymakers argue, for example, over the perceived failure of Africa to develop, the impact of the colonial experience on contemporary events, the influence of global economic structures on African development patterns, and the role of the state versus the private and nonprofit sectors in the development process.

Closely related to the externalist/internalist debate is one concerning the most appropriate approach to development in Africa. The structuralists evoke dependency theory and world systems theory to suggest that, even though the colonial era has ended, historical patterns of resource extraction persist. In many instances, African nations are supplying cheap commodities (minerals, oil, lumber, cotton, coffee, cocoa, tea, and sugar) to Europe and North America in exchange for relatively expensive manufactured goods. As such, the participation of African nations in the global economy under current conditions leads to a process of underdevelopment. The best way to avoid this trap, according to the structuralists, is to diversify the national economy by producing imported goods at home, an approach also known as import substitution.

The general failure of import substitution, and the related Third World debt crisis, led the international financial institutions (especially the World Bank and International Monetary Fund) to begin pushing for free-market reforms (known as structural adjustment) of African economies beginning in the early 1980s. With the basic aim of balancing the national budget and spurring economic growth, these programs obliged African government to privatize state-owned enterprises, devalue local currencies, cut government programs, and expand exports. The uneven success and social costs of these free-market policies have led to bitter debates in academic and policy circles. Many contend that the role of the state in the development process has been underestimated by neoclassical economists.

In addition to the above, the African development scene has also been swirling with a number of other fascinating policy debates in recent years. Although cell phone usage and Internet access lag behind that of other regions, the growth in the use of these devices has just exploded on the continent in recent years. The dramatic proliferation of these devices (and the innovative way in which they are being used, such as Internet banking) has revived an older debate about ability of some technologies to allow countries and regions to circumvent more traditional and costly paths to development. This optimistic narrative of the cell phone and Internet as "leapfrog" technologies has been countered by a more somber assessment, which sees these as reinforcing longstanding power inequities.

There are also ongoing contestations about the ability of foreign aid to foster development on the continent. Pessimistic commentators assert that aid is actually undermining or underdeveloping African countries. This argument is countered by those who believe that we must move beyond this old debate and focus on new forms and approaches to assistance. Some have looked to China as a major new player on the development scene in Africa, asserting that the "Chinese model" is substantially different than "Western" forms of engagement on the continent. Others see China as just another colonial power pursuing its own interests.

Agriculture, Food, and the Environment

Since the global media focused attention on large-scale droughts in the early 1970s and mid-1980s, famine and environmental destruction in Africa have loomed large in the public imagination. Key debates have centered on whether or not climate change will substantially alter African farming practice, if the continent's population is growing too quickly for its agricultural base, whether or not a new "African Green Revolution" is needed, and if community-based approaches to wildlife management make more sense than the traditional park strategy.

Climate variability and drought have long been a reality for African farmers. Scientists have studied climate variability on the continent, from major shifts in paleoclimates (e.g., we know that the Sahara desert was once much wetter) to intensive study of annual and decadal rainfall patterns in semiarid regions such as the Sahel. Now the concern is that the global climate change, when downscaled to Africa and various African subregions, will be especially problematic for the continent. A major concern is how African farmers will adapt to these changes. Others point out that the continent's farmers are not only dealing with environmental change, but also with significant economic shifts (a problem sometimes referred to as "double-exposure"). In the end, these commentators suggest that economic change will lead to more substantial shifts in African agriculture than environmental change.

A fundamental debate persists as to whether or not famine and food insecurity in Africa are the result of underproduction or the maldistribution of food. Many development assistance programs, including the international network of crop development institutes under the aegis of the Consultative Group on International Agricultural Research (CGIAR), are predicated on the assumption that Africa's food problems will be resolved by increases in food production. Economist and Nobel laureate Amartya Sen has argued that famine rarely results from an absolute shortage of food, but the inability of poor households to access available supplies.

Related to the aforementioned debate are two long-standing discussions. The first concerns the relationship between food production and population growth. The eighteenth-century parson Thomas Malthus asserted that human populations would inevitably grow more quickly than food production, ultimately leading to famine. Contemporary neo-Malthusians, such as Paul Ehrlich, have similarly argued that urgent measures are needed to control population in order to avert disaster. Ester Boserup has countered that Malthus had it all wrong because it is population density that controls the level of food production, rather than food production quantities setting population thresholds. In recent years, some of the best case studies (both pro-Boserupian and pro-Malthusian) on the relationship between population growth and food production have come out of Africa. The second debate concerns a recent push for a new African Green Revolution, an attempt to increase yields through improved seeds and inputs. This initiative has received prominent support from former UN secretary general Kofi Annan and Columbia University Economist Jeffrey Sachs. Critics are skeptical on two fronts. First, and as discussed above, they wonder if hunger in Africa is really a supply-side problem that improving yields can resolve. They argue for more locally appropriate technologies.

Second, they are concerned about repeating the costly social and environmental errors of the green revolution in Asia where packages of improved seeds and inputs led to pesticide problems and increasing economic differentiation in rural areas.

The "tragedy of the commons" is another prominent paradigm that has been used to explain resource problems in Africa. According to this theory, commonly held natural resources will tend to be overexploited by individuals seeking to maximize personal gain. The solution advocated by economists is to privatize common resources, as it is believed that the private owner will more carefully husband environmental resources over the long term. The commons is actually a misnomer in the African context because many African communities have effectively managed commonly held natural resources through traditional mechanisms of control and enforcement. It is when these traditional mechanisms break down, and the commons become "open-access resources" (or resources where there is no effective management authority), that problems develop. The loss of wildlife in Africa has been characterized as an open-access resource problem. Environmentalists argue that drastic measures are needed to protect the continent's dwindling wildlife resources from poachers and overuse by poor households. As such, the park model has been the traditional approach to conservation. Critics suggest that the park model is deeply flawed because it disenfranchises local people, often taking a well-managed community resource and turning it into something which is exclusively and poorly managed by the state. They suggest that reengaging communities in the management of local wildlife resources is the only way to move forward.

Social Issues

Perhaps more than any other set of contested African issues, those pertaining to the social sphere tend to provoke deep-seated emotional responses. Different aspects of the AIDS crisis in Africa, female genital cutting, the position of women in African societies, and debt relief all have attracted considerable media attention and scrutiny in recent years.

Many of these issues get at a deeper debate between those who assert there are certain universal rights and wrongs and cultural pluralists who believe we need to evaluate practices within their own cultural context. Advocates of the universality of norms disparage defenders of certain African practices as cultural relativists. Cultural relativism is cast as problematic because it may be used as an excuse to say that anything goes. In contrast, cultural pluralists assert that there are separate and valid cultural and moral systems that may involve social mores that are not easily reconcilable with one another. Cultural pluralists are not necessarily cultural relativists as many would argue that everything does not go, for example, murder is wrong. The challenge for cultural pluralists is to determine if a practice violates a universal norm when it is viewed in its proper cultural context (rather than in the cultural context of another). The result of this deep philosophical divide is that we often see Western feminists pitted against multiculturalists (two groups that frequently function as intellectual allies in the North American context) over some of the issues addressed in this section of the book. No issue better exemplifies this divide than the debate over female genital cutting.

If one does want to work for change, how should this be approached? In the case of empowering women, some argue that fundamental legal changes must be made, while others suggest that the creation of economic opportunities is the key.

Certain aspects of the HIV/AIDS policy debate in Africa also involve the question of human rights. Should all people be eligible for more expensive HIV/AIDS treatments because it is their basic human right, or do we need to consider the fact that some rationing may need to occur so that governments may adequately invest in more cost-effective prevention strategies? Foreign debt may also be considered a human rights issue because it, and related structural adjustment policies, reduces the state's ability to address discrimination, vulnerability, and inequality. Debt relief could help reduce poverty and promote health. Others argue that such relief may only promote irresponsibility and stop any attempts to address the factors that created the debt in the first place.

Politics, Governance, and Conflict Resolution

The terrain of politics, governance, and conflict resolution is simultaneously one of the most hopeful and distressing realms in contemporary African studies. Although more contested elections have been held in the last 20 years than at any other time in the postcolonial period, and popular uprisings in Egypt and Tunisia have led to the downfall of dictators, the African continent suffers from more instances of civil strife than other world regions. Key debates have focused on the success or nonsuccess of multiparty democracy in the African context, the impact of increased female participation on African politics, reasons for corruption among African officials, and the role of African states in their own peacekeeping efforts.

An underlying theme related to several of these questions concerns the most appropriate form of governance in the African context. Proponents of multiparty democracy assert that this form of government will promote economic growth and minimize ethnic tensions. They also believe that a healthy civil society will serve as a check on corruption and other government excesses. Increasingly, foreign assistance for a variety of projects is conditional upon certain types of democratic reform.

Other scholars and African leaders see the imposition of democracy in Africa as a form of neoimperialism. They suggest that the problem with multiparty democracy in many African countries is that it leads to the formation of too many political parties, each with a regional or ethnic outlook, and none representing the interests of the country as a whole. Furthermore, some Afrocentrists assert that the one-party state is more consistent with traditional consensus decision making that occurs at the village level. They maintain that the process of competitive elections is a foreign notion that is divisive in the African context. They have also argued that democracy may actually inhibit economic growth. According to this argument, the problem with democracy is that it does not allow leaders to make tough economic decisions, such as the austerity measures required under structural adjustment.

The flourishing female participation in African politics has been a sign of hope on the continent. From the election of a female president in Liberia to strong female representations in the parliaments of Rwanda and Burundi, many see women as a force that will fundamentally change African politics. Others are more sanguine and see female politicians as constrained by a variety of internal and external factors. Finally, what is the role of African peacekeeping forces on the continent? Some argue that international peacekeepers are critical for resolving some ethnic conflicts, while others suggest that Africans must play a central role in resolving their own conflicts.

Internet References . . .

H-Africa

H-Africa encourages discussion of Africa's history and culture. It offers a variety of listservs related to specific themes and regions of the continent. The site also allows one to search past discussions, link to other sites that deal with continent, and examine reviews of books on the region.

http://www.h-net.org/~africa/

African Studies Association

The African Studies Association is the largest African studies organization in North America. It coordinates a meeting each year where scholarly presentation takes place, and publishes several Africa-related journals.

http://www.africanstudies.org/

Columbia University's African Studies Internet Resources

A catalog of links to African studies programs, resources by region and country, electronic journals and newspapers, and resources by topic. It also has a directory of African studies scholars.

http://www.columbia.edu/cu/lweb/indiv/africa/cuvl/

The University of Pennsylvania's African Studies Center, History Section

The history page of this Web site offers links to several other reputable sites dealing with African history and archival resources.

http://www.africa.upenn.edu/About_African/ww_hist.html

History

*I*t is important for students of Africa to place the continent in its historical and global context. Sub-Saharan Africa has long-standing connections with other areas of the world. Debates rage over the nature of these connections, their persistence and impact on the future of Africa, why and how Africa was colonized, and how African ideas influenced development in other regions of the world.

- Did the Trans-Atlantic Slave Trade Underdevelop Africa?
- Have the Contributions of Africans Been Recognized for Developing New World Agriculture?
- Is European Subjugation of Africans Ultimately Explained by Differences in Land, Plant, and Animal Resources?
- Did Colonialism Distort Contemporary African Development?

ISSUE 1

Did the Trans-Atlantic Slave Trade Underdevelop Africa?

YES: Paul E. Lovejoy, from "The Impact of the Atlantic Slave Trade on Africa: A Review of the Literature," *Journal of African History* (1989)

NO: John Thornton, from *Africa and the Africans in the Making of the Atlantic World, 1400–1680* (Cambridge University Press, 1992)

ISSUE SUMMARY

YES: Paul E. Lovejoy, professor of history at York University, argues that the trans-Atlantic slave trade significantly transformed African society. It led to an absolute loss of population on the continent and a large increase in the enslaved population that was retained in Africa. The economic advantages of exporting slaves did not offset the social and political costs of participation, there were disastrous demographic impacts, and Africa's relative position in world trade declined.

NO: John Thornton is a professor of history at Boston University. He notes that slavery was widespread and indigenous in African society. Europeans simply worked with this existing market and African merchants, who were not dominated by Europeans, responded by providing more slaves. African leaders who allowed the slave trade to continue were neither forced to do so against their will, nor did they make irrational decisions. As such, the preexisting institution of slavery in Africa is as much responsible as any external force for the development of the trans-Atlantic slave trade.

The debate about the impact of the trans-Atlantic slave trade on Africa, and related discussions concerning the role that African elites played in this process dates back to at least the time period leading up to the abolition of the slave trade within the British Empire in 1807. During this time frame, some were arguing that the majority of Africans were already slaves in Africa, and consequently their situation would not be improved by ending the trans-Atlantic slave trade.

Apologists seeking to justify slavery in the American South also pointed to the existence of slavery in Africa and the role of African elites in the trans-Atlantic trade. Many academics (who would not have considered themselves apologists) writing in the 1950s and 1960s also made similar observations. Walter Rodney is probably the scholar most widely known for challenging these views about the nature of slavery in Africa prior to the arrival of Europeans. For an example of Rodney's writing in this vein, see his 1966 article in the *Journal of African History* entitled "African Slavery and Other Forms of Social Oppression on the Upper Guinea Coast in the Context of the Atlantic Slave Trade." In this article, Rodney essentially argued that the institution of slavery in Africa was really quite different from, and a more benign form of, what came to be practiced in the Americas. Furthermore, the trans-Atlantic slave trade left deep and lasting scars on societies in Africa.

Paul Lovejoy, professor of history at York University, is writing from a perspective similar to that described for Walter Rodney above. He argues that the trans-Atlantic slave trade significantly transformed African society. It led to an absolute loss of population on the continent and a large increase in the enslaved population that was retained in Africa. The economic advantages of exporting slaves did not offset the social and political costs of participation, there were disastrous demographic impacts, and Africa's relative position in world trade declined. Lovejoy, therefore, supports the "transformation thesis," which holds that the external slave trade dramatically reshaped slavery and society in Africa.

The contrarian view in this issue, presented by John Thornton, professor of history at Boston University, was written in the early 1990s when Walter Rodney's and Paul Lovejoy's views, and related structuralist scholarship, were more dominant. He notes that slavery was widespread and indigenous in African society. Most importantly for the time period when he was writing, he argues that African elites had considerable agency (or power) in determining the shape and character of the trans-Atlantic slave trade. He asserts that Europeans worked with existing African slave markets and that African merchants, who were not dominated by Europeans, responded by providing more slaves. African leaders who allowed the slave trade to continue were neither forced to do so against their will, nor did they make irrational decisions. As such, the preexisting institution of slavery in Africa is as much responsible as any external force for the development of the trans-Atlantic slave trade. While Thornton's views are still hotly contested, he did successfully tap into a broader theme in post-structuralist literature on Africa, that is, the notion that Africans were not helpless pawns in a world dominated by Europeans, but active and influential participants.

The debate about the impact of trans-Atlantic slavery on Africa, and the role of Africans in this process, was reignited by African American scholar Henry Louis Gates in his 1999 PBS series entitled *The Wonders of the African World*. In the second episode of this series he suggested that contemporary Africans bear a collective guilt for what he referred to as "the black-on-black Holocaust" that occurred during the era of the trans-Atlantic slave trade. The then president of the African Studies Association, Lansiné Kaba, responded to Gates in a November 2000 address that subsequently was published in the April 2001 issue of the *African Studies Review* as "The Atlantic Slave Trade Was Not a 'Black-on-Black Holocaust.'"

YES

Paul E. Lovejoy

The Impact of the Atlantic Slave Trade on Africa: A Review of the Literature

African History and the Atlantic Slave Trade

The significance of the Atlantic slave trade for African history has been the subject of considerable discussion among historians and merits attempts from time to time to review the literature. The present such attempt addresses several, but not all, the key issues that have emerged in recent years. These are, in order of discussion here: What was the volume of the Atlantic slave trade? More specifically, what were the demographic trends of the trade with respect to regional origins, ethnicity, gender and age? Finally, what was the impact of the slave trade on Africa? In brief, what is the state of the debate over the slave trade?

My own position in the debate is clear: the European slave trade across the Atlantic marked a radical break in the history of Africa, most especially because it was a major influence in transforming African society.

> The history of slavery involved the interaction between enslavement, the slave trade, and the domestic use of slaves within Africa. An examination of this interaction demonstrates the emergence of a system of slavery that was basic to the political economy of many parts of the continent. This system expanded until the last decades of the nineteenth century. The process of enslavement increased; the trade grew in response to new and larger markets, and the use of slaves in Africa became more common. Related to the articulation of this system, with its structural links to other parts of the world, was the consolidation within Africa of a political and social structure that relied extensively on slavery.

The transformation thesis identifies slavery as a central feature of African history over the past millennium. The Atlantic trade was only one, although a major, influence on the transformation of society. The Muslim slave trade was also important, and other internal African developments strongly influenced social change as well. According to this interpretation, the task of the historian is to weigh the relative importance of the various factors that incorporated Africa

From *Journal of African History,* vol. 30, issue 3, November 1989, pp. 365–367, 386–393.

into an "international system of slavery" that included Africa, the Americas, western Europe and the Islamic world.

David Eltis has challenged this interpretation. On the basis of his study of the nineteenth-century Atlantic trade and an analysis of the value of the Atlantic trade between the 1680s and 1860s, Eltis has concluded that neither the scale nor the value of the Atlantic trade was sufficiently large to have had more than a marginal influence on the course of African history. According to Eltis,

> The slave trade for most regions and most periods was not a critically important influence over the course of African history. At the very least, those who would place the slave trade as central to West African and west-central African history should be able to point to stronger common threads, if not themes, across African regions than have so far come to light.

With respect to slavery, he claims that "whatever the origins or nature of structural changes in African slavery, it is unlikely that external influences could have been very great."

The contribution which Eltis makes is two-fold, it seems to me. First, he brings more precision to an analysis of the volume and direction of the Atlantic trade in the nineteenth century. His study of nutritional trends, age and gender, and mortality are particularly significant. Secondly, he has articulated a model of economic development for the pre-colonial era that must be taken seriously, although I disagree with his conclusions. Did exports determine the extent of economic change, as measured by standard economic indicators? He concludes that an export-led model of economic development has little to offer in interpreting African history. Climate and human genius, according to Eltis, were far more important than the export sector.

Eltis bases his startling conclusions on an analysis of the relative importance of the Atlantic trade on Africa, as determined on a *per capita* basis. In a study undertaken jointly with Lawrence C. Jennings, it is claimed that the annual average *per capita* trade of those parts of Africa involved in the Atlantic slave trade was significantly less than in other parts of the Atlantic basin and that the African share in world trade declined in relative terms from the 1680s to the 1860s. They conclude that neither the absolute nor the relative value of Atlantic trade was very great; in general, foreign trade had only a weak influence on African economies. According to Eltis,

> . . . on the assumption that the improbably low figure of 15 million people lived in West Africa [in the 1780s] at subsistence levels, then imports from Atlantic trade may be taken at about 9 per cent of West African incomes in the 1780s. With assumptions that are more in accord with reality (i.e. a population of 25 million or more and domestic production in excess of subsistence), then imports decline in importance to well below 5 per cent. For other decades in the century when both slave prices and exports were lower, imports would have been much less significant. For west-central Africa, population densities were much lower but import/income ratios could not have been much greater.

Indeed, "The majority of Africans . . . would have been about as well off, and would have been performing the same tasks in the same socioeconomic environment, if there had been no trading contact" with Europe. Eltis even advances the astonishing conclusion for Asante that ". . . the ratio of the level of exports, either before or after 1807, to any plausible population estimates of Asante suggests that the slave trade can never have been important."

The rise of commodity exports in the nineteenth century had virtually no impact on Africa either. According to Eltis,

> . . . the slave and commodity trades together formed such a small percentage of total African economic activity that either could expand without there being any impact on the growth path of the other. . . . [I]n the mid-nineteenth century neither the slave nor the commodity traders were large enough to have to face the problem of inelastic supplies of the factors of production.

In short, neither the Atlantic slave trade nor its suppression had much influence on African history.

The following review of the recent literature on the Atlantic slave trade provides a context in which to assess the revisionist interpretation of Eltis (and Jennings). I begin with the new studies on the volume of the slave trade, in which a consensus seems to have emerged. I then consider the analytical refinements in the regional and ethnic origins of the exported slave population. The demographic data allow a closer examination of the gender and age profile of the trade, which is the subject of the next section of this article. Finally, I return to an assessment of the arguments of Eltis, particularly with regard to the demographic impact of the trade on Africa. A number of important themes are not considered, including the economic significance of slavery in Africa and the importance of imported commodities on African society and economy. Nonetheless, I believe that I can demonstrate that Eltis' provocative conclusions are seriously flawed. . . .

Impact of the Atlantic Slave Trade on Africa

The discussion of the volume of the trade, the regional and ethnic origins of the exported population, and the sex and age profiles of slaves should indicate that much of Eltis' revisionist interpretation cannot be accepted. Otherwise, these factors would not matter to African history, although they are crucial to the history of slavery in the Americas. While the slight modifications in the volume and direction of the Atlantic slave trade do not affect the argument of Eltis (and Jennings), the issues remain: did the slave trade have a dramatic impact on exporting regions? Did the suppression of the Atlantic trade in the nineteenth century have a significant effect on the course of slavery? My informed opinion is that both the trade and it suppression were major factors in African history, and to show this I will examine, in order, the following issues: (1) the economic impact of the trade; (2) its demographic implications; and (3) the incidence of slavery in Africa. There are certainly other issues, but these will have to suffice.

One of the principal conclusions of Eltis and Jennings seems likely but only modifies my analysis: Africa's share of world trade from the late seventeenth until the mid-nineteenth century was relatively small in comparison with other parts of the Atlantic world, and Africa's share of that trade declined in relative terms during the period of the slave trade. Eltis and Jennings use an estimate of £0.8–£1.1 for *per capita* incomes in western Africa for the 1780s. They calculate that the export trade amounted to £0.1 per person each year. The proportion of exports to total income certainly appears to be very low by comparison with other parts of the Atlantic basin, although the room for error in measuring *per capita* income and the value of the export trade are enormous.

This observation on the value of the export trade can be accepted, but the implications that Eltis and Jennings draw from it cannot. The ratio of the value of the external trade to *per capita* income is not an accurate indicator of the impact of the slave trade on Africa. *Per capita* income in western Africa was certainly very low by the standards of other parts of the Atlantic basin. Africa was very poor. Almost any incremental increase over subsistence would have had a disproportionate impact on the economy. Eltis and Jennings quantify the relative poverty of Africa, but they are wrong to conclude that the lack of prosperity was an accurate gauge of the degree of isolation from the impact of the slave trade.

The simulation model developed by Patrick Manning provides one way to establish that the slave trade had a significant, indeed devastating, impact on Africa. Manning's model is a statistical means of measuring demographic change under conditions of enslavement, slave trade and slave exports. His analysis is based on the demography of the Atlantic slave trade and certain broad assumptions about demographic change that establish the parameters of the historical possibilities. Manning estimates that the population of those areas of West and west-central Africa that provided slaves for export was in the order of 22–25 million in the early eighteenth century. He projects a growth rate for that population during the eighteenth and early nineteenth centuries of 0–5 per cent, which he considers the maximum possibility. His simulation "model suggests that no growth rate of less than one per cent could have counterbalanced the loss of slaves in the late eighteenth-century." He uses a counterfactual argument to make the same point: "With a growth rate of 0.5%, the 1700 populations . . . [of] 22 to 25 million would have led to 1850 populations of from 46 million to 53 million, more than double the actual 1850 population." Manning concludes that "the simulation of demographic impact of the Atlantic slave trade provides support for the hypothesis of African population decline through the agency of the slave trade."

Furthermore, the simulation model suggests that the incidence of slavery increased in Africa. According to Manning, "As an accompaniment to the estimated nine million slaves landed in the New World [between 1700 and 1850] . . . , some twenty-one million persons were captured in Africa, seven million of whom were brought into domestic slavery, and five million of whom suffered death within a year of capture." As the discussion of the sex profile of the export trade makes clear, more women were retained in Africa than men. Not only did the slave population increase, therefore, but the incidence of polygyny increased as well. Indeed, the two phenomena were closely related.

By 1770 the Atlantic trade resulted in a slave population in the Americas of approximately 2,340,000. Manning's simulation suggests that the slave population in West and west-central Africa could not have been much different. It is safe to say that the slave population was at least 10 per cent of the total population of 22–25 million and that the proportion of slaves was rising. Manning concludes that there were 3 million slaves in those parts of Africa that serviced the Atlantic trade at the turn of the nineteenth century, virtually the same number as in the Americas.

The dramatic growth in the African slave population is the transformation that I highlighted initially in 1979 and more fully in 1983. The transformation was the result of a dialectical relationship between slavery in the Americas and the enslavement, trade and use of slaves in Africa. The production of slaves for the Americas also produced slaves for Africa. It is difficult to prove that the Atlantic slave trade *caused* the transformation of slavery in Africa, but it is likely. The simulation model, as well as the thesis that I advanced in *Transformations in Slavery,* consider that enslavement, trade and the use of slaves were interrelated, across the Atlantic and across the Sahara. African, European, and Muslim merchants wanted slaves, and African, European and Muslim slave-owners used slaves. The low value of exports and imports apparently confused Eltis and Jennings so that they did not perceive the importance of this inter-relationship, but an examination of any part of Africa that was a supplier of the export trade reveals the dialectic.

Miller's study of Angola provides the most dramatic example of impact of the slave trade on a region in Africa. As the export data make clear, approximately 40 per cent of all slaves in the Americas came from west-central Africa, and Miller estimates that deaths in Africa related to capture and enslavement roughly equalled the number of slaves exported, that is 50,000–60,000 per year in the last half of the eighteenth century. In addition, "fully as many more people [were] seized as slaves but left to reside in other parts of western central Africa." Total population displacement would have been in the order of 100,000–120,000 per year. Admittedly, Miller paints this as a worst-possible scenario, but even if demographic change was less severe it must have been dramatic. No one has argued as much, but it may be that matrilineality and the export trade were interrelated. They certainly reinforced each other.

Miller's analysis confirms the gender and age structure of the trade. In pursuing a discovery made earlier by John Thornton, Miller shows that the sex ratio of the population in those areas most heavily involved in the export trade was strongly skewed towards women and girls. Polygyny was a central institution of wealth and political power, and slaves (females) were most heavily concentrated around the principal courts of the region. Miller refers to the centralization that was associated with the slave trade as the "great transformation," which is solace to the theorist. On the basis of Miller's analysis, it is impossible to conclude that the slave trade had a marginal impact on west-central Africa.

Eltis is on shaky ground in suggesting that the suppression of the slave trade was not a decisive event for Africa. According to his interpretation, the increased incidence of slavery in the nineteenth century was unconnected with the collapse of the Atlantic trade. Instead, increased demand for slaves arose from "rejuvenated Islam" and late in the century from European demand for primary products from Africa. Indeed there was a dramatic increase in slavery as a result of the *jihads* and Muslim commercial expansion in East Africa. Although Eltis does not explicitly argue the point, his proposed revision is one of timing and concentration, not of substance, and he presents no new data. I remain unconvinced. Certainly there was an expansion in production, based on slave labour, late in the nineteenth century, but that development is also part of the transformation thesis.

Eltis disputes the thesis that the slaving frontier continued to move inland during the early nineteenth century and that the number of slaves in Africa increased dramatically in these decades. He bases this conclusion on an analysis of slave prices:

> . . . because the price of all slaves declined, it seems clear that although domestic [African] demand increased, it did not increase sufficiently to offset the decline in trans-Atlantic demand. As a consequence, the number of slaves traded as well as the price of those slaves declined during the century . . . ; accordingly, suppression must have meant some reduction in the enslavement of Africans.

Eltis is correct that the price of slaves dropped between the 1790s and the 1820s and continued at a depressed level for most of the continent for the rest of the century. Slaves were cheap, and in real terms prices may have continued to fall in many of the major slave markets, but this does not mean that enslavement decreased in intensity. The cheapness of slaves reflects two factors, the generally low prices of basic commodities in Africa and the glut of the slave market.

Before the suppression of the trans-Atlantic trade, according to Eltis, the demand for slaves was divided into two sectors: one within Africa and the second in the Americas. (In fact, there was a third sector—the external Muslim market.) The collapse of American demand inevitably depressed prices in Africa, He concludes that the African market did not increase sufficiently to offset the loss of American sales; according to Eltis demand and supply declined, although he only provides slave prices as evidence. Prices did not rebound after the decline of the 1790–1820 period, it is true, but the reason was a combination of factors. Indeed Eltis shows that American demand did not decline, in the aggregate, in this period but regained its former heights. Decline only began in the 1850s, well after the period that is crucial to his argument.

Demand is an expression of price, so if prices were generally low, then demand might appear to have been depressed. In fact, the contrary could have been true, if more were known about the price structure of African economies. African demand for slaves was determined by the value of what slaves could produce, and this marginal revenue product was primarily a function of the value of food consumption, housing costs, social requirements, taxation and similar expenditures that were not much affected by international markets.

Any qualitative assessment of the eighteenth and nineteenth centuries would judge Africa to have been comparatively poor, as the ratio of exports to *per capita* income reveals. Prices in general were low, and the price structure would have influenced the cost of slaves accordingly.

Eltis and Jennings, in concentrating on export-led economic development, have raised an important consideration: more needs to be known about internal African price structures. Until such information is available, however, it is hard to justify their conclusions, and there is sufficient evidence available to argue the contrary.

Eltis' conclusion that enslavement declined in the early nineteenth century contradicts the research of most Africanist historians who have written on the period. In almost every part of Africa for most of the century, enslavement was rampant. Slaves were generated on a scale previously unknown, as can be attested by the following examples: the wars of the *mfecane* in southern and central Africa, the activities of Arabs, Swahili, Yao, Nyamwezi, Chikunda and others in eastern and south-eastern Africa, and the raiding of Muslim slavers in the southern Sudan and north-central Africa. None of these cases have much, if anything, to do with the suppression of the Atlantic slave trade, and hence could be dismissed by those who favour the Eltis thesis. But what about the phenomenal levels of enslavement during the *jihads,* often in areas that once did or could have fed the Atlantic trade? How are the collapse of the Lunda states and the havoc of enslavement instigated by the Cokwe to be explained? Can the insecurity of Igbo country in the nineteenth century and the enslavement resulting during the Yoruba wars be easily dismissed? The combined impact of these phenomena was to maintain a glutted market and hence a depressed price for slaves almost everywhere. The "transformation thesis" holds that the external slave trade, particularly the trans-Atlantic sector but also the Islamic market, shaped slavery and society in Africa, and that internal factors intensified slavery as the external trade contracted.

The enslavement of people and the growth of the slave population in Africa continued apace for the whole of the nineteenth century, despite local variations. As yet there are few estimates of the scale of the African slave population, but some insights can be gained by a comparison of certain parts of West Africa in *c.* 1900 with the Americas on the eve of emancipation there.

The slave population of the Americas rose from 2,340,000 in *c.* 1770 and peaked at 2,968,000 by the end of the century. The revolt of St Domingue reduced this total considerably; St Domingue had a slave population of 480,000 in 1791. The independence of mainland Spanish America after *c.* 1820, with its slave population of a couple hundred thousand, and the emancipation of 674,000 British slaves in 1834 reduced the total further, but the number of slaves continued to expand in the Spanish Caribbean, the U.S.A. and Brazil, reaching a peak just before the emancipation of U.S.A. slaves in the early 1860s. In 1860, there were almost 4 million slaves in the U.S.A. and another 1.5–1.8 million slaves in Brazil and the Spanish Caribbean, for an estimated total of 5.5–5.8 million slaves. With the freeing of slaves in the U.S.A., the slave population declined considerably to a level well below two million. Puerto Rico had 47,000 slaves in 1867; Cuba 288,000 slaves in 1871, and Brazil 1,511,000 slaves in 1872. With the final emancipation of slaves in Cuba in 1880 and Brazil in 1888, slavery came to an end in the Americas.

We may compare the American figures with those of the Western Sudan that have been assembled by Martin Klein. Various estimates between 1905 and 1913 put the slave population of Haut-Sénégal-Niger at about 702,000, or 18 per cent of the total population of 3,942,000. The slave population of French Guinea was 490,000, or 34.6 per cent of the total population (1,418,000). For the French Sudan as a whole, there were approximately 1,192,000 slaves in a total population of 5,134,000, but these estimates were made *after* the slave exodus that occurred during and immediately after the French conquest. That exodus reached a climax in 1905–06, by which time hundreds of thousands of slaves had fled. Before the exodus, the slave population was probably in the order of 1.5–2 million.

The extent of slavery in the Sokoto Caliphate was comparable. J. S. Hogendorn and I have calculated that the number of slaves in the Caliphate probably represented one-quarter of the total population of 10 million in 1900. Both the percentage of slaves and the scale of the population are intended as conservative estimates. While there is a slight overlap between Klein's figures and our own, these are not significant. Eight of the thirty emirates in the Caliphate came under French rule, but only two are included in Klein's sample, and they were both small emirates. Whether Hogendorn's and my figures are accepted or not, there can be little doubt that the Sokoto Caliphate may well have been the second or third largest slave society in modern history. Only the United States in 1860 (and maybe Brazil as well) had more slaves than the Caliphate did in 1900.

About a decade after the final emancipation of slaves in the Americas, there were at least twice as many slaves in Islamic West Africa as there had been in Brazil and Cuba in 1870 and at least as many as in the U.S.A. at the start of its Civil War. These comparisons are striking evidence that slavery in Africa has to be taken seriously by historians of both Africa and the Americas. As should be obvious, no attempt is made here to estimate the slave populations of other parts of Africa, particularly areas that fed the Atlantic slave trade, but it is known that the percentage of slaves in Asante, the Yoruba states, the Igbo country and elsewhere was high.

Conclusion

The economic costs of the slave trade in African economies and societies were severe, despite Eltis' interpretation to the contrary. First, the low *per capita* income from the trade indicates that the economic advantages of exporting slaves were nowhere near large enough to offset the social and political costs of participation. Secondly, the size of the trade, including enslavement, related deaths, social dislocation and exports, was sufficient to have had a disastrous demographic impact. Thirdly, because the rise of produce exports started at such a low base and at a time when slave exports were becoming less important, western Africa suffered a relative decline in its position in world trade.

There were other heavy costs which Eltis has failed to appreciate. It is possible to calculate the gross barter terms of trade and *per capita* income from the slave trade and compare western Africa with other parts of the world. But it is difficult to assess the full costs of "producing" slaves because of the nature of enslavement. In an economic sense, as Robert Paul Thomas and Richard Bean have demonstrated, slaves were a "free good," like fish, as far as those doing the enslaving were concerned. There were costs associated with "production," but the real cost in human terms included the loss of life from enslavement, subsequent famines and disease. Furthermore, the destruction of property during wars and raids also represented a loss. Manning's simulation model has attempted to account for some of these costs, although it will never be possible to do the kind of analysis that is possible in measuring the volume and direction of the trans-Atlantic trade itself. Miller has come closest to demonstrating the effects of this impact on a particular region, but his analysis, too, is based on a considerable degree of conjecture.

When the indirect costs of enslavement and trading are taken into consideration, the insights of Eltis and Jennings take on a new meaning. Rather than demonstrating that the Atlantic slave trade had virtually no impact on western Africa, it can be concluded that the impact was in fact strongly negative, although profound.

John Thornton  **NO**

Africa and the Africans in the Making of the Atlantic World

If Africans were experienced traders and were not somehow dominated by European merchants due to European market control or some superiority in manufacturing or trading techniques, then we can say confidently that Africa's commercial relationship with Europe was not unlike international trade any-where in the world of the period. But historians have balked at this conclusion because they believe that the slave trade, which was an important branch of Afro-European commerce from the beginning, should not be viewed as a simple commodity exchange. After all, slaves are also a source of labor, and at least to some extent, removal from Africa represented a major loss to Africa. The sale of slaves must therefore have been harmful to Africa, and African decisions to sell must have been forced or involuntary for one or more reasons.

The idea of the slave trade as a harmful commerce is especially supported by the work of historical demographers. Most who have studied the question of the demographic consequences of the trade have reached broad agreement that the trade was demographically damaging from fairly early period, especially when examined from a local or regional (as opposed to a continental) perspective. In addition to the net demographic drain, which began early in some areas (like Angola), the loss of adult males had potentially damaging impacts on sex ratios, dependency rates, and perhaps the sexual division of labor.

In addition to these demographic effects, historians interested in social and political history have followed Walter Rodney in arguing that the slave trade caused social disruption (such as increasing warfare and related military damage), adversely altered judicial systems, or increased inequality. Moreover, Rodney argued that the slave trade increased the numbers of slaves being held in Africa and intensified their exploitation, a position that Paul Lovejoy, its most recent advocate, calls the "transformation thesis." Because of this per-ception of a widespread negative impact, many scholars have argued that the slave trade, if not other forms of commerce, must have been forced on unwill-ing African participants, perhaps through the type of commercial inequities that we have already discussed or perhaps through some sort of military pres-sure (to be discussed in a subsequent chapter).

When Rodney presented his conclusions on the negative impact and hence special status of the slave trade as a branch of trade, it was quickly contested by J. D. Fage, and more recently, the transformation thesis has been attacked by David Eltis. As these scholars see it, slavery was widespread and indigenous in African society, as was, naturally enough, a commerce in slaves. Europeans simply tapped this existing market, and Africans responded to the increased demand over the centuries by providing more slaves. The demographic impact, although important, was local and difficult to disentangle from losses due to internal wars and slave trading on the domestic African market. In any case, the decision makers who allowed the trade to continue, whether merchants or political leaders, did not personally suffer the larger-scale losses and were able to maintain their operations. Consequently, one need not accept that they were forced into participation against their will or made decisions irrationally.

The evidence for the period before 1680 generally supports this second position. Slavery was widespread in Africa, and its growth and development were largely independent of the Atlantic trade, except that insofar as the Atlantic commerce stimulated internal commerce and development it also led to more widespread holding of slaves. The Atlantic slave trade was the outgrowth of this internal slavery. Its demographic impact, however, even in the early stages was significant, but the people adversely affected by this impact were not the ones making the decisions about participation.

In order to understand this position it is critical to correctly comprehend the place of the institution of slavery in Africa and furthermore to understand why the structure of African societies gave slavery a different meaning than it had in Europe or the colonial Americas. The same analysis explains the reasons for slavery's extension (if indeed it was extended) during the period of the Atlantic trade and its correlation with commercial and economic growth.

Thus, as we will see in this chapter and the next, the slave trade (and the Atlantic trade in general) should not be seen as an "impact" brought in from outside and functioning as some sort of autonomous factor in African history. Instead, it grew out of and was rationalized by the African societies who participated in it and had complete control over it until the slaves were loaded onto European ships for transfer to Atlantic societies.

The reason that slavery was widespread in Africa was not, as some have asserted, because Africa was an economically underdeveloped region in which forced labor had not yet been replaced by free labor. Instead, slavery was rooted in deep-seated legal and institutional structures of African societies, and it functioned quite differently from the way it functioned in European societies.

Slavery was widespread in Atlantic Africa because slaves were the only form of private, revenue-producing property recognized in African law. By contrast, in European legal systems, land was the primary form of private, revenue-producing property, and slavery was relatively minor. Indeed, ownership of land was usually a precondition in Europe to making productive use of slaves, at least in agriculture. Because of this legal feature, slavery was in many ways the functional equivalent of the landlord–tenant relationship in Europe and was perhaps as widespread.

Thus, it was the absence of landed private property—or, to be more precise, it was the corporate ownership of land—that made slavery so pervasive an aspect of African society. Anthropologists have noted this feature among modern Africans, or those living in the so-called ethnographic present or traditional societies. Anthropologists have regarded the absence of private or personal ownership of landed property as unusual, because it departs from the European pattern and from the home cultural experience of most anthropological observers, and has therefore seemed to require an explanation. . . .

At first glance, this corporatist social structure seems to allow no one to acquire sources of income beyond what they could produce by their own labor or trade if they were not granted a revenue assignment by the state. Modern commentators on Africa have occasionally noted this, and precolonial African societies have sometimes been characterized as unprogressive because the overdeveloped role of the state inhibited private initiative by limiting secure wealth. In particular, these commentators believed that the absence of any form of private wealth other than through the state greatly inhibited the growth of capitalism and, ultimately, progress in Africa.

It is precisely here, however, that slavery is so important in Africa, and why it played such a large role there. If Africans did not have private ownership of one factor of production (land), they could still own another, labor (the third factor, capital, was relatively unimportant before the Industrial Revolution). Private ownership of labor therefore provided the African entrepreneur with secure and reproducing wealth. This ownership or control over labor might be developed through the lineage, where junior members were subordinate to the senior members, though this is less visible in older documentation.

Another important institution of dependency was marriage, where wives were generally subordinate to their husbands. Sometimes women might be used on a large scale as a labor force. For example, in Warri, Bonaventura da Firenze noted in 1656 that the ruler had a substantial harem of wives who produced cloth for sale. Similarly, the king of Whydah's wives, reputed to number over a thousand, were employed constantly in making a special cloth that was exported. Such examples give weight to the often-repeated assertion that African wealth was measured in wives, in the sense both that polygamy was indicative of prestige and that such wives were often labor forces.

Of course, the concept of ownership of labor also constituted slavery, and slavery was possibly the most important avenue for private, reproducing wealth available to Africans. Therefore, it is hardly surprising that it should be so widespread and, moreover, be a good indicator of the most dynamic segments of African society, where private initiate was operating most freely.

The significance of African slavery can be understood by comparing it briefly with slavery in Europe. Both societies possessed the institution, and both tended to define slaves in the same way—as subordinate family members, in some ways equivalent to permanent children. This is precisely how slaves are dealt with in *Siete partidas,* following a precedent that goes all the way back to Aristotle, if not before. Modern research clearly reveals that this is also how Africans defined slavery in the late precolonial and early colonial period.

Seventeenth-century African data do not deal with the legal technicalities, though we have little reason to believe that they differed from those uncovered by modern anthropological research. For Kongo, where the remarkable documentation allows glimpses of the underlying ideology, the term for a slave, *nleke,* was the same as for a child, suggesting the family idiom prevailed there.

Where the differences can be found is not in the legal technicalities but in the way slaves were used. In theory, there have been no differences in this respect either, but in practice, African slaves served in a much wider variety of ways than did European or Euro-American slaves. In Europe, if people acquired some wealth that they wished to invest in secure, reproducing form, they were likely to buy land. Of course, land did not produce wealth by itself, but usually the land was let out to tenants in exchange for rents or was worked under the owner's supervision by hired workers. In neither case would such people have to have recourse to slaves to acquire a work force.

From what we know of slave labor in Europe in this period, it would appear that they were employed in work for which no hired worker or tenant could be found or at least was willing to undertake the work under the conditions that the landowner wished. As we shall see, this lay behind most of the employment of slaves in the New World as well. Consequently, slaves typically had difficult, demanding, and degrading work, and they were often mistreated by exploitative masters who were anxious to maximize profits. Even in the case of slaves with apparently good jobs, such as domestic servants, often the institution allowed highly talented or unusual persons to be retained at a lower cost than free people of similar qualifications.

This was not necessarily the case in Africa, however. People wishing to invest wealth in reproducing form could not buy land, for there was no landed property. Hence, their only recourse was to purchase slaves, which as their personal property could be inherited and could generate wealth for them. They would have no trouble in obtaining land to put these slaves into agricultural production, for African law made land available to whoever would cultivate it, free or slave, as long as no previous cultivator was actively using it.

Consequently, African slaves were often treated no differently from peasant cultivators, as indeed they were the functional equivalent of free tenants and hired workers in Europe. This situation, the result of the institutional differences between Europe and Africa, has given rise to the idea that African slaves were well treated, or at least better treated than European slaves. Giacinto Brugiotti da Vetralla described slaves in central Africa as "slaves in name only" by virtue of their relative freedom and the wide variety of employments to which they were put. Likewise, as we shall see, slaves were often employed as administrators, soldiers, and even royal advisors, thus enjoying great freedom of movement and elite lifestyles.

This did not mean, of course, that slaves never received the same sort of difficult, dangerous, or degrading work that slaves in Europe might have done, although in Africa often such work might just as easily have been done by free people doing labor service for the state. In any case, Valentim Fernandes's description of slave labor in Senegambia around 1500, one of the few explicit texts on the nature of slave labor, shows that slaves working in agricultural production worked one day a week for their own account and the rest for their master, a work regime that was identical for slaves serving in Portuguese sugar mills on the island colony of São Tomé in the same period. Slaves employed in mining in Africa may have suffered under conditions similar to those of slaves in European mining operations, though the evidence is less certain.

On the whole, however, African slavery need not have been degrading or the labor performed by slaves done under any more coercion (or involving any more resistance) than that of free laborers or tenants in Europe. Therefore, the idea that African dependence on slave labor led to the development of a reluctant work force or inhibited innovation is probably overdone.

For Europe and the European colonies in America, the distinction between the productivity of slave and free labor may have validity (though even there it is a matter of intense debate); in Africa the distinction is probably less applicable. The exact nature of the labor regime, rather than the legal status of the workers, is more relevant to a description of African economic history, and in this instance, different legal structures led Africans and Europeans to develop the institution of slavery in substantially different ways. Consequently, the conventional wisdom concerning slavery developed from the study of European or colonial American societies with landed private property simply cannot apply in Africa.

African slaves were typically used in two different ways. First of all, slaves became the preeminent form of private investment and the manifestation of private wealth—a secure form of reproducing wealth equivalent to landowning in Europe. Second, slaves were used by state officials as a dependent and loyal group, both for the production of revenue and for performing administrative and military service in the struggle between kinds or executives who wished to centralize their states and other elite parties who sought to control royal absolutism.

The private employment of slaves as heritable, wealth-producing dependents was perhaps the most striking African use. Dapper, in describing private wealth in Kongo, noted that although the households of the nobility were not wealthy in ready cash, nor did they possess much in the way of luxury goods, they were wealthy in slaves. This, he believed, was the main form of wealth in central Africa.

Likewise, slaves represented the way to achieve wealth for ambitious commoners in the Gold Coast states, and the state did attempt to regulate their acquisition. According to de Marees, a commoner who had become wealthy through trade might be able to attain noble status by sponsoring an expensive ceremony in which nobility was conferred upon him. Although the ceremony was ruinously expensive, the noble-to-be was willing to undertake it because it allowed him to acquire slaves, which, as Dapper noted a few years later, would make it possible for him to recover the expenses of the ceremony, for "as soon as he gets some goods he bestows them on slaves, for that is what their wealth consists of."

Recently, several historians have followed the careers of some prominent Gold Coast merchants who rose from relative obscurity to become great economic and political actors on the coast, using documentation from the records of the Dutch, English, and Danish commercial houses. In all these accounts, the acquisition of slaves to carry goods, cultivate lands, protect the household, and assist in trading figures prominently as an essential step. Indeed, the careers can in some ways be seen as parallel to that of the European commoners who invested first in land and then in titles of nobility, though in Africa, of course, the investment was in slaves and then in nobility.

Slaves as reproducing wealth figured prominently among the Julas and other Moslem commercial groups of the western Sudan and Senegambia. Richard Jobson, the English gold trader who spent considerable time traveling up the Gambia deep into the Sudan in the 1620s, noted that the Julas ("Juliettos") had constructed a chain of villages worked by their slaves, who provided them with provisions and served as carriers on their commercial expeditions. The heads of these villages acquired special rights (in some ways equivalent to those of nobility) from the rulers of the state in which they settled. Philip Curtin's detailed study of the Julas and other Moslem commercial groups in the late seventeenth and early eighteenth centuries emphasizes their extent and organization. Yves Person, focusing on a later period, compared them to the French bourgeoisie rising from common to noble status or seizing power if thwarted: hence he compared a series of "Dyula Revolutions" from the late eighteenth century onward to the French Revolution.

Thus, in Africa the development of commerce and the social mobility based on commerce was intimately linked to the growth of slavery, for slaves in villages performing agricultural work or carrying goods in caravans or working in mines under private supervision were essential to private commercial development.

This last point is significant to consider in transformation thesis. Both Rodney and Lovejoy, advocates of the idea that the development of the Atlantic slave trade extended slavery and resulted in larger numbers of people being enslaved and being worse treated, see this as a direct external input, foreign to African political economy. Yet, the development and extension of slavery, if it did take place (and this point is never proved by either author), might just as well be seen as the result of economic growth in Africa, perhaps stimulated by commercial opportunities from overseas, perhaps by a growing domestic economy. Even the increased incidence of maltreatment (another point that is completely without proof) may indicate only more aggressive use of the labor force by entrepreneurs, just as the European work force faced increased exploitation during the early stages of the Industrial Revolution.

The use of slaves by private people to increase and maintain their wealth was just one of the ways in which slaves were utilized in African societies. Another one, almost of equal importance, was their use by the political elite to increase their power. Slaves employed by the political elite might be used as a form of wealth-generating property, just as they were in private hands, or they might be used to create dependent administrations or armies. In this latter capacity, Africa created many wealthy and powerful slaves.

Most large African states were collections of smaller ones that had been joined through alliance and conquest, and typically the rulers of these smaller constituent states continued to exercise local authority, and the ruler of the large state found his power checked by them. Developing private resources that would answer only to themselves was an important way in which African rulers could overcome such checks and create hierarchical authority centered on their own thrones. Slaves, who could be the private property of a king or his family or might also be the property of the state, were an ideal form of loyal workers, soldiers, and retainers.

The powerful Sudanese empires relied heavily on slave armies and slave administrators to keep a fractious and locally ascendant nobility in check. These nobles were often descendants of the rulers of the constituent smaller states; this was probably the status of the territorial rulers of Mali in al-'Umaris's fourteenth-century description. These constituent states were called *civitas* by Antonio Malafante, Genonese traveler who left a description of the empire of Songhay and its neighbors in 1477, a term that implies both subordination and self-government in Latin. An anonymous description of the "Empire of Great Fulo" written about 1600 states that it dominated the whole Senegal valley and was composed of some twenty smaller units. In accounts of the western provinces of Mali during the late fifteenth and early sixteenth centuries, Portuguese travelers describe local "kings" (heads of constituent states) as virtually sovereign in their local rights and government, yet simultaneously describe Mali as a powerful overlord. These descriptions and later ones of Mali and Kaabu, a state that based its authority on being a province of Mali, all reveal apparent local sovereignty coexisting with the apparently strong rights of the overlord, who at least extracted tribute and obedience and might even intervene in local affairs.

One can observe the same with regard to a somewhat shadowy kingdom of "Kquoja" that dominated Sierra Leone from a capital near Cape Mount in the late sixteenth and early seventeenth centuries. Although Alvares de Almada noted that the Kquoja kings collected regular tribute and taxes from local rulers, historians have generally not seen it as a unitary state.

In some cases, perhaps including both Kaabu and sixteenth-century Mali, the strength of local states did cut in on revenue and authority exercised by their overlords, but slaves often offered a way around this Alvise da Mosto's description of Jolof in the mid-fifteenth century provides a good example. Here, according to da Mosto, the king was beholden to three or four other powerful nobles, each of whom controlled a region (clearly the constituent states), gave him revenue when they chose, and moreover exercised the right to elect him. But the king was able to obtain revenue of his own by distributing slaves in villages to each of his several wives; this income belonged to him. Not only did this give him independent support, but it allowed him to develop a large retinue of dependents who carried out his administrative tasks, numbering some 200 people in all. Unfortunately, the fact that at least one of his subordinates, "Budomel" (title of the ruler of Kajorr [Kayor]), was doing the same thing at the local level may well have ultimately limited his capacity to develop more central power.

Sixteenth- and seventeenth-century evidence from the *Tarikh al-Fettash*, a locally composed source on Songhay, shows quite clearly how the development of an army and administration of slaves helped that empire to become centralized. Tymowski has analyzed this text and showed that rice plantations worked by slaves, as well as villages of slaves settled throughout the country, supported an army of slaves and a bureaucracy of slaves through which the emperors conducted their business, neglecting whatever obligations they may have had to the local nobles.

Slavery probably also aided centralizing monarchs in central Africa as well as West Africa. Kongo seems to have originally been a federation of states, at least as sixteenth- and seventeenth-century tradition and law described it. The original kings of the federation owed their election to the votes of several electors, who were the heads of the member states. But collecting slaves into a central place gave the Kongo kings great power—the capital city of Mbanza Kongo and its surrounding area formed a great agricultural center already in 1491, and probably had ten times the population density of rural areas a century later. The slaves, many of whom occupied estates around the capital, provided Kongo with both the wealth and the demographic resources to centralize. As early as 1526, documents from Kongo show that the provinces (constituent states) were in the hands of royally appointed people (mostly kinsmen), and by the mid-seventeenth century local power and election were regarded more as a curse than a blessing.

Slavery played a role in the centralization of nearby Ndongo as well. Like Kongo, Ndongo rulers may have benefited from the concentration of slaves in their capital—for Kabasa, Ndongo's capital, was also described as a large town in a densely populated area. In addition, the ruler had villages of slaves who paid revenue to him scattered around in his domain. These villages were called *kijiko* (which actually means "slave" in Kimbundu), which a document of 1612 rendered as "populated places whose residents are slaves of the said king." Perhaps more significant in Ndongo, however, was the use that the kings made of slaves as administrators, for the ruler had officials, the *tendala* and the *ngolambole* (judicial and military officials, respectively), who supervised subordinates and collected tax and tribute from his slaves.

We have already seen that African rulers were sometimes limited in the amount of absolute power they could exercise. Some societies had rules of election that allowed officials to choose a weak ruler. In some of the smaller states, the use of slaves may have helped rulers develop more autocratic systems of government. In dealing with the states of the eastern Gold Coast, for example, Dapper noted that they were all quite strongly centralized and, moreover, had an abundance of slaves. Similarly, Alvaro Velho noted that among the smaller states of Sierra Leone, the income that rulers obtained from their slaves was their only steady source of income.

Thus slaves could be found in all parts of Atlantic Africa, performing all sorts of duties. When Europeans came to Africa and offered to buy slaves, it is hardly surprising that they were almost immediately accepted. Not only were slaves found widely in Africa, but the area had a well-developed slave trade, as evidenced by the numbers of slaves in private hands. Anyone who had the wherewithal could obtain slaves from the domestic market, though sometimes it required royal or state permission, as in the Gold Coast. Europeans could tap this market just as any African could.

Moreover, the most likely owners of slaves—wealthy merchants and state officials or rulers—were exactly the people with whom European traders came into contact. Because merchants selling gold, ivory products, mats, copper bracelets, pepper, or any other trade commodity in Africa would also be interested in the buying and selling of slaves, European merchants could readily find sources.

This was not so much because Africans were inveterate slave dealers, as it was because the legal basis for wealth in Africa lay in the idea of transferring ownership of people. This legal structure made slavery and slave marketing widespread and created secondary legal mechanisms for securing and regulating the sale of slaves, which Europeans could use as well as Africans.

The significance of African slavery in the development of the slave trade can be clearly seen in the remarkable speed with which the continent began exporting slaves. As soon as the Portuguese had reached the Senegal region and abandoned their early strategy of raiding for commerce 700–1,000 slaves were exported per year, first with caravans bound for the Sahara (after 1448). After Diogo Gomes's diplomatic mission to the West African rulers in 1456, which opened markets north of the Gambia, exports took a dramatic turn upward, reaching as many as 1,200–2,500 slaves per year by the end of the century.

Thus, from 1450 onward, even before their ships actually reached the Senegal River, Portuguese merchants were buying slaves from northward-bound caravans from the post at Arguim, tapping a long-standing trans-Sahara trade. It is not surprising that Avelino Teixeira da Mota has been able to document the diversion of the Saharan slave trade from North Africa to the Atlantic coast in the same period. The reason that such dramatic numbers were reached immediately may indicate nothing more, therefore, than that a preexisting engagement with foreign markets was transferred to Atlantic ones. Most of the early European slave trading with West Africa, even that with such relatively remote regions as Benin and the Niger delta, known in the sixteenth century as the "River of Slaves," was simply an internal trade diverted to the Atlantic. Pacheco Pereira mentioned that the country of "Opuu," probably the Jukun kingdom on the Benue River, was a major source of slaves for the region.

The slave trade of the Benin coast shows another interesting aspect of African slavery and the export slave trade. The Saharan trade was mainly an export trade, but it also involved some internal trade. This is demonstrated by the fact that the Portuguese resold a large number of the Benin coast slaves to the Gold Coast. We know that such slaves were not simply used in the coastal mines (though we can be sure that many were) because the king of Portugal ordered this trade to cease (unsuccessfully, as it turns out) to prevent them from being sold to Moslems. These Moslems had to be northern Jula merchants who also visited the coastal goldfields, and thus these slaves may well have been employed in goldfields located quite far in the interior.

That existing internal use and commerce in slaves lay behind the export trade is even more strongly suggested by the trade of central Africa. Unlike the West African trade, which drew on an ancient slave trade with North Africa and might thus have already been affected by external contacts, the central African region had no such external links. Nevertheless, the king of Portugal regarded Kongo as sufficiently important a potential exporter of slaves that he granted settlers in São Tomé privileges to engage in the slave trade in 1493, just a few years after the development of official trade there. Kongo indeed became an important source of slaves for the Tomistas by 1502. Unfortunately we possess no early statistics for the volume of this trade, but Valentim Fernandes noted that around 1507, in addition to some 2,000 slaves working on sugar plantations, the island held 5,000–6,000 slaves awaiting reexport. Presumably these slaves were recent imports who had probably arrived within the last year, and certainly half, but probably the majority, originated in central Africa. When the books of the royal factor on the island were inspected by Bernardo da Segura in 1516, they showed annual imports, mostly from Kongo, of nearly 4,500 slaves.

Slaves from central Africa were so numerous that they soon exceeded the capacity of São Tomé and the Mina trade to absorb them, and so they began the long journey to European markets. Although most of the slaves available in the port towns of Lisbon, Balencia, and Seville in the 1470s and 1480s came from western West Africa, Jolof in particular, by 1512 "Manicongos" were arriving in Seville, and Portuguese reports of 1513 mention a whole ship from Kongo making delivery in Europe.

Thus, at some point, probably with twenty years of first contact, central Africa was able to supply exports of slaves equal to the entire exports of West Africa. Clearly this sort of volume could not simply have been the occasional export of odd misfits. Nor have we any reason to believe that the Portuguese were able to either acquire the slaves themselves (except as clients of the Kongo kings) or force the Kongo to obtain the export slaves against their will. Instead, the growth of Kongo's trade had to draw on a well-developed system of slavery, slave marketing, and slave delivery that preexisted any European contact.

We must therefore conclude that the Atlantic slave trade and African participation in it had solid origins in African societies and legal systems. The institution of slavery was widespread in Africa and accepted in all the exporting regions, and the capture, purchase, transport, and sale of slaves was a regular feature of African society. This preexisting social arrangement was thus as much responsible as any external force for the development of the Atlantic slave trade.

POSTSCRIPT

Did the Trans-Atlantic Slave Trade Underdevelop Africa?

Today, John Thornton is not alone in his assertion that African elites had considerable influence on the continent's relationship with Europeans during the slave trading era. David Eltis (see *The Rise of African Slavery in the Americas,* Cambridge University Press, 2000) and Herbert Klein (see *The Atlantic Slave Trade,* Cambridge University Press, 1999) also have arrived at similar conclusions. Nonetheless, John Thornton's views remain quite controversial. While few question his motives or the quality of his scholarship, many disagree with his conclusions. It is significant that Thornton ends his analysis before the advent of the eighteenth century at which point important changes in global trade and technology drastically transform the relationship between Africa and Europe. As Patrick Manning notes in a 1993 review of Thorton's book in the *American Historical Review,* "I question whether African institutions of slavery, and especially of enslavement, could have been as extensive in the sixteenth century, when some five thousand slaves were exported annually, as in the eighteenth century, when ten times that many slaves left Africa each year."

Thornton's text, while controversial, has become a departure point for new scholars who have been inspired to put his assertions to the test. One example of this is Walter Hawthorne's book *Planting Rice and Harvesting Slaves: Transformations Along the Guinea-Bissau Coast, 1400–1900* (Heinemann, 2003). Here, Hawthorne examines the Balanta of the Upper Guinea Coast and demonstrates that those living outside of more centralized African states were not mere victims of enslavement. In response to outside pressure, the Balanta developed a new rice production system and often found ways to produce slaves themselves.

A final point worth touching on is Thornton's argument that slavery developed in Africa prior to European arrival because African institutions did not permit private ownership of land, but did permit ownership of labor. It is true that labor has generally been a relatively scarce factor in African farming systems and control of labor certainly would have been a key factor in amassing wealth. Labor control in this context, however, does not necessarily correlate to the type of slavery that evolved alongside capitalism in the Atlantic world. For a good discussion of the latter phenomenon, see Eric Williams' *Capitalism and Slavery* (The University of North Carolina Press, 1994).

For an example of Paul Lovejoy's more recent work on this issue, see his book *Transformations in Slavery: A History of Slavery in Africa* (Cambridge University Press, 2000). While he makes many of the same arguments that he made in his selection for this issue, his argument is bolstered by new evidence on the numbers of slaves traded.

ISSUE 2

Have the Contributions of Africans Been Recognized for Developing New World Agriculture?

YES: Duncan Clinch Heyward, from *Seed from Madagascar* (University of North Carolina Press, 1937)

NO: Judith Carney, from "Agroenvironments and Slave Strategies in the Diffusion of Rice Culture to the Americas," in Karl S. Zimmerer and Thomas J. Bassett, eds., *Political Ecology: An Integrative Approach to Geography and Environment-Development Studies* (Guilford Press, 2003)

ISSUE SUMMARY

YES: Duncan Clinch Heyward, a former Carolina rice planter writing in the middle of the last century, represents the mainstream view that Europeans were primarily responsible for developing South Carolina's remarkable rice plantations in the eighteenth century. In his own accounting of the rise of rice cultivation in the Carolinas, Duncan suggests that the techniques and approaches must have been derived from those observed in China.

NO: Judith Carney, a professor of geography at UCLA, explains that slaves from rice-producing areas in West Africa have only recently been recognized for their intellectual contributions to the development of rice cultivation in the New World. Carney describes how her work, and that of others, challenged the view that slaves were mere field hands, "showing that they contributed agronomic expertise as well as skilled labor to the emergent plantation economy."

$\mathbf{A}$s Judith Carney describes in her selection for this issue, it had long been thought that African slaves contributed little more than labor to the New World's burgeoning plantation economies of the seventeenth, eighteenth, and nineteenth centuries. But the pioneering work of Wood (*Black Majority: Negroes in Colonial South Carolina from 1670 Through the Stone Rebellion,* Knopf, 1974), Littlefield (*Rice and Slaves: Ethnicity and the Slave Trade in Colonial South Carolina,* Louisiana State University Press, 1981) and Carney (*Black Rice: The African Origins of Rice Cultivation in the Americas,* Harvard University Press, 2001) demonstrated Africans' knowledge of rice production was critical to the success of New World agriculture.

The aforementioned studies also form part of a growing literature on the Atlantic World, a global region centered on the Atlantic basin. Scholars working in this context often focus on the Atlantic Ocean as a conduit for the transmission of people and ideas. Other than the titles just mentioned, other examples of scholarship in this arena include: H. Taylor's *Circling Dixie: Contemporary Southern Culture Through a Transatlantic Lens* (Rutgers University Press, 2001) and Carney and Voeks' 2003 article in *Progress in Human Geography,* entitled "Landscape Legacies of the African Diaspora in Brazil."

In this issue, Judith Carney, a professor of geography at UCLA, explains that slaves from rice-producing areas in West Africa have only recently been recognized for their intellectual contributions to the development of rice cultivation in the New World. Carney describes how her work, and that of others, challenged the view that slaves were mere field hands, "showing that they contributed agronomic expertise as well as skilled labor to the emergent plantation economy."

Carney largely was reacting to an older body of literature that celebrated the contributions of southern planters in establishing successful rice plantations in the Carolinas in the eighteenth century. The selection by Duncan Heyward, a former Carolina rice planter writing in the middle of the last century, is an example of one such tract. According to Heyward, British functionaries identified rice as a suitable crop for the Carolinas and had rice seed sent to the colony. Heyward further suggested that the techniques and approaches for such cultivation must have been observed in China and brought to the Carolinas via European settlers. Those familiar with the Reconstruction and post-Reconstruction periods in U.S. history would not be errant in associating the work of Heyward and others with a larger body of Southern scholarship that tended to glorify the antebellum period. Very interestingly, however, there are a few cases of more contemporary works that highlight the contributions of Southern rice planters. One recent example is Richard Schulze's *Carolina Gold Rice: The Ebb and Flow History of a Low-Country Cash Crop* (History Press, 2005). The author is the descendant of a former rice plantation owner and is reviving Carolina Gold rice as an heirloom variety.

Carolina Gold Rice

Often during my years as a planter, when the rice industry on our South Atlantic coast was rapidly being abandoned, I have sat under a great cypress, growing on my river bank, and, looking across the broad expanse of my rice fields, have thought of their strange and remarkable history. There would come to my mind the great and fundamental changes, racial and social as well as economic, which have taken place in a short space of time. And I have wished that the old tree above me could tell the tragic story of those fields, recounting events of which it had been a silent witness.

If the tree could only have spoken, I know its story would have begun at a time when the swamp, on whose edge it grew, was the favorite hunting ground of the red man. It would then have told of the coming of the white man, who drove the red man far away and took from him his lands. Next it would have told how the black man came, brought from far across the sea, how he felled the trees in the swamps and cleared them of their dense undergrowth, letting the sunshine in; then, how he drained the lowlands and grew crops of golden grain; and finally it would have told of the emancipation of the black man, who, after years of servitude, worked on faithfully as a freedman.

The rest of the story of my rice fields would for me have needed no telling. It would have dealt with the years when the white man was compelled, by conditions beyond his control, to give up planting, and the black man moved away seeking employment elsewhere, leaving fertile lands, the only naturally irrigated ones in this country, to revert to their former state.

The story of our former rice fields begins more than two centuries ago.

Carolina Gold rice, world renowned because of its superior quality as compared with all other varieties of rice throughout the world, was grown from seed brought to the province of Carolina about the year 1685. This rice had been raised in Madagascar, and a brigantine sailing from that distant island happened, in distress, to put into the port of Charles Town. While his vessel was being repaired, its captain, John Thurber, made the acquaintance of some of the leading citizens of that town. Among them was Dr. Henry Woodward, probably its best known citizen, for he had the distinction of being the first English settler in the province. He had accompanied Sandford on his first exploring expedition and had volunteered to remain alone at Port Royal in order that he might study the language of the Indians and familiarize himself with the country, in the interests of the Lords Proprietors.

To Woodward Captain Thurber gave a small quantity of rice—less, we are told, than a bushel—which happened to be on his ship. "The gentleman of the name of Woodward," to quote the earliest account of this occurrence, "himself planted some of it, and gave some to a few of his friends to plant."

Thus it came about that I, on the distaff side a descendant of Dr. Woodward, seem to have been destined to spend the best years of my life in seeking to revive an industry in the pursuit of which four generations of my family had been successful.

Three states bordering on the South Atlantic, North Carolina, South Carolina, and Georgia, are the only states in this country where for upwards of two centuries rice was grown. Nearly all of it, however, was produced in South Carolina and Georgia, partly because of a slightly warmer climate, but mainly because of the numerous tidal rivers which flow through these states and empty into the ocean. Along the Cape Fear River in North Carolina, it must be admitted, the finest quality of rice was grown, and for many years seed raised there was sold to the planters farther south, in order to preserve the quality of their rice.

The principal rivers in South Carolina, along which rice was planted, were the Waccamaw, the PeeDee, the Santee, the Cooper, the Edisto, and the Combahee. There were also many large rice plantations on the Savannah River, which separates South Carolina and Georgia. Farther south in Georgia were the Ogeechee, the Altamaha, and the Satilla rivers, the last near the Florida line. Some of these rivers are long, having their sources in the mountains, while others are much shorter. All of them are affected for a number of miles by the rise and fall of the tide, the result being that the fresh water they contain is backed up in the rivers and then drawn down again as the water in the ocean rises and falls. Great salt-water marshes lie on either side of the rivers as they approach the ocean, while higher up they were originally bordered by dense cypress, gum, and cedar swamps where the water was fresh, though rising and falling with the tide. It was in these fresh-water swamps that rice was successfully grown for the longest period of years.

As early as August 31, 1663, Lord Albemarle, one of the Lords Proprietors, wrote to the Governor of the Barbadoes, advising the planters there to settle in the proposed province of Carolina. Among the inducements offered he suggested the growing of rice. In his letter he said, "The commodyties I meane are wine, oyle, reasons, currents, rice, silk, etc." Of these, rice alone was destined to be successfully produced over a long period of years.

The plans of the Lords Proprietors regarding the growing of rice in the new province took practical shape in 1672, for in that year they had a barrel of rice sent to Charles Town in a vessel named the "William and Ralph;" and one must assume that this rice was intended for seed. By 1690 some headway had been made in the growing of rice, for the leading men of the province petitioned Governor Sothell to arrange with the Lords Proprietors that the people be allowed to "pay their quit rents in the most valuable and merchantable produce of their land," among which they included rice. These products, they reported, were "naturally produced here."

Also during this period the General Assembly of the province had ratified acts to protect those who should perfect labor-saving machinery for the purpose of husking and cleaning rice. A few years later the Assembly protested against an export duty on rice. How different was this attitude of the early rice planters from that of the planters of my day! We wanted an import duty. Though Southerners, we favored a tariff on rice, for we sorely needed protection. However, we never went so far as to claim that ours was an "infant industry."

By the year 1700 there was being produced in the province more rice "than we had ships to transport," according to the Governor and Council. Edward Randolph, collector of customs for the Southern department of North America that year, wrote: "They have now found out the true way of raising and husking rice. There has been above three hundred tons shipped this year to England besides about thirty tons to the Islands." This progress must have caused the Lords Proprietors to be exceedingly optimistic as to the future of rice cultivation in Carolina, for they congratulated themselves upon "what a staple the province of Carolina may be capable of furnishing Europe withall," and added that "the grocers do assure us it's better than any foreign rice by at least 8s the hundred weight."

At any rate, for the praise so soon bestowed by the grocers of London on the quality of the rice exported from the province of Carolina, and for the demand from abroad for this variety of rice, and also for the success which for upwards of two centuries attended the growing of rice in South Carolina and Georgia, we are principally indebted to Captain Thurber. Had it not been for him, the once celebrated Carolina Gold rice probably would never have been planted in America.

Another variety of rice later planted in South Carolina was known as Carolina White rice. This rice made a beautiful sample when prepared for market, and could scarcely be distinguished from the Gold rice, but its tendency to shatter when being harvested, if slightly over-ripe, caused it to be planted only to a limited extent.

It has never been known definitely from what country this Carolina White rice was first imported, but many of our rice planters believed it had been brought from China, where for unnumbered centuries rice has been grown and where there are today numerous varieties. Were I to hazard a guess as to who was responsible for bringing this white rice to the province, I should name Robert Rowand, born in Glasgow in 1738, who, when a boy, came to Charles Town. He later purchased an inland swamp plantation, on Rantowles Creek, about twenty-five miles southwest of Charles Town. He must have succeeded as a planter, for seemingly he was a man of means and traveled extensively. There is every reason to believe that he visited China, and that, while there, he had certain attractive pictures painted, illustrating the way the Chinese planted rice in those days, and showing the implements used in the process.

A few years ago in the town of Summerville, South Carolina, at the home of a friend, the late Mr. S. Lewis Simons, I saw these pictures. (They have since been destroyed by fire.) There were quite a number of them, and I am sure there are no others like them anywhere in this country. In size, they were about eighteen by fourteen inches, exceedingly well executed in bright water colors. They showed in great detail the growing of rice in China, from the preparation of the ground to the gathering and pounding of the grain. The pictures were evidently painted by a Chinese artist. The figures of men and women, the landscape, with its green trees here and there, blue mountains and hills rising out of the level plains intersected by canals, were entirely Oriental. On the front of the cover was written, "Painted in China prior to the Revolutionary War for one of the first South Carolina Rice Planters, Illustrating the Chinese Method of Cultivating Rice." The rice planter referred to was undoubtedly Robert Rowand, and the pictures had come into the possession of Mr. Simons through his wife, a Miss Mayrant, the Mayrant family having been rice planters for several generations, Mrs. Simons was a lineal descendant of Robert Rowand on her maternal side through the Drayton family.

The first of these Chinese pictures shows the plowing and harrowing of the soil of the seed-beds, both processes being done under water, the Chinaman and his black "water buffalo" nearly up to their knees, and the latter looking as if he did not trust his footing and was anxious to turn back. Then follows the sowing of the seed broadcast on the water, the transplanting of the rice by hand in the fields, the cultivating of the growing crop, the harvesting, until finally the rice is carried to the barnyard, where it is threshed and pounded. Anyone at all familiar with the methods used by the early planters in South Carolina cannot fail to be struck by their similarity to the methods shown in the old Chinese pictures, and especially is this true of the implements used. With the exception of sowing the rice in the water and transplanting it, nearly every picture recalled our system of planting. Many of the implements were almost identical with ours. There were the flail-sticks, being used in exactly the same way; the sickle, only a little straighter than ours, with which the rice was cut; the mortar and pestle and hand-fans; the "boards," as the Negroes used to call them, with which, indoors on a floor, the rice was pushed from place to place; and also the baskets in which the rice was carried.

These paintings of Robert Rowand's convinced me that our early methods of rice culture were adopted largely from the Chinese. For if, instead of the Chinese settings of the pictures, if in the place of the men with queues and black slanting eyes, dressed in bright-colored costumes, there could have been substituted the scenery of our rice fields and the Low Country Negroes at work, I could readily have believed the scenes were laid in our Black Border instead of in that far Eastern land.

Judith Carney  **NO**

Agroenvironments and Slave Strategies in the Diffusion of Rice Culture to the Americas

By the mid-1700s a distinct cultivation system, based on rice, shouldered the American and African Atlantic. One locus of rice cultivation extended inland from West Africas Upper Guinea Coast, another flourished along the coastal plain of South Carolina and Georgia, and a third developed in the corridor between Brazil's Northeast and the eastern Amazon region (Figure 1). This tidal rice cultivation system depended upon enslaved African labor. In West Africa farmers planted rice as a subsistence crop on small holdings, with surpluses occasionally marketed, while in South Carolina, Georgia, and Brazil, cultivation depended on a plantation system and West African slaves to produce a crop destined for international markets.

While rice cultivation continues in West Africa today, its demise in South Carolina and Georgia swiftly followed the abolition of slavery. Brazil's experiment in tidal rice cultivation, modeled after that of Carolina, did not withstand its competition in the 19th century. Yet the U.S. South's most lucrative plantation economy continued to inspire nostalgia well into the 20th century when the crop and the princely fortunes it delivered remained no more than a vestige of the coastal landscape. Numerous commentaries documented the lifeways of European American planters, their achievements, and their ingenuity in shaping a profitable landscape from malarial swamps. These accounts have never presented African Americans as having contributed anything but their unskilled labor. The planter-biased rendition of the origins of American rice cultivation prevailed until 1974 when the historian Peter Wood carefully examined the role of slaves in the Carolina plantation system during the colonial period. His scholarship recast the prevalent view of slaves as mere field hands to one that showed that they contributed agronomic expertise as well as skilled labor to the emergent plantation economy. Littlefield built upon Wood's path-breaking thesis by discussing the antiquity of African rice-farming practices and by revealing that more than 40% of South Carolina's slaves during the colonial period originated in West Africa's rice cultivation zone.

While this research has resulted in a revised view of the rice plantation economy as a fusion of both European and African cultures, the agency of

From *Political Ecology: An Integrative Approach to Geography and Environment-Development Studies* Karl S. Zimmerer and Thomas J. Bassett, eds. (Guilford Publications, 2003). Copyright © 2003 by Guilford Publications. Adapted from *Political Geography,* vol. 31, no. 1, 1998, Johns Hopkins University Press. Reprinted by permission.

Figure 1

Rice cultivation along the Atlantic Basin, 1760–1860

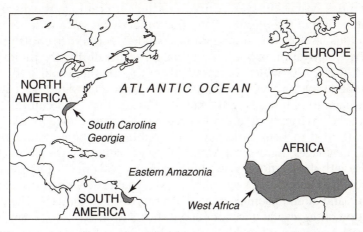

African slaves in its evolution is still debated. Current formulations question whether planters recruited slaves from West Africa's rice coast to help them develop a crop whose potential they independently discovered, or whether African-born slaves initiated rice planting in South Carolina by teaching planters to grow a preferred food crop. The absence of archival materials that would document a tutorial role for African slaves is not surprising given the paucity of records available in general for the early colonial period, and because racism over time institutionalized white denial of the intellectual capacity of bondsmen. An understanding of the potential role of slaves demands other forms of historical enquiry.

This study is situated in a growing trend in scholarship that integrates detailed ethnographic and ecological investigation, particularly of agroenvironments, with social and environmental history. Such an integration invites the use of multiple and diverse tools, including analysis of archival materials and historical documents (e.g., travelers' narratives, colonial accounts, maps), oral histories, agroecological methods, and ethnographic inquiry. The following discussion of West African rice-farming technology and culture is complemented by extensive field-work conducted on rice systems in Senegambia by the author, by Olga R. Linares in Senegal, and by Paul Richards in Sierra Leone. Similarities between today's West African rice culture and that of the antebellum U.S. South do not prove in themselves the case for rice technology transfer by African slaves. Field-based geographical studies, however, can substantiate and elaborate upon often-sparse observations in the archival record; provide theoretical frameworks, inspired by political ecology, for understanding past human–environment relations; and produce a richer portrait of past landscapes.

This [selection] combines geographical and historical perspectives to examine the likely contributions of African-born slaves to the colonial rice economy. The approach identifies and describes the principal West African microenvironments planted to rice in the first section as the basis for examining, in the one following, the systems that emerged in South Carolina during the colonial period. While archival documentation, albeit fragmentary, exists on rice systems in West Africa from the 14th century, the discussion in the first section of these systems and their underlying soil and water management principles is based on modern field studies. Focus then shifts to the history of rice cultivation in South Carolina, especially during the hundred years from 1670 to 1770, which is crucial since it spans the initial settlement by planters and slaves as well as the expansion of tidal (tidewater) rice cultivation into Georgia. In emphasizing the complex nexus that links culture, technology, and the environment, attention is directed to the indigenous knowledge systems formed in West Africa by ethnic groups speaking Mande and West Atlantic languages. Across the Middle Passage of slavery they brought their expertise with them, and then established rice as a favored dietary staple in the Americas. This knowledge system, moreover, enabled enslaved rice growers to negotiate and alter, to some extent, the terms of their bondage. The concluding section raises several questions about the issue of technology development and transfer, suggests a lingering Eurocentric bias in historical reconstructions of the agricultural development of the Americas, and discusses the scholarly implications of this research.

The Agronomic and Technological Basis of West African Rice Systems

Some 4,000 years ago, West Africans domesticated rice along the floodplain and inland delta of the upper and middle Niger River in Mali. The species of rice originally planted in this primary center of domestication, *Oryza glaberrima*, differs from Asian rice, *Orvza sativa*. While both species are currently planted in West Africa, the indigenous African center extends along the coast from Senegal to Côte d'lvoire and into the Sahelian interior along riverbanks, inland swamps, and lake margins. Within this diverse geographic and climatic setting two secondary centers of *glaberrima* domestication emerged: one, on floodplains of the Gambia River and its tributaries; and another, farther south in the forested Guinea highlands where rainfall reaches 2,000 millimeters/year. By the end of the 17th century rice had crossed the Atlantic Basin to the United States, appearing first as a rainfed crop in South Carolina before diffusing along river floodplains and into Georgia from the 1750s.

Many similarities characterized rice production on both sides of the Atlantic Basin. In both West Africa and South Carolina the most productive system developed along floodplains. Precipitation in each region follows a marked seasonal pattern, with rains generally occurring during the months from May/June to September/October. Rice cultivation flourished in South Carolina and Georgia under annual precipitation averages of 1,200–1,400 millimeters, a figure that represents the midrange of a more diverse rainfall pattern influencing West African rice cultivation, where precipitation increases dramatically over short distances from north to south. Accordingly, in the Malian primary center and the Gambian secondary center of rice domestication semi-arid (900 millimeters/year) conditions prevail, while southward in Guinea-Bissau and Sierra Leone precipitation exceeds 1,500 millimeters per annum.

The topography of the rice-growing region on both sides of the Atlantic presents a similar visual field. Coastlines are irregularly shaped and formed from alluvial deposits that also create estuarine islands. Tidal regimes on the American coast differ somewhat from those of Africa. The steep descent from the piedmont in South Carolina and Georgia delivers freshwater tidal flows to floodplains just 10 miles from the Atlantic coast. But the less pronounced gradient of rivers in West Africa's rice region means that freshwater tides meet marine water much farther upstream from the coast; on the Gambia River, salinity permanently affects the lower 80 kilometers but intrudes seasonally more than 200 kilometers upstream. Rice cultivation is adapted to the annual retreat and advance of the saline corridor. Even coastal estuaries inundated by ocean tides served as a basis for West African rice experimentation. South of the Gambia River, where precipitation exceeds 1,500 millimeters/year, West Africans learned to plant rice in marine estuaries, by enclosing plots and allowing rainfall to flush out accumulated salts. Water saturation is the key to planting such soils, as it prevents oxidization to an acidic condition that would preclude further cultivation. An elaborate network of embankments provides a barrier to seawater intrusion, while dikes, canals, and sluice gates enable field drainage of rainwater used for desalination. . . .

African Rice and American Continuities

By 1860, rice cultivation extended over 100,000 acres along the Eastern Seaboard from North Carolina's Cape Fear River to Florida's St. John's River, and inland for some 35 miles along tidal waterways. The initial stage of the rice plantation economy dated to the first hundred years of South Carolina's settlement (1670–1770) and, especially, the decades prior to the 1739 Stono slave rebellion. Rice cultivation systems analogous to those in West Africa, as well as identical principles and devices for water control and milling, were already evident in this period. Dramatic increases in slave imports during the 18th century facilitated the evolution of the Carolina rice plantation economy. Technology development unfolded in tandem with the appearance of the task labor system that regulated work on coastal rice plantations. As the crop grew in economic importance, agroenvironments favored for colonial rice production shifted from uplands to inland swamps and, from the 1730s, to the tidal (tidewater) cultivation system that led Carolina rice to global prominence.

This section presents an overview of the historical and geographical circumstances under which rice became a plantation crop in South Carolina. The technical changes marking the evolution of the colonial rice economy illuminate three issues that bear on comparative studies of technology and culture: first, the need to examine the technical components of production as parts of integrated systems of knowledge and not merely as isolated elements; second, the significance of cultural funds of knowledge for technology transfer; and third, in situations of unequal power relations, the extent to which claims for technological ingenuity can rest on cultural dispossession and appropriation of knowledge.

Slaves accompanied the first settlers to South Carolina in 1670; within 2 years they formed one-fourth of the colony's population; and as early as 1708 they outnumbered whites in the colony. Rice cultivation appears early in the colonial record, with planting already underway in the 1690s. By 1695 South Carolina recorded its first shipment of rice: one and one-quarter barrels to Jamaica. In 1699 exports reached 330 tons, and by the 1720s rice had emerged as the leading trade item. Years later, in 1748, South Carolina governor James Glen drew attention to the significance of rice experimentation during the 1690s for development of the colony's economy.

By the 1740s, documents firmly establish the presence of the upland, inland swamp, and tidal floodplain production systems. But rice cultivation in these areas is implied even earlier in the comments of one plantation manager, John Stewart, who claimed in the 1690s to have successfully sown rice in 22 different locations. One major point distinguished patterns of land use in West African and South Carolinian rice systems. In West Africa, subsistence security shaped the crop's production, thereby favoring cultivation in numerous microenvironments along a landscape gradient. Rice cultivation in colonial Carolina began as a subsistence crop, planted similarly in diverse environments, but as it became a plantation crop emphasis shifted to specific agroenvironments along the landscape continuum—from rainfed, to inland swamp, to tidal production—to maximize yields and returns on capital and labor.

Upland rice production received initial attention because it complemented the early colony's economic emphasis on stock raising and extraction of forest products. Slave labor buttressed this agropastoral system, which involved clearing forests, producing naval stores (pine pitch, tar, and resin), cattle herding, and subsistence farming. Export of salted beef, deerskins, and naval stores in turn generated capital to purchase additional slaves. The number of enslaved Africans imported to the colony dramatically increased from 3,000 in 1703 to nearly 12,000 by 1720, which enabled a shift in rice cultivation to the more productive inland swamp system.

The higher yielding inland swamp system initiated the first attempts at water control in Carolina's rice fields. After clearing swamp forests, slaves developed the network of berms and sluices necessary for converting plots into reservoirs. Like its counterpart in West Africa, the inland swamp system impounded water from rainfall, subterranean springs, high water tables, or creeks to saturate the soil. The objective was to drown unwanted weeds and thereby reduce the labor spent on weeding, as in West Africa. West African principles also guided the cultivation of rice in coastal marshes. Rice was grown in saltwater marshes near the terminus of freshwater streams in soils influenced by the Atlantic Ocean. The conversion of a saline marsh to a rice field depended upon soil desalination, a process not as easily achieved with South Carolina's annual precipitation regime (1,100–1,200 millimeters), which is lower than the average 1,500 millimeters per year that regulate the West African mangrove system. However, by diverting an adjoining freshwater creek or stream, salts could be rinsed from the field. The principle of canalizing water for controlled flooding also extended to other settings, such as in places where subterranean springs flowed near the soil surface. Detailed knowledge of landscape topography and hydrological conditions thus enabled the proliferation of rice growing in diverse inland swamp microenvironments.

During the 18th century rice cultivation in such areas innovated to more elaborate systems of sluices that released reserve water on demand for controlled flooding at critical stages of the cropping cycle. This inland swamp system flourished where the landscape gradient sloped from rain-fed farming to the inner edge of a tidal swamp. Enclosure of a swamp with earthen embankments created a reservoir for storing rainwater, the system's principal source of irrigation. The reservoir fed water, through a sluice gate and canal, by gravity flow to the inland rice field, while a drainage canal and sluice placed at the lower end of the rice field emptied excess water into a nearby stream, creek, or river. Whereas the principle of constructing a reservoir for controlled field flooding is identical to the West African mangrove rice system, the innovative changes that subsequently developed may well provide an instance of what geographer Paul Richards calls "agrarian creolization." The term refers to the convergence of different knowledge systems and their recombination into new hybridized forms, spearheading the process of innovation.

By the mid-18th century the emphasis on rice had shifted from inland swamps to tidal river floodplains, first in South Carolina, and subsequently in Georgia. The swelling number of slaves directly entering South Carolina from West Africa between the 1730s and the 1770s proved crucial in this spatial relocation of the rice economy. Some 35,000 slaves were imported into the colony during the first half of the century and over 58,000 between 1750 and 1775, making South Carolina the largest importer of enslaved Africans on the North American mainland between 1706 and 1775. The share of slaves brought directly from the West African rice coast grew during these crucial decades of tidewater rice development from 12% (1730s) to 54% (1749–1765), and then to 64% (1769–1774) (Richardson, 1991). This pattern is illuminated in a typical handbill from the colonial period, which announces the sale in Charlestown (Charleston) on July 24, 1769 of enslaved men, women, and children from Sierra Leone.

Tidewater cultivation occurred on floodplains along tidal rivers where, similar to its mangrove rice counterpart in West Africa, the diurnal variation in sea level facilitated field flooding and drainage. Preparation of a tidal floodplain for rice cultivation followed principles already outlined for the mangrove rice system. The rice field was embanked at sufficient height to prevent tidal spillover, while the earth removed in the process created adjacent canals. Sluices built into the embankment and field sections operated as valves for flooding and drainage. The next step involved dividing the area into plots (in South Carolina these were termed "quarter sections," of some 10–30 acres), with river water delivered through secondary ditches. This elaborate system of water control enabled the adjustment of land units to labor demands and allowed slaves to directly sow rice along the floodplain. Then the rice was planted directly in the floodplain, as it is in African floodplain cultivation.

Tidewater cultivation required considerable landscape modification and ever-greater numbers of enslaved laborers than rain-fed and inland swamp cultivation. Leland Ferguson, a historical archaeologist, vividly captures the staggering human effort involved in transforming Carolina's tidal swamps to rice fields:

> These fields are surrounded by more than a mile of earthen dikes or "banks" as they were called. Built by slaves, these banks . . . were taller than a person and up to 15 feet wide. By the turn of the eighteenth century, rice banks on the 12½ mile stretch of the East Branch of Cooper River measured more than 55 miles long and contained more than 6.4 million cubic feet of earth. By 1850, aided only by hand-held tools, slaves in the Carolina rice zone had built earthworks "nearly three times the volume of Cheops, the world's largest pyramid."

While such landscape change placed considerable demands on slave labor for construction and maintenance, it reduced the need for manual weeding, one of the most labor-intensive tasks in rice production. The systematic lifting and lowering of water was achieved by sluices, known as "trunks," located in the embankment and secondary dikes; by the late colonial period these devices had evolved into hanging floodgates. With full control of an adjacent tidal river, the rice field could be flooded on demand for irrigation and weeding and to renew the soil annually with alluvial deposits. Because of this increasing reliance on water control technology, in tidal cultivation one slave could manage 5 acres of rice, as opposed to just 2 acres in the inland rice system.

The history of the term "trunk" for sluice gate in Carolina also suggests evidence for technology transfer from the West African rice coast. While the hanging gate technology likely provides another example of agrarian creolization, its name refers to an earlier device. Even when the hanging gate replaced earlier forms, Carolina planters continued to call sluice gates "trunks." In the 1930s planter descendant David Doar stumbled upon the term's origin: the earliest sluice gates were formed from hollowed-out cypress trunks. The original Carolina sluice system looked and functioned exactly like its African counterpart. Reference to them as "trunks" throughout the antebellum period suggests that the technological expertise of Africans again proved significant in Carolina rice history.

Tidewater rice cultivation led South Carolina to global economic prominence in the 18th century. It was made possible by the expertise of enslaved Africans in cultivating as well as processing the crop. Their experience with planting a whole range of interconnected environments along a landscape gradient likely permitted the sequence of adaptations that marked the growth of the South Carolina rice industry.

African contributions to Carolina rice history were not limited to cultivation practices. They also extended to the method by which the grain was processed for consumption. Until suitable mechanical mills were developed around the time of the American Revolution, the entire export crop was milled by hand, in the traditional method long used by African women, with a mortar and pestle. Even the fanner baskets used to winnow rice on Carolina and Georgia plantations display a likely African origin, as anthropologist Dale Rosengarten suggests. She links the coiled-basket-weaving tradition in fanner baskets to the Senegambian rice region, as it did not exist among Native Americans of the Southeast region.

Enslaved Africans thus contributed significantly more than physical labor to colonial rice production. They provided the critical expertise in establishing rice cultivation in South Carolina, even though accounts by planters and their descendants have long claimed for their forebears the ingenuity in developing the system. Planter accounts, however, fail to explain how they learned to transform wetland landscapes by careful observance of tidal dynamics, soils, microenvironments, and water regimes. Such systems of planting cereals in standing water did not exist in England at the time, yet surfaced within two decades of Carolina's settlement. Wetland rice represents a far more complex farming system than the rainfall cultivation practiced by the English. Enslaved Africans from the African rice region were thus the only settlers present in the Carolina colony who possessed this knowledge system.

The delayed recognition of this significant contribution to the history of the Americas stems, in part, from the way that rice has been examined by scholars. By emphasizing rice as a cereal, as a grain consumed and traded internationally, studies have failed to place it within its proper agroecological context for studies of diffusion and technology transfer. An examination of rice as a landscape of microenvironments brings into focus the underlying soil and water management regimes that inform its cultivation as well as its cultural origins. Recovery of the African contribution to American rice history also involves considering consumption as well as production, processing as well as cultivation, and the types of labor and knowledge that mediated the spatial movement of rice from field to kitchen. This requires sensitivity to ecological as well as gendered forms of knowledge in technology transfer.

While this overview of rice beginnings in South Carolina argues that planters reaped the benefits of a rice-farming system perfected by West Africans over millennia, an important question remains. Why would enslaved West Africans transfer a sophisticated rice system to plantation owners when the result spelled endless and often lethal toil in malarial swamps?

The answer is perhaps revealed in the appearance by 1712 of the task labor system that characterized coastal rice plantations. It was distinguished from the more pervasive "gang" form of work typifying plantation slavery. In the gang system, "the laborer was compelled to work the entire day," while "under the task system the slave was assigned a certain amount of work for the day, and after completing the task he could use his time as he pleased." Without underestimating the real toil involved in the two systems, the task system did set normative limits to daily work demands. Such seemingly minor differences between the two systems could deliver tangible improvements in slave nutrition and health, as Johan Bolzius implied in 1751 with his observation: "If the Negroes are Skillful and industrious, they plant something for themselves after the day's work."

The task labor system appeared at the crucial juncture of the evolution of rice as a plantation crop in the Carolina colony and the shift to the more productive, but labor-demanding, inland swamp system. A similar system of limiting demands placed on enslaved labor was already in existence along Africa's Rice Coast. The appearance of the task labor system in Carolina's fledgling rice economy may well represent the outcome of negotiation and struggle between master and slave over knowledge of rice culture and the labor process to implement it. In providing crucial technological acumen, slaves perhaps discovered a mechanism to negotiate improved conditions of bondage. But by the 19th century such gains had eroded, as the frontier for slave escape closed and slavery appeared to be a permanent feature of Southern agriculture. The task labor system became little different than the gang form of slavery.

Conclusion

"What skill they displayed and engineering ability they showed when they laid out these thousands of fields and tens of thousands of banks and ditches in order to suit their purpose and attain their ends! As one views this vast hydraulic work, he is amazed to learn that all of this was accomplished in face of seemingly insuperable difficulties by every-day planters who had as tools only the axe, the spade, and the hoe, in the hands of intractable negro men and women, but lately brought from the jungles of Africa." In 1936, when David Doar, descendant of Carolina planters, echoed the prevailing view that slaves contributed little besides labor to the evolution of the South Carolina rice economy, no historical research suggested otherwise. While recent research challenges such unquestioned assumptions, a bias nonetheless endures against considering West Africans as the originators of rice culture in the Americas.

Even authoritative texts on rice cultivation, such as that of D. H. Grist, express such a bias. In reference to a type of paddy rice cultivation found in British Guiana (now Guyana) and the neighboring former Dutch colony of Surinam, Grist describes the "empoldering" technique as "a method of restricting floods and thus securing adjacent areas from submergence." While this technique is strikingly similar to that employed in mangrove rice production along the West African rice coast, he attributes the system to 18th-century Dutch colonizers. In that era, however, Surinam possessed one of the highest ratios of Africans to Europeans of any New World plantation society (65:1 in Surinam compared to Jamaica's 10:1).

More recent work in Brazilian rice history repeats this perspective. Even though rice was not grown in Portugal during the colonization of the Americas, Pereira attributes its 16th-century introduction in Brazil's eastern Amazon and Northeast to migrants from the Azores and Portugal. There is no discussion of how the African mortar and pestle (a device not used in Portugal) came to be the sole technology for milling rice in Brazil until the mid-18th century, when it was replaced by water mills that successfully removed the hulls without grain breakage. Nor is there any acknowledgment of the possibility that rice culture became established in Brazil, as in South Carolina, because African rice routinely provisioned slave ships, providing the enslaved an opportunity to grow their food staple for subsistence.

Evidence from the American and African Atlantic thus suggests that slaves from West Africa's indigenous rice area established rice culture. A crop initially planted for subsistence became, in 18th-century South Carolina, the first cereal globally traded as a plantation export crop. This complex indigenous knowledge system guided the transformation of Carolina's swamps while serving as a source of technological innovation in rice culture, captured in the notion of agrarian creolization. These achievements in tidal rice diffused southward in the 18th century along rivers in Georgia and Florida, and, from midcentury, overseas to similar environments in Brazils Northeast and eastern Amazon regions. While rice systems of the Americas would eventually bear the imprimatur of both African and European influences, its appearance initially is linked to a knowledge system developed in West Africa and carried across the Middle Passage by slavery's victims.

Thus, as Europeans and Africans faced each other in a new territory under dramatically altered and unequal power relations, the enslaved established a subsistence crop long valued in West Africa. With the abolition of slavery, rice history led to cultural dispossession and appropriation by descendants of slave owners who credited the beginnings of rice farming to European ingenuity and presumed mastery of technology. A careful reading of the archival and historical record—one attuned to agroenvironments, power relations, ecological principles, and social history—reveals a dramatically different narrative.

The ending of this story is not yet settled: this cross-cultural, social–environmental history of transatlantic rice culture also cuts across current political and cultural debates. In the U.S. South, historical preservation of antebellum landscapes is charged with the controversy of memory politics. While many descendants of white planters view preservation of plantation landscapes as a source of pride, many African Americans perceive these as sites of toil, terror, and shame. This discussion of rice technology transfer demonstrates that despite the brutality of bondage, African slaves were active and ingenious shapers of antebellum agroenvironments rather than mere physical laborers. Historical preservation and ecological restoration efforts in the South should acknowledge the rich hybrid nature of these cultural landscapes in a way similar to that currently underway in the reconsideration of southern family histories by black and white descendants of slave owners.

This study also has implications for the hotly debated issue of intellectual property rights over agricultural seeds. Today, cultivars and their germ plasm are increasingly engineered, patented, and privatized. This market logic reduces seeds to their mere biology. As this investigation of rice cultivation on the Atlantic Rim implies, however, seeds cannot be so easily separated from their political and social context. The introduction of African rice in the New World was not simply the movement of seeds from one environment to another, but rather the transfer and transformation of a *rice culture,* with attendant continuities and changes in technology, labor organization, social structures, and cultural meanings.

POSTSCRIPT

Have the Contributions of Africans Been Recognized for Developing New World Agriculture?

Despite the wide acceptance of Carney's scholarship, certain aspects of her work have been critiqued. In a review of Carney's book on this subject, *Black Rice* (summarized in Carney's selection in this issue), in a 2002 issue of the *William and Mary Quarterly,* Philip Morgan raises several questions about her treatment of the slave trade. He disputes her claim that Carolina planters imported a greater percentage of female slaves than did Caribbean planters. He questions whether Carolina planters paid more for female slaves than other New World slave markets (but cannot prove otherwise). He also questions whether South Carolineans were sourcing relatively more slaves from rice-producing areas in West Africa at critical junctures in the development of the rice economy. New scholarship by Max Edelson (see *Plantation Enterprise in Colonial South Carolina,* Harvard University Press, 2006) also suggests that European colonists played a much larger role in the development of a Carolina rice economy than Carney admits.

Another question to ponder, and this is raised by Carney in her selection, is why Africans would have been willing to share their rice-growing expertise with their white oppressors. Carney's answer is that they used this knowledge to bargain for a "task labor system" rather than the more oppressive "gang system," which forced slaves to work the entire day. Under the task system, a slave could use his or her time for their own activities after completing a certain quantity of work. As Carney notes, this negotiated approach was eventually lost in the nineteenth century.

A final issue to consider is the somewhat subjective nature of regions as units of analysis. What the burgeoning scholarship on the Atlantic World demonstrates is that historical transfers of peoples and ideas may bind together the two areas (i.e., the Atlantic sides of the Americas and Africa) that traditionally have been studied separately. Furthermore, it is important to note that Carney was originally trained as an Africanist and undertook most of her scholarship in West Africa. What Carney's scholarship demonstrates is that she was able to use her deep understanding of the African context to shed light on the situation in the Americas. For more reflections on comparative regional studies, see William Moseley's May 2005 article in the *Southeastern Geographer* entitled "Regional Geographies of the U.S. Southeast and Sub-Saharan Africa: The Potential for Comparative Insights."

ISSUE 3

Is European Subjugation of Africans Ultimately Explained by Differences in Land, Plant, and Animal Resources?

YES: Jared Diamond, from "Why Europeans Were the Ones to Colonize Sub-Saharan Africa," in *Guns, Germs, and Steel: The Fates of Human Societies* (W. W. Norton & Company, 1999)

NO: Lucy Jarosz, from "A Human Geographer's Response to *Guns, Germs, and Steel:* The Case of Agrarian Development and Change in Madagascar," *Antipode* (2003)

ISSUE SUMMARY

YES: Jared Diamond, professor of physiology and biogeography at UCLA, argues that Europeans were able to colonize Africa (rather than vice versa) because of the advantages of guns, widespread literacy, and political organization. These advantages stem ultimately from different historical trajectories that are linked to "differences in real estate" (or differences in physical geography).

NO: Lucy Jarosz, associate professor of geography at the University of Washington, is troubled by Diamond's narrow conception of geography and asserts that explaining differences in wealth and power between regions must also take account of social, political, and economic connections. She focuses on the specific case of Madagascar and argues that the intentions of the colonizer are as or more important than their military power for determining the nature of the colonial relationship.

Many in the social sciences at the turn of the last century were heavily influenced by a school of thought known as Social Darwinism. Social Darwinists took the ideas of Charles Darwin and used these concepts to explain perceived differences between human societies and races. One such prominent thinker was the German philosopher Fredrich Ratzel (1844–1904). Ratzel argued that differences in climate largely explained (what he perceived to be) Europeans' advanced state of development relative to that of the various peoples of the tropical regions.

Ratzel and others reasoned that the cooler temperatures, changing seasons, and long winters of the temperate regions led to more forethought, energy, and planning, whereas the heat, absence of a harsh winter, and relative abundance of the tropics resulted in contentment, sloth, and a lack of foresight.

This particular perspective came to be known as the geographic factor, environmentalism (different than the contemporary definition of this term), or environmental determinism, and was, not surprisingly, very popular among the colonial powers (e.g., France, England) for explaining or rationalizing their subjugation of African peoples. Many were beginning to doubt environmental determinism by the 1930s, and this blatantly racist perspective eventually fell out of favor. People questioned these ideas because there were too many exceptions. How does one explain the Egyptian pyramids, Mayan temples, or the Great Zimbabwe if people in the tropics are supposed to be less advanced?

Although environmental determinism has long since fallen out of favor, some social scientists are beginning to reassert a milder version of environmental determinism (sometimes referred to as environmental possibilism). Within the discipline of economics, a new subfield known as "new economic geography" has flourished in recent years by explaining lagging development in terms of differences in geography or real estate. Economists writing in this new genre include Jeffrey Sachs and Ricardo Hausmann among others. Europe and North America were twice blessed. They had the most productive economic systems reinforced, and partly caused, by their favorable geographic conditions. In contrast, the tropics and great land-based empires were twice cursed, with weaker economic institutions, and as a rule, poorer geographical endowments" (*The Economist*, 1997).

In this issue, Jared Diamond, trained in physiology, and now a professor of biogeography at UCLA, argues that Europeans were able to colonize Africa because of the advantages of guns, widespread literacy, and political organization. These advantages arose from different historical trajectories that were linked to "differences in real estate." This selection is taken from Diamond's Pulitzer Prize–winning book, *Guns, Germs, and Steel*. In the opening chapter of this text, Diamond asserts that he is not an environmental determinist. He further asserts that his theory is anything but racist because he explains varying levels of development in terms of natural endowments and geography rather than mental ability.

Lucy Jarosz, an associate professor of geography at the University of Washington, asserts that explaining differences in wealth and power between regions must also take account of social, political, and economic connections. She focuses on the specific case of Madagascar and argues that the intentions of the colonizer are as or more important than their military power for determining the nature of the colonial relationship. She describes how Madagascar had previously been peopled and colonized by immigrants from South and Southeast Asia, East Africa, and the Arabic Peninsula. These early waves of immigration and colonization led to settled agriculture and international trade, as well as the development of a local elite. According to Jarosz, the colonial experience with the French was quite different because this power reshaped Madagascar's economy for its own benefit.

YES

Jared Diamond

Why Europeans Were the Ones to Colonize Sub-Saharan Africa

Africa's diverse peoples resulted from its diverse geography and its long prehistory. Africa is the only continent to extend from the northern to the southern temperate zone, while also encompassing some of the world's driest deserts, largest tropical rain forests, and highest equatorial mountains. Humans have lived in Africa far longer than anywhere else: our remote ancestors originated there around 7 million years ago, and anatomically modern *Homo sapiens* may have arisen there since then. The long interactions between Africa's many peoples generated its fascinating prehistory, including two of the most dramatic population movements of the past 5,000 years—the Bantu expansion and the Indonesian colonization of Madagascar. All of those past interactions continue to have heavy consequences, because the details of who arrived where before whom are shaping Africa today. . . .

Now let's turn to the remaining question in our puzzle of African prehistory: why Europeans were the ones to colonize sub-Saharan Africa. That it was not the other way around is especially surprising, because Africa was the sole cradle of human evolution for millions of years, as well as perhaps the homeland of anatomically modern *Homo sapiens*. To these advantages of Africa's enormous head start were added those of highly diverse climates and habitats and of the world's highest human diversity. An extraterrestrial visiting Earth 10,000 years ago might have been forgiven for predicting that Europe would end up as a set of vassal states of a sub-Saharan African empire.

The proximate reasons behind the outcome of Africa's collision with Europe are clear. Just as in their encounter with Native Americans, Europeans entering Africa enjoyed the triple advantage of guns and other technology, widespread literacy, and the political organization necessary to sustain expensive programs of exploration and conquest. Those advantages manifested themselves almost as soon as the collisions started: barely four years after Vasco da Gama first reached the East African coast, in 1498, he returned with a fleet bristling with cannons to compel the surrender of East Africa's most important port, Kilwa, which controlled the Zimbabwe gold trade. But why did Europeans develop those three advantages before sub-Saharan Africans could?

As we have discussed, all three arose historically from the development of food production. But food production was delayed in sub-Saharan Africa

From *Guns, Germs, and Steel: The Fates of Human Societies* by Jared Diamond (W. W. Norton, 1997). Copyright © 1997 by Jared Diamond. Reprinted by permission of W. W. Norton & Company, Inc., Random House Group Ltd., and Brockman Agency, Inc.

(compared with Eurasia) by Africa's paucity of domesticable native animal and plant species, its much smaller area suitable for indigenous food production, and its north–south axis, which retarded the spread of food production and inventions. Let's examine how those factors operated.

First, as regards domestic animals, we've already seen that those of sub-Saharan Africa came from Eurasia, with the possible exception of a few from North Africa. As a result, domestic animals did not reach sub-Saharan Africa until thousands of years after they began to be utilized by emerging Eurasian civilizations. That's initially surprising, because we think of Africa as *the* continent of big wild mammals. But . . . a wild animal, to be domesticated, must be sufficiently docile, submissive to humans, cheap to feed, and immune to diseases and must grow rapidly and breed well in captivity. Eurasia's native cows, sheep, goats, horses, and pigs were among the world's few large wild animal species to pass all those tests. Their African equivalents—such as the African buffalo, zebra, bush pig, rhino, and hippopotamus—have never been domesticated, not even in modern times.

It's true, of course, that some large African animals have occasionally been *tamed*. Hannibal enlisted tamed African elephants in his unsuccessful war against Rome, and ancient Egyptians may have tamed giraffes and other species. But none of those tamed animals was actually domesticated— that is, selectively bred in captivity and genetically modified so as to become more useful to humans. Had Africa's rhinos and hippos been domesticated and ridden, they would not only have fed armies but also have provided an unstoppable cavalry to cut through the ranks of European horsemen. Rhino-mounted Bantu shock troops could have overthrown the Roman Empire. It never happened.

A second factor is a corresponding, though less extreme, disparity between sub-Saharan Africa and Eurasia in domesticable plants. The Sahel, Ethiopia, and West Africa did yield indigenous crops, but many fewer varieties than grew in Eurasia. Because of the limited variety of wild starting material suitable for plant domestication, even Africa's earliest agriculture may have begun several thousand years later than that of the Fertile Crescent.

Thus, as far as plant and animal domestication was concerned, the head start and high diversity lay with Eurasia, not with Africa. A third factor is that Africa's area is only about half that of Eurasia. Furthermore, only about one-third of its area falls within the sub-Saharan zone north of the equator that was occupied by farmers and herders before 1000 B.C. Today, the total population of Africa is less than 700 million, compared with 4 billion for Eurasia. But, all other things being equal, more land and more people mean more competing societies and inventions, hence a faster pace of development.

The remaining factor behind Africa's slower rate of post-Pleistocene development compared with Eurasia's is the different orientation of the main axes of these continents. Like that of the Americas, Africa's major axis is north–south, whereas Eurasia's is east–west. As one moves along a north–south axis, one traverses zones differing greatly in climate, habitat, rainfall, day length, and diseases of crops and livestock. Hence crops and animals domesticated or acquired in one part of Africa bad great difficulty in moving to other parts.

In contrast, crops and animals moved easily between Eurasian societies thousands of miles apart but at the same latitude and sharing similar climates and day lengths.

The slow passage or complete halt of crops and livestock along Africa's north–south axis had important consequences. For example, the Mediterranean crops that became Egypt's staples require winter rains and seasonal variation in day length for their germination. Those crops were unable to spread south of the Sudan, beyond which they encountered summer rains and little or no seasonal variation in daylight. Egypt's wheat and barley never reached the Mediterranean climate at the Cape of Good Hope until European colonists brought them in 1652, and the Khoisan never developed agriculture. Similarly, the Sahel crops adapted to summer rain and to little or no seasonal variation in day length were brought by the Bantu into southern Africa but could not grow at the Cape itself, thereby halting the advance of Bantu agriculture. Bananas and other tropical Asian crops for which Africa's climate is eminently suitable, and which today are among the most productive staples of tropical African agriculture, were unable to reach Africa by land routes. They apparently did not arrive until the first millennium A.D., long after their domestication in Asia, because they had to wait for large-scale boat traffic across the Indian Ocean.

Africa's north–south axis also seriously impeded the spread of livestock, Equatorial Africa's tsetse flies, carrying trypanosomes to which native African wild mammals are resistant, proved devastating to introduced Eurasian and North African species of livestock. The cows that the Bantu acquired from the tsetse-free Sahel zone failed to survive the Bantu expansion through the equatorial forest. Although horses had already reached Egypt around 1800 B.C. and transformed North African warfare soon thereafter, they did not cross the Sahara to drive the rise of cavalry-mounted West African kingdoms until the first millennium A.D., and they never spread south through the tsetse fly zone. While cattle, sheep, and goats had already reached the northern edge of the Serengeti in the third millennium B.C., it took more than 2,000 years beyond that for livestock to cross the Serengeti and reach southern Africa.

Similarly slow in spreading down Africa's north–south axis was human technology. Pottery, recorded in the Sudan and Sahara around 8000 B.C., did not reach the Cape until around A.D. 1. Although writing developed in Egypt by 3000 B.C. and spread in an alphabetized form to the Nubian kingdom of Meroë, and although alphabetic writing reached Ethiopia (possibly from Arabia), writing did not arise independently in the rest of Africa, where it was instead brought in from the outside by Arabs and Europeans.

In short, Europe's colonization of Africa had nothing to do with differences between European and African peoples themselves, as white racists assume. Rather, it was due to accidents of geography and biogeography—in particular, to the continents' different areas, axes, and suites of wild plant and animal species. That is, the different historical trajectories of Africa and Europe stem ultimately from differences in real estate.

Lucy Jarosz

A Human Geographer's Response to *Guns, Germs, and Steel:* The Case of Agrarian Development and Change in Madagascar

In his book, *Guns, Germs, and Steel* (1999; hereafter *GGS*), Jared Diamond asks important questions about why wealth and power are distributed so unequally throughout the world and how this pattern originated: "For instance, why weren't Native Americans, Africans, and Aboriginal Australians the ones who decimated, subjugated, or exterminated Europeans and Asians?" (p 15). He identifies the possession of guns, germs, and steel as the critical explanation, tracing their arrival in European imperial hands through a 13,000-year history of agriculture and technology. According to Diamond's argument, Europe's colonization of Africa was due to "accidents of geography and biogeography—in particular, to the continents' different areas, axes, and suites of wild plant and animal species. That is, the different historical trajectories of Africa and Europe stem ultimately from differences in real estate" (p 401).

I appreciate Diamond's emphasis upon the role of geography within the broad sweep of human history, but as a human geographer, I am concerned by the narrow definition of geography that this widely read book puts forward. Geography encompasses the realms of the humanities and the social sciences in its examination and explanation of society-environment relations. Explaining differences in wealth and power among and within world regions and diverse societies through time must take social, economic, and political connections into account, along with their relationship to environmental attributes and ecological change, in an attempt to link global, regional, and local scales of analysis. These explanations must also consider the influence of the human imagination and sociocultural processes in shaping the diversity of agrarian and ecological landscapes.

As an example of the limits to purely environmental explanations of contemporary global disparities, I offer an overview of the political economy of food and agriculture in Madagascar. This island was subject to two distinct forms of colonization, first from East Africa, the Middle East, and Southeast Asia and subsequently from Western Europe. The first wave did not devastate

From *Antipode*, September 2003, pp. 823–827. Copyright © 2003 by John Wiley & Sons. Reprinted by permission via Rightslink.

the food-production economy in the way that the second wave did. The reasons for this lie with issues of politics, ideology, and power. I argue that beyond environment and rates of technological diffusion, human relationships are critical to explaining the dynamics of agrarian development and change, as well as the origins of inequality.

Food and Agriculture in Pre-European Madagascar

Records of Madagascar's aboriginal populations are sparse and based upon archaeological sources and oral history. The Vazimba, the earliest human inhabitants, are believed to be of Indonesian and East African descent and settled the island between the fifth and seventh centuries. Later immigrants and sociocultural influences arrived from South and Southeast Asia, East Africa, and the Arabic peninsula between the 11th and 16th centuries (Kottak 1980). Agricultural practices on the island varied both with regional ecological attributes and with the technologies and sociocultural perspectives of local populations. Shifting cultivation was practiced throughout the island, while in the most densely inhabited area of Madagascar—the central plateau—the Merina and Betsileo people cultivated irrigated rice. Early settlers domesticated livestock in the south, which subsequently diffused to the west. This situation affirms Diamond's thesis about the multiple influences of early immigration and colonization in Africa as well as the importance of settled agriculture to the formation and advancement of sophisticated nation-states. However, the development of agriculture in Madagascar's central plateau did not depend upon exceptional environmental differences. Most of the plateau is made up of poor lateritic clay soils, and only a very small percentage—less than 5% of the total area—is able to support irrigated rice fields. This area's geology is very similar to areas of Zimbabwe, which does not support nearly the population density. This comparison prompted Gourou (1956:346) to remark,

> The geographer would be wrong if he considered the natural environment as independent from man. The natural environment in two ways depends on man: first, man has transformed profoundly the environment . . . ; second, man has interpreted the environment in terms of his techniques. The same natural environment will result in different human landscapes when interpreted (and transformed) by traditional European peasant civilization, by Chinese civilization, and by modern American civilization. The Merina of Madagascar and the Bemba of Rhodesia have created very different human landscapes in similar physical environments. No progress in understanding the human aspects of the landscape is possible if these aspects are simply considered to react to the physical elements of the landscape. Explanation does not progress if the human groups are considered to be compelled by the natural environment to adopt such-and-such techniques.

Food and Agriculture Following European Colonization

When Diego Diaz "discovered" Madagascar in 1500, a Merina monarchy headed a centralized state based upon a caste system of nobles, commoners, and slaves and managed to unite the island under a single rule based upon livestock and rice cultivation. A commercial class was conducting international trade in rice, beef, and slaves. The mixing of various groups of people on the island led to both innovation and conflict. The Merinas' hold on the island was tenuous, as war and slave-raiding among various ethnic groups were commonplace. This led the Merina monarchy to develop political and economic relations with both Britain and France in order to help consolidate its power on the island, and to hold the influence of both imperial powers in tension throughout the 1800s. These changing dynamics of sociopolitical relationships and cultural processes shaped agricultural landscapes and production and triggered imperial interest in the island.

Relationships with Britain and France were used by competing political units on the island and by aspiring dominant groups within the Merina elite to advance their political and economic agendas. Beginning in 1817, British officers helped train the Malagasy army, and a treaty with Britain recognized the Merina claim to control of the whole island. In 1863, the Merina signed trade and diplomatic treaties with both the United States and England concerning Malagasy sovereignty, commerce, and freedom of religion. The French had claimed jurisdiction over all or part of the island since the beginning of the 19th century, and these claims were pressed with growing insistence in response to the development of imperialist competition in Europe. The Merina monarchy aligned itself with Britain, hoping to protect itself from French conquest. This strategy ultimately failed. In 1890, Britain and France signed a treaty exchanging French recognition of British control over Zanzibar for British acceptance of the French claim to Madagascar. In 1896, the French claimed the island as one of its colonies, abolished the Merina monarchy, and ended trade with Britain and the United States, while a system of French monopolies and oligopolies dominated trade and credit.

This brief colonial history of Madagascar illustrates some intriguing parallels and significant discrepancies with Diamond's archetypal European conquest, the overthrow of the Inca Empire at Cajamarca. At the eve of European contact, both societies were highly organized and stratified, practiced intensive agricultural production, and engaged in long-distance trade. Diamond focuses attention on the moment of military conflict to emphasize the importance of guns, swords, horses, and smallpox in Atahuallpa's capture and the eventual conquest of the Inca Empire by Spain. These are the "proximate factors"—the means of conquest—the acquisition which Diamond sees as reflecting the "ultimate factors" of continental size and axes. While the colonial outcome in Madagascar was similar, the definitive factors in this process were not. As in the rest of Africa, germs—in the forms of malaria and dengue fever—delayed and impeded European contact and colonization. And while military power was certainly present, conquest, colonization, and impoverishment were effected by economic power through integration into the global market via taxation, slavery, and forced labor, the establishment of export crops, and resource appropriation and control. This suggests a different set of ultimate factors: the scale and intentions of the societies employing the proximal controls.

Power, Resource Extraction, and Capitalism: From Local to Global Scales

The first wave of Madagascar's colonization established settled forms of agriculture and developed international trade in agricultural products to build and maintain the wealth and power of local, regionally based elites. In the second wave, the goal was to reshape the economy and commerce so as to reorient and extract wealth and profit for French companies and creditors, rather than for the further enrichment of the Malagasy living on the island. Coffee and peas were introduced as export crops, timber was extracted, and environmental disruption and degradation ensued as taxation and forced labor laws spurred waves of regional migration and increased shifting cultivation as a crucial way to address food insecurity. Rice and beef were requisitioned to feed French troops and the slaves on sugar plantations on Mauritius. Famine and chronic food shortages were two key outcomes of the second wave of colonization.

European colonizers were similarly eager to extract and control the Incan empire's labor, land and gold. This was realized through the destruction of the political and economic power of the ruling elite and the subsequent restructuring of the empire's economy to serve the development and commercial interests of the colonizer within the global economy. I suggest that a focus upon the shifting political and economic dimensions of resource control and extraction highlights the primacy of politics and economics in European colonization as vital "ultimate factors." A focus upon the proximal military battles obscures this context and these connections in both the Inca and the Merina examples.

Environmental differences matter in contemporary global inequalities, as Diamond suggests, but so do the processes of colonization. While I agree with Diamond that it is wrong to attribute inequality and processes of colonization to any sort of racial superiority, it is also wrong to ignore the role of racism within the processes of colonization. Malagasy people were treated as animals and slaves in forced labor gangs during European colonization. The image of African people as less than human has consistently permeated racist discourse throughout the colonial and postcolonial periods; it is this aspect of colonization that Diamond challenges. However, the emphasis upon environmental difference in *GGS* lets us off the hook in terms of thinking deeply about geopolitical and economic relationships and the contributions of human ingenuity, imagination, and even cruelty to agricultural development and change. In the case of Africa, European colonization often replaced or appropriated the indigenous systems of agricultural production, resulting in the emergence of food shortages, chronic poverty, and environmental marginalization.

Environmental history and differences alone do not explain why today Madagascar is one of the poorest countries in the world. The importance of geopolitical relationships, the development of capitalism, and dynamics of regional, national, and global food networks within specific environments are critical components of an accurate understanding of inequality and poverty. It is not enough to focus on environmental difference at the millennial scale, since the forms and processes of contemporary disparity are maintained by relationships established in the recent colonial past and reworked and deepened in the postcolonial period. We must not neglect the complex linkages and relationships among European and African societies and environments, as well as the realities of imperialism, power, and racism, when explaining the harsh realities of inequality and poverty that surround us.

POSTSCRIPT

Is European Subjugation of Africans Ultimately Explained by Differences in Land, Plant, and Animal Resources?

A couple of interesting points arise from this issue. The first concerns whether or not the ideas that Diamond and others are espousing may be characterized as environmental determinism. To be fair, Diamond's argument differs from those of early twentieth-century environmental determinists in several respects. First, he does not suggest that climatic conditions directly result in a certain level of social organization and development. Rather, he asserts that settled agriculture and animal husbandry were able to first develop in those places that possessed wild plants and animals that were amenable to domestication. This is important, according to Diamond, because those societies that had densely settled agricultural populations first were more likely to develop: (1) centralized political structures, (2) writing and metallurgy, and (3) immunity to certain communicable diseases after a period of affliction. As such, resource endowments ultimately allow for certain possibilities, but they do not directly determine levels of development. Second, Diamond also attaches considerable importance to the location of a group of people vis à vis other human populations. Those groups that were in contact with other human societies and possessed similar environmental conditions were more likely to benefit from the diffusion of ideas. As such, Europeans were able to benefit from several innovations that diffused from the Middle East and Asia, even though they did make such advances independently.

Some would be more comfortable characterizing Diamond as an environmental possibilist, rather than an environmental determinist. Environmental possibilists argue that the environment presents humans societies with a range of options from which they may pick and choose. Environmental conditions are important because they constitute an initial range of possibilities. However, technology may help groups overcome certain environmental constraints. As such, humans adapt to certain environmental constraints or conditions, but the process of human adaptation is very dynamic. Yet, even allowing for such distinctions or nuances, many remain critical of Diamond (and those who share his views) for his lack of attention to the pernicious effects for Africa of the global capitalist system that slowly emerges from the age of exploration (fifteenth century) forward.

A second point of note concerns the sordid history of environmental determinism. While it is up for debate if Diamond's perspective really constitutes an environmental determinist view, it is worth underscoring that many in the sciences and social sciences in the previous era played a role in and were influenced by colonialism. Theories like environmental determinism helped justify the behavior of colonialists, by supporting the notion that people from the temperate regions (Europe) were superior to people from the tropics. Even though we may think of science as the neutral pursuit of knowledge, we should also recognize that science may be influenced by culture and politics.

For other examples of readings that partially explain varying levels of development in terms of differences in geography, see Jeffrey Sachs's 1997 article in the *Economist* entitled "Nature, Nurture and Growth" or Ricardo Hausmann's 2001 article in *Foreign Policy* entitled "Prisoners of Geography." For a text that questions this view, see *Late Victorian Holocausts* by historian Mike Davis (Verso Press, 2002).

ISSUE 4

Did Colonialism Distort Contemporary African Development?

YES: Marcus Colchester, from "Slave and Enclave: Towards a Political Ecology of Equatorial Africa," *The Ecologist* (September/ October 1993)

NO: Robin M. Grier, from "Colonial Legacies and Economic Growth," *Public Choice* (March 1999)

ISSUE SUMMARY

YES: Marcus Colchester, director of the Forest Peoples Programme of the World Rainforest Movement, argues that rural communities in equatorial Africa are today on the point of collapse because they have been weakened by centuries of outside intervention. In Gabon, the Congo, and the Central African Republic, an enduring colonial legacy of the French are lands and forests controlled by state institutions that operate as patron-client networks to enrich indigenous elite and outside commercial interests.

NO: Robin M. Grier, assistant professor of economics at the University of Oklahoma, contends that African colonies that were held for longer periods of time tend to have performed better, on average, after independence.

T he degree to which the colonial experience has impacted contemporary patterns of development in Africa is a major issue of discussion. Prior to 1880 roughly 90 percent of sub-Saharan Africa was still ruled by Africans. Of course slavery had had a profound impact on the continent, but the European presence in Africa was limited largely to coastal enclaves before this time. At the famous Berlin Conference of 1884–1885 (at which Africans were not represented), the European powers established ground rules to divide up the continent. Two decades later the only uncolonized states were Ethiopia and Liberia. The principal colonial powers in Africa were Great Britain and France, followed by Portugal, Belgium, Germany, and Spain. Italy would also control some African countries in the late colonial period.

What were the objectives of European colonial endeavors in Africa? Of course the colonizers themselves suggested, or rationalized, that they were agents of progress who had a civilizing influence on the African people. The reality, widely accepted by scholars today, is that African lands and peoples were colonized to provide key raw materials for the European powers. While accepting resource extraction as the principal goal of colonialism, some scholars assert that the experience had some positive benefits, namely the development of infrastructure (roads, railroads, etc.) and educational and medical systems that benefit the African peoples today. Others suggest that the influence of colonialism has been overblown, and that African leaders merely use the colonial experience as a scapegoat for their own mismanagement of African affairs.

Those who suggest that colonialism has had an enduring negative influence on Africa point out that the infrastructure, as well as the educational systems left behind, nonsensical national borders, and the political legacy of colonial governance, all serve to distort, rather than to facilitate, contemporary economic and political development. For example, the roads and railways built during the colonial era often extend from the capital or port city to interior regions rich in resources, rather than connecting a country in a fashion that would promote national unity. Educational systems emphasized rote learning as they were developed to train low-level civil servants, rather than managers and directors. Enduring and nonsensical national borders have compromised the economic viability of many African nations (there are a large number of landlocked countries in Africa) and aggravated ethnic tensions (because they often do not respect ethnic boundaries). Finally, colonial governance was anything but democratic.

In the following selections, Marcus Colchester and Robin M. Grier present contrasting views on the enduring legacy of colonialism in Africa. Colchester examines contemporary patterns of resource extraction in equatorial Africa and draws a link between these and colonial practices. In Gabon, the Congo, and the Central African Republic, an enduring colonial legacy of the French are lands and forests controlled by state institutions that operate as patron-client networks to enrich indigenous elite and outside commercial interests. In many ways, Colchester is suggesting that colonial patterns of resource extraction never really ended; we simply have gone from overt colonial control to neocolonial regimes of extraction in the postindependence era. In contrast, Grier argues that on average, African colonies that were held for longer periods of time tend to perform better after independence. Furthermore, she asserts that the level of education at the time of independence helps explain the development gap between former British and French colonies in Africa.

YES

Marcus Colchester

Slave and Enclave: Towards a Political Ecology of Equatorial Africa

The three countries of Equatorial Africa—Gabon, the Congo and the Central African Republic (CAR)—are among the most urbanized in Africa, largely as a result of the resettling of rural communities in colonial times. Of a total population of five million, only some 40 per cent live in rural areas, comprising about one hundred, linguistically closely-related peoples who are described in Western anthropology as the "Western Bantu," as well as some 120,000 "pygmies." Those remaining in the forests are reliant on self-provisioning economies, based on shifting cultivation, treecropping, hunting and fishing.

Some 47 million hectares of closed tropical forests in the three countries, combined with those of neighbouring Angola, Cameroon, Zaire and Equatorial Guinea, make up the second largest area of closed tropical forests in the world after Amazonia. These forests are some of the most diverse in Africa, and contain an abundance of wildlife, including forest elephants, lowland gorillas, various kinds of chimpanzees, forest antelopes and a wide variety of birds.

The social and political structures of Equatorial Africa have been markedly transformed by the European slave trade, the French colonial era and by the subsequent interventions of commercial interests and the new African states. Despite the current political liberalization taking place in Gabon, the Congo and the Central African Republic after three decades of single party politics, the forests and the peoples who rely on them are still being sidelined.

The Slave Trade

Long before Bantu society came into contact with colonial Europeans, its egalitarian traditions—a reflection of mobile, decentralized settlements which included an unease towards the accretion of power—had been progressively overlain during several centuries by more hierarchical forms of social organization, resulting from warfare and competition for land. This process had gradually reduced the accountability of leaders to their people with disastrous social and, later, ecological consequences. The European slave trade, however, is the most obvious example of this phenomenon, during which several millions of Africans died and millions more were transported overseas. The Portuguese

From *The Ecologist,* vol. 23, no. 5, September/October 1993, pp. 166–171. Copyright © 1993 by The Ecologist. Reprinted by permission.

began the trade in slaves on the coasts of Equatorial Africa around 1580, but it only became vigorous after about 1640. The European slavers themselves, however, were minimally engaged in raiding, never going far inland. Capture was carried out by Africans which not only intensified raiding and war between local African communities but also transformed previous systems of bondage and servile working conditions into those of absolute slavery.

This meant that slaving had a profound impact not only on those communities whose members were captured but also on those engaged in the trade. To gain control of the trading network, dispersed and differentiated social groups merged their numbers and identities to increase their power and domain. These groups were dominated by "trading firms" often comprising a chief and his sons. The increasing importance of inheritable wealth, capital accumulation and the corresponding need to resist redistributive customs led many matrilineal societies of the middle Congo to become patrilineal. Trading firms swelled their numbers by recruiting male and female slaves as labour and as wives to produce pliable heirs devoid of inheritance rights through the maternal line.

Colonial Repression and Resistance

Direct colonial rule of Equatorial Africa by the French from the 1880s onwards further exacerbated the tendency towards more hierarchical forms of social organization. As the French government was unable and unwilling to administer directly the vast area of more than 700,000 square kilometres, it allocated 80 per cent of the region to some 40 companies.

Within these vast concessions—the area controlled by the Compagnie Française du Haut Congo encompassed 3.6 million hectares—the companies had almost sovereign control including the right to their own police force and legislation, slavery, violence, killing and inhuman punishment were widely documented. Their express aim was to extract the natural resources—chiefly wild rubber, timber, ivory, and later coffee, cocoa and palm oil—as cheaply and as quickly as possible.

Labour shortages, rather than shortages of land, were the main constraint of these enclave economies. To extract labour for the concession areas, plantations, road-building, portage and, in the 1920s and 1930s, for building the Congo-Ocean railroad, "man hunts" became fundamental to the colonial economy. The military was used to support concessionaires who had insufficient labour. In the rubber regions of Ubangi (now the Central African Republic), for example, villagers who did not manage to flee from the troops sent in to ensure their "participation" were tied together and brought naked to the forests to tap the rubber vines. They lived in the open and ate whatever they could find. A French missionary who witnessed the scene wrote: "The population was reduced to the darkest misery . . . never had they lived through such times, not even in the worst days of the Arab [slaving] invasions."

The colonial authorities also introduced various taxes and levies to oblige local peoples to enter the cash economy; in practice, this meant to work for the concessionaires. Along the coasts, logging rapidly replaced the trade in non-timber products, becoming the foundation for the political economy of the region. The extraction was carried out with axes and handsaws, the huge logs being dragged by hand to the rivers and floated out to waiting trading ships. By the 1930s, an estimated third to one half of men from the interior villages of what is now Gabon had been brought down to the logging camps on the coast, where they worked for minimal wages and in appalling conditions for months at a time. Disease was rife, alcoholism rampant and thousands died. Meanwhile, the workload of the women and children who remained in the villages increased correspondingly and their health too was undermined by the numerous diseases—influenza, yellow fever, sleeping sickness and venereal infections—which returning labourers brought back.

The conscience of the colonial authorities was obviously pricked by the all too evident degradation and exploitation of the local people. As Governor General Reste noted in 1937:

> "The logging camps are great devourers of men . . . Everything has been subordinated to the exploitation of the forest. The forests have sterilized Gabon, smacking down the men and taking off with the women. This is the image of Gabon: a land without roads, without social programmes, without economic organization, the exploitation of forests having sapped all the living force from the country . . . There is not a single indigenous teacher, doctor or vet, agricultural officer or public works agent."

Those making the profits—some two billion francs between 1927 and 1938—were a few French companies in whose hands the logging concessions were concentrated. By 1939, of the one million hectares of Gabon under concessions, 66 per cent was controlled by just seven companies with only some 84 others controlling the rest. These companies were to play a key role in the transition to Independence.

Resistance, however, was widespread throughout French Equatorial Africa; villages flared up against demands to yield labour and forest products to the concessionaires and taxation to the French administration. The result was a protracted and brutal war in which the colonial regime sought to bring the whole interior of the colony under its control to facilitate its commercial exploitation. The taking of hostages, including women and children, pillage, arbitrary imprisonment, executions and massacres, the torching of settlements and the sacking of whole communities were commonplace. By the 1930s, when the last areas of resistance were being quashed in what is now the west of the Central African Republic, the French were using planes to spot villagers hiding out in the forests before sending in the army to bring them out.

Regroupement des Villages

Resettlement of dispersed and shifting African communities into larger, permanent villages on roads and portage trails became a central plank of French administrative policy throughout its colony. One of the main aims of this *regroupement des villages* was to control the local people—to oblige them to render tax and labour and to prevent further rebellions.

Regroupement had devastating impacts on the local peoples. One mission-ary reported to André Gide that the local people:

> "prefer anything, even death to portage . . . Dispersion of the tribes has been going on for more than a year. Villages are breaking up, families are scattering, everyone abandons his tribes, village, family and plot to live in the bush like wild animals to escape being recruited. No more cultivation, no more food . . ."

In the early days of colonial rule, it was carried out with little if any con-sideration of customary land rights, causing conflicts over land between dif-ferent social groups. Traditional institutions, residence patterns and ties with the land were overturned while overcrowding in the new villages led to declin-ing standards of nutrition and exposed people to epidemics. Sleeping sickness increased throughout the region in the first half of the 20th century, causing a massive decline in population.

Even after Independence in 1960, Gabon and the Congo pursued the policy right into the 1970s. As a regional *préfèt* noted in Gabon in 1963:

> "We have had enough of these isolated hamlets, which, being so numer-ous and of eccentric location, lost in the vastness of the Gabonese for-est, have never allowed the Gabonese government to have control of their populations or permanent contact with them and have thereby prevented an improvement of their standard of living."

Convinced that "no family head can cut himself off from the duty to modernize," the post-colonial government continued to tear villages away from their crops without any compensation, overturning residence patterns that reflected local ways of life and throwing into disarray, at least temporarily, traditional systems for allocating rights to land. In some respects, *regroupement* under the independent administration was more onerous than in the colonial era: houses in the new settlements had to be laid out in regular rows with defined sizes of houseplots and pathways.

The Indigenous Elite

The striking continuity between the colonial and independence policies was assured by an élite of French-educated Africans. Under French law, most Afri-cans were subject to the *indigenat,* a set of laws which ascribed an inferior sta-tus to local people: "persons under the *indigenat* were subject to penalties and taxation without the legal protection afforded 'citizens.'" However, a small number of Frenchified indigenous people were considered as *évolué* and thus accorded "citizen" status, becoming a local "élite attuned to the French pres-ence and subservient to its interests."

Cooption of the indigenous leadership extended down to the commu-nity level through a hierarchy of *chefs du canton and chefs du village,* chosen to act as intermediaries between the villagers and the authorities. Dependent on the colonial administration for their positions and often resented and even secretly ridiculed by the villagers themselves, these leaders became ready tools in the colonialists' hands and assisted with the unjust exploitation in the concessions. Favoured ethnic groups emerged, considered to be more "evolved" and less "backward;" groups which had been intermediaries in pre-colonial trade now became an important support base for the administration. This structure and practice still persist; local leaders and chiefs, and favoured ethnic groups, continue to owe their primary allegiance to the urban élites and to the administration, not to the villagers or to the remoter, more traditional, rural communities.

Independence: *Plus Ça Change . . .*

After Independence from France in 1960, French Equatorial Africa was divided into the three countries of Gabon, the Congo and the Central African Republic, a division based on the previous colonial administrative divisions. The crucial concern of the departing colonial power was to ensure that the new "independent" governments supported French interests; President de Gaulle warned African states that "France would intervene if it considered its interests in jeopardy." In both Gabon and the Congo, France's principal aim was to guarantee that French logging companies were assured continued access to the forests. Maintaining access to strategic minerals, notably manganese and uranium, the latter being of critical importance to France's civil and military nuclear programme, was also central to French policy. In the Central African Republic, France's main preoccupation was to protect its cotton, coffee and diamond interests.

Gabon

French interests were decisive in selecting the future leadership in Gabon after Independence; French logging interests poured funds into the successful election campaign of Leon Mba, an *évolué* from the coastal region. After Mba's accession to power, the press was suppressed, political demonstrations banned, freedom of expression curtailed, other political parties gradually excluded from power and the Constitution changed along French lines to vest power in the Presidency, a post that Mba assumed himself. However, when Mba dissolved the National Assembly in January 1964 to institute one-party rule, an army coup sought to oust him from power and restore parliamentary democracy.

The extent to which Mba's dictatorial regime was synonymous with "French interests" then became blatantly apparent. Within twenty-four hours of the coup, French paratroops flew in to restore Mba to power. After a few days of fighting, the coup was over and the opposition imprisoned, despite widespread protests and riots. The French government was unperturbed by international condemnation of the intervention; the paratroops still remain in the Camp de Gaulle on the outskirts of Gabon's capital, Libreville, to this day, where they share a hilltop with the presidential palace, an unforgettable symbol of the coincidence of interests between the French and the ruling indigenous élite.

With the establishment of a one-party state, abuse of power and office became the norm. Wealth became more concentrated in the hands of the ruling élite, and the network of patronage became further removed from the concerns of ordinary citizens. The Presidency passed smoothly from Mba to Omar Bongo, "the choice of a powerful group of Frenchmen whose influence in Gabon continued after independence." Bongo and his cronies have since amassed substantial fortunes, having "transformed Gabon into their private preserve, handsomely enriching themselves in the process." They have transferred *billions* of French francs annually to Swiss and French banks. Opponents of the regime have been arbitrarily imprisoned and tortured, among other human rights violations.

Central African Republic

Post-Independence politics in the Central African Republic were not dissimilar. After the independent-minded first President Barthelemy Boganda died in an aeroplane accident, his successor, David Dacko, received strong French support. But Dacko's repressive policies and lack of effective economic reforms made him unpopular locally; he was widely perceived as a "French puppet caring only about cultivating French interests."

When Jean-Bedel Bokassa replaced Dacko on New Year's Day 1966, it seemed that the country, which had been virtually bankrupted by Dacko's regime, might be given a chance to recover. However, Bokassa's capricious and violent rule became synonymous with the worst excesses of African dictatorship—"the systematic perversion of the state into a predatory instrument of its ruler." Massive corruption was the norm, and Bokassa himself appeared to make no distinction between the revenues to the state treasury and his personal income.

France only withdrew its support for the regime in 1979, when it was revealed that Bokassa visited prisons personally to torture and kill those who had stood up to his whims. As in Gabon, French paratroops were sent in and Dacko restored to power, to be replaced on his death in 1981 by army strongman General Andre Kolingba, the current President.

Congo

Independence in the Congo pursued a different course. Initially, the post-Independence regime was modelled in the neo-colonial mould—servile to French political and economic interests. However, "it was swiftly corroded by venality and became an embarrassment not only to its internal supporters but also to its French sponsors." After a street uprising in 1963, the regime was overthrown and a Marxist-Leninist government assumed power. The French did not intervene.

French influence within the Congo remained strong, however, and, despite the Congolese government's rhetoric to the contrary, the role of foreign capital was scarcely diminished. Although the state created marketing monopolies for agriculture and forestry, and nationalized some other sectors—including the petroleum distribution network—the timber concessions, some of the oil palm plantations and many other import-export concerns remained in foreign hands. Foreign oil companies, too, were assured a satisfactory cut.

Logging Enclave Perpetuated

In all three countries of Equatorial Africa, logging has intensified since Independence. Mechanized logging extended the area of extraction during the 1970s and 1980s, up to the remote forests of the northern Congo and the south of the CAR; the rate of extraction increased some six-fold between 1950 and 1970. The industry has remained an enclave of foreign companies who enjoy the patronage of the governing élites. In the Congo, foreign companies, or joint operations dominated by foreign capital, produce the vast bulk of the timber—nearly 80 per cent of the sawlogs, 90 per cent of the sawn wood and 92 per cent of the plywood. In common with both the Gabon and the CAR, the Congolese government "largely lacks adequate technical and economic competence to control and rationally manage its forests."

During the 1970s, logging in the Congo was widely used as a fraudulent mechanism for capital flight, through false declarations of the quantity and type of timber being exported and through transfer pricing. The government attempted to curb this by creating a state monopoly to market timber, but the inefficient and ineffective agency ran at a loss due to collusion between loggers and officials.

A leaked report, prepared for the World Bank, reveals that the logging industry in the Congo is still swindling the government of millions of dollars. Unpaid taxes, stamp duty, transport and stumpage (duty payable on each tree cut) fees are estimated to exceed US$12 million on declared production alone, while huge quantities of timber are slipping across the border illegally into neighbouring Cameroon and Central African Republic. According to the report, "almost all the companies in the forestry sector are 'outside the law'" and "forestry administration is nonexistent." As a result, "the forest is left to the mercy of the loggers who do what they like without being accountable to anyone."

Companies are taking maximum advantage of this lack of supervision. For example, the French company, Forestière Nord Congo, has exclusive rights over 10 years to log some 187,000 hectares in the north of the country. Its contract obliges the company to process 60 per cent of the logs on site and to establish a major sawmill and wood-processing works constructed out of new, imported materials. In exchange, Forestière Nord Congo has received generous benefits—substantially reduced import duties and a five-year tax holiday on wood production, including company taxes, property tax and stumpage fees. Yet, in complete violation of the agreement, the company bought a nonfunctioning second-hand mill, has not processed any timber, has exported sawlogs for six years through its tax loophole, and has not paid even the tax it should have rendered. In total, Forestière Nord Congo has cost the Congo some US$2.9 million in lost revenue.

Elsewhere, the Société Congolais Bois de Ouesso (whose board includes the Congolese President himself) received foreign "aid" in the late 1980s and early 1990s to install a highly-sophisticated saw mill and veneer producing works. With technical advice from the Finnish company, Jaako Poyry Oy, the World Bank backed the project with some US$12 million. But costs soon rocketed from an estimated US$39 million to US$63 million. Further loans were incurred from several African banks, but the mill was never completed; today only a small sawmill is working. The World Bank report notes "the situation is catastrophic and no further activity within the present arrangement is possible." The company's failure is attributed "quite simply to the overvaluation of the project which has allowed some vultures to enrich themselves immeasurably at the Congo's expense."

In the Central African Republic, where concessions have been granted in 48 per cent of the "exploitable" forests, the accountability of the logging industry is more lax. Illegal cross-border logging into the forests of north Congo was observed by FAO [Food and Agriculture Organization] technicians in the mid-1970s, and smuggling of timber down the Oubangi and Congo rivers to the Congolese capital, Brazzaville, on the Zairean border was normal. Today it is common knowledge in Bangui, the capital of the Central African Republic, that nearly all the forest concessions are being illegally logged, the regulations are flouted and much of the timber is clandestinely leaving the country via new road connections with Cameroon. CAR President Kolingba himself is alleged to be closely linked to businessman M. Kamash who owns SCADS, the company which processes timber illegally slipped across the border from north Congo. In common with all the concessionaires in the country, "SCAD carries out no management and employs no foresters—they are just timber merchants who mine the forests."

Since the Gabonese forestry service is funded from the central government budget rather than through stumpage fees and other tariffs on logging itself, there are few incentives for foresters to impose the many rules and regulations to control logging, while the loggers themselves make sure the foresters are provided with suitable incentives not to apply them.

Rural Stagnation

For the rural communities of Equatorial Africa, the long history of exploitative, extractive and enclavistic development has meant marginalization and poverty. Despite the statistically high per capita incomes of Gabon and the Congo relative to other Sub-Saharan countries, the rural people are poor and getting poorer.

The explicit aim of most government efforts concerning agriculture has been to replace itinerant, family-based, labour-intensive agriculture with fixed, capital-intensive, mechanized agriculture serviced by wage labour. Heavily-subsidized agribusiness schemes to promote large-scale farming—cattle ranches, sugar plantations, battery farms of poultry, rice schemes, rubber and oil palm estates, and banana plantations—have undercut small farmers, destroying the last elements of a cohesive rural society.

A century of neglect and disruption of small-holder agriculture has had inevitable consequences. In the Central African Republic, only one per cent of the country is farmed. In Gabon, agriculture accounts for only eight per cent of the GDP [gross domestic product], occupies only 0.5 per cent of the land area and supplies only 10–15 per cent of the country's food needs, the remaining 85 per cent being imported; even traditional peasant crops such as taro, yams, mangoes, avocados and vegetables are imported from neighbouring states, particularly Cameroon.

In the Congo, agriculture yields only 5.9 per cent of GDP. Since Independence, there has been a massive migration to the cities. By 1990, 52 per cent of the total population, and 85 per cent of men aged between 25 and 29 years, lived in two cities alone, Brazzaville and Pointe Noire, although the Congo barely has any industrial base. As in colonial times, the lack of young men in the countryside means that 70 per cent of Congolese farms are managed by women. Education, preferentially given to young men, exacerbates this trend, resulting in the towns being considered the domain of men and the countryside that of women.

The marginalization of smallholder agriculture has also transformed local political institutions. Increasing mobility has weakened community ties and diminished customs which favour the redistribution of wealth and land. Echoing the political shifts which took place during the European slave trade, matrilineal groups have become more patrilineal and marriages with women classified as "slaves" (that is, without lineage and therefore without kin to make demands on agnatic inheritance) have been favoured. These internal trends have been reinforced by imposed national laws which favour cognatic succession, all tendencies which have further weakened the status and security of women.

Robin M. Grier

Colonial Legacies and Economic Growth

Introduction

Development theorists have long hypothesized whether the identity of the colonial power mattered for subsequent growth and development. Many authors . . . once concluded that the colonial experience was insignificantly different under the major colonial powers. Recent research though has shown that colonialism did significantly affect development patterns. . . . In this [selection], I examine whether the duration of colonization has a significant effect on later development and growth and whether human or physical capital in place at the time of independence can help to explain why British colonies perform significantly better than French ones.

In the first empirical application of the [selection], I pool data from 63 former colonies and find that the length of colonization is positively and significantly correlated with economic growth over the 1961–1990 period. While the direction of causality is not conclusive, I find no evidence to support exploitation theory. Given that a country was colonized, since it would be impossible to test the counterfactual hypothesis, the longer it was held by the mother country, the better it did economically in the post-colonial era.

In the second empirical application, I reduce my sample to 24 countries in Africa. . . . I find that the length of colonization is still positively and significantly related to economic growth and that former British colonies still outperform their French counterparts. . . . I find that the newly independent British colonies were significantly more educated than the French ones. . . . I find that the inclusion of education at independence can explain the development gap between the former British and French colonies and the positive relationship between length of colonization and growth. . . .

Data and Variables

My empirical work addresses two questions. First, does the length of the colonization period matter for subsequent growth and development? Second, can education levels at the time of independence explain why British ex-colonies perform significantly better than their French counterparts?

In the first empirical application of the [selection], I perform a cross national study of 63 ex-colonial states and test to see whether the length of

From *Public Choice*, March 1999, excerpts pp. 317–318, 321–326, 328. Copyright © 1999 by Springer Science and Business Media. Reprinted by permission via Rightslink.

colonization is correlated with subsequent growth rates. . . . In the second empirical application, I reduce the sample to British and French Africa and test whether the duration of colonization matters for later development and whether human or physical capital levels at the time of independence can help to explain why British ex-colonies perform better on average than French ones. . . .

Econometric Results

Does the Identity of the Colonizing Power Matter for Economic Growth?

Five year averages are calculated on data from 63 countries for the years 1961–1990, resulting in 6 observations for each country and a sample size of 378 data points. . . .

Because of the vast differences in the length of colonization for different ex-colonial states, I add a variable called TIME to determine the effect of the duration of colonialism on subsequent growth, where TIME is the year of independence less the year of arrival. The most common argument against colonialism is that it exploited the native population, causing dependency and instability in the colonial states. The results . . . show that TIME is positively and significantly related to subsequent economic growth. I do not claim that longer colonization causes higher economic growth, but my findings do reject a crude form of the exploitation theory. While it is possible that countries in my sample might have had higher growth rates if they had not been subjected to colonialism, the results do show that colonies that were held for longer periods of time than other colonies have had more economic success in the post-colonial era. . . .

Does Colonialism Matter for African Development?

. . . I found that former French colonies perform significantly worse on average than British ones. It is possible that an African effect is exaggerating the development gap between the two. Because all of the French colonies were African (except Haiti, whose population is primarily African), and Africa has been characterized by poverty and underdevelopment in the last century, it may be that an African effect is biasing the earlier results. Limiting the sample to Africa also gets rid of the high-performing British outliers, like the United States and Canada, and helps to better distinguish the differences between French and British non-settlement colonies. . . .

French ex-colonies perform 1.38 percentage points worse on average than their British counterparts. That is, even when the sample is limited to Africa, the former French colonies still lag significantly behind the British ones. . . .

The results also show TIME to be positively and significantly related to economic growth, which implies that the length of colonization is positively related to growth and development. The next section, which examines the institutional legacies of colonialism, tries to determine why TIME is significantly related to growth and why the French ex-colonies perform worse than British ones, even when the sample is restricted to Africa.

The Colonial Legacy

To determine why ex-British colonies perform better than French ones, I look at the level of human and physical capital the colonizing power left at independence. . . . [E]ducation is an important component of growth and development. The British and French had contrasting philosophies of education, which translated into very different types of colonial education. The largest difference between colonial education policies is that the British made a conscious effort to avoid alienating the native culture, by teaching in the vernacular languages and training teachers from the indigenous tribes. While most of the teachers in French Africa were imported from France, the Advisory Committee on Native Education in the British Tropical African Dependencies recommended that, "Teachers for village schools should, when possible, be selected from pupils belonging to the tribe and district who are familiar with its language, tradition, and customs. . . ."

Students in British Africa were, for the most part, taught in their native language and in their tribal villages, which significantly eased the learning process. In contrast, in the French system, most students were boarded and only able to go home for the summertime vacation. Students were required to speak French, and all vernacular languages were forbidden. . . .

Conclusion

The literature on colonialism and underdevelopment is mostly theoretical and anecdotal, and has, for the most part, failed to take advantage of the more formal empirical work being done in new growth theory. This essay has tried to close that gap by presenting some empirical tests of oft-debated questions in the literature.

I find that the identity of the colonizing power has a significant and permanent effect on subsequent growth and development, which would deny the validity of a crude exploitation hypothesis. Colonies that were held for longer periods of time than other countries tend to perform better, on average, after independence. This finding holds up even when the sample is reduced to British and French Africa.

I also find that the level of education at the time of independence can help to explain much of the development gap between the former British and French colonies in Africa. Even correcting for the length of colonization, which has a positive influence on education levels and subsequent growth, I find support for a separate British effect on education. That is, the data imply that the British were more successful in educating their dependents than were the French.

POSTSCRIPT

Did Colonialism Distort Contemporary African Development?

Colchester describes a long process wherein relatively egalitarian and decentralized rural Bantu settlements in equatorial Africa were slowly transformed into more hierarchical forms of social organization. This process began with warfare and competition among Africans, was exacerbated by the slave trade and French colonial rule (for an excellent discussion of contemporary academic debates regarding the involvement of Africans in the trans-Atlantic slave trade, see Lassina Kaba's 2000 article in the *African Studies Review,* entitled "The Atlantic Slave Trade Was Not a 'Black-on-Black' Holocaust"), and continues today as the indigenous elite are more responsive to outside commercial interests than their own people. The limited accountability of leaders to their own people has led to exploitative logging with disastrous social and ecological consequences. As such, colonialism is not solely responsible for the lack of accountability among some African leaders in equatorial Africa today, but Colchester would argue that it definitely facilitated such behavior and created a set of relationships between French and African elites that persist today.

The process by which African elites were co-opted is worth highlighting. Colchester describes a Frenchified indigenous elite who were considered "advanced" enough to be granted French-citizen status. Through a process of acculturation, this elite came to share in and promote the interests of the French in their colonies. This approach, adopted by the French in their African colonies, is often referred to as the *policy of assimilation*. The policy of assimilation held out the carrot to all Africans (in French colonies) of potentially becoming French citizens. Although seemingly egalitarian on the one hand, it was a terribly ethnocentric policy on the other because it asserted that Africans only became civilized by adopting French culture. In many ways, the French policy of assimilation was an effective strategy for developing the group of "dependent elites" conceptualized by Andre Gunder Frank in his dependency theory.

The French policy of assimilation may be contrasted with the British policy of indirect rule. In nonsettler colonies such as Ghana and Nigeria, the policy of indirect rule meant that the British administered a country via its traditional leaders (up to a certain level in the colonial hierarchy). This difference in policy approaches is often used to explain why the British invested more heavily in education (because they relied on more Africans to run the colonial civil service). This is also a difference highlighted by Grier in her article for this issue. Grier further notes that the British encouraged instruction, at least at the lower grade levels, in local languages, whereas the French insisted that local people be educated in French (an approach consistent with their policy of assimilation).

Of course the French and the British were not the only colonial powers in Africa, but their approaches receive the most attention among academics because they had the largest number of colonies. For further information on the legacy of British and French colonialism, see a book chapter by the well-known Africanist Crawford Young, entitled "The Heritage of Colonialism," in John W. Harbeson and Donald Rothchild, eds., *Africa in World Politics: The African State System in Flux* (Westview Press, 2000). The Belgians and Portuguese have a reputation for having been particularly brutal colonial powers. See Adam Hochschild, *King Leopold's Ghost: A Story of Greed, Terror, and Heroism in Colonial Africa* (Houghton Mifflin, 1998) for a gripping account of conditions in the Belgian Congo, or Allen Isaacman, *Cotton Is the Mother of Poverty: Peasants, Work, and Rural Struggle in Colonial Mozambique, 1931–1961* (Heinemann, 1996), regarding the situation under the Portuguese in Mozambique.

Internet References . . .

World Bank Group for Sub-Saharan Africa

World Bank Group for Sub-Saharan Africa includes annual reports, publications, speeches, and other sources of information about issues including rural development, education, and the incorporation of indigenous knowledge into development.

http://web.worldbank.org/WBSITE/EXTERNAL/COUNTRIES/
AFRICAEXT/0,,menuPK:258649~pagePK:158889
~piPK:146815~theSitePK:258644,00.html

United Nations Development Programme

The United Nations Development Programme provides information on UN development work in Africa. A wide range of topics is covered, including poverty and globalization.

http://www.undp.org/africa/

Economic Commission for Africa

The United Nations' Economic Commission for Africa provides information on the regional integration of African economies, including information on regional economic organizations. The organization is attempting to reform and integrate African economies to improve the quality of life on the continent.

http://www.uneca.org/index.htm

United States Agency for International Development (USAID) in Africa

USAID in Africa details African development from the perspective of USAID, the bilateral development assistance arm of the U.S. government. Information is available by country, topic, or date and ranges from short articles and press releases to full publications.

http://www.usaid.gov/locations/sub-saharan_africa/

Development

What constitutes "development" in the African context very much remains an open question. Since the majority of African nations gained independence in the 1960s, how best to pursue this process has been an often-debated question. The role of the state, the commercial sector, international financial institutions, and donors in development is highly contested. Controversies have also raged over the extent to which Africa's progress on the development front has been influenced by its position in the global economic system, its relationship to major economic powers, and emerging technology clusters.

- Have Free-Market Policies Worked for Africa?
- Do Cell Phones and the Internet Foster "Leapfrog" Development in Africa?
- Is Increasing Chinese Investment Good for African Development?
- Does Foreign Aid Undermine Development in Africa?

ISSUE 5

Have Free-Market Policies Worked for Africa?

YES: **Fudzai Pamacheche and Baboucarr Koma**, from "Privatization in Sub-Saharan Africa—An Essential Route to Poverty Alleviation," *Africa Integration Review* (2007)

NO: **Thandika Mkandawire**, from "The Global Economic Context." In: B. Wisner, C. Toulmin, and R. Chitiga, eds., *Towards a New Map of Africa* (Earthscan, 2005)

ISSUE SUMMARY

YES: Fudzai Pamacheche, of the Southern African Development Community Directorate on Trade, Industry, Finance, and Investment, and Baboucarr Koma, of the Private Sector Development, Investment and Resource Mobilization Division of the African Union Commission, argue that privatization is an example of a free-market policy that has benefited African people. They suggest that the privatization of public enterprises creates efficiency gains, stable and reduced prices, lower government subsidies and the redirection of scarce resources elsewhere, the payment of dividends to government, and increased employment. They further argue that privatization programs, if well designed, are crucial for poverty alleviation and economic integration regionally and globally.

NO: Thandika Mkandawire, director of the UN Research Institute for Social Development, counters that while African governments have reshaped domestic policies to make their economies more open, growth has faltered. Mkandawire assesses structural adjustment from a developmental perspective, judging its effects on economic development and the eradication of poverty. He suggests that structural adjustment policies designed to integrate Africa into the global economy have failed because "they have completely sidestepped the developmental needs of the continent and the strategic questions on the form of integration appropriate to addressing these needs."

In order to understand the past 30 years of free-market policies in Africa, it is important to take a step back to understand the institutions and particular histories which informed this approach. Two critical players are the World Bank and the International Monetary Fund (IMF), which are sometimes referred to as Bretton Woods institutions (along with the World Trade Organization) because they were established during a conference of the major economic powers at Bretton Woods, New Hampshire, in 1944. The World Bank, or International Bank for Reconstruction and Development (IBRD), was established to rebuild Europe and Japan in the aftermath of the war and the IMF was created to provide loans to help countries resolve short-term balance of payment problems. The IMF and the World Bank have persisted as major multilateral economic development institutions that are particularly powerful in Africa. These institutions are influential in Africa because much of the debt incurred by national governments is public rather than private (i.e., loans from bilateral or multilateral development agencies rather than loans from commercial banks). In Africa, the IMF tends to focus on lending to resolve short-term problems (e.g., controlling inflation), whereas the World Bank has a slightly longer "developmental" view.

Following the Third World "debt crisis" of the 1970s, the World Bank and the IMF initiated a form of free market–oriented, policy-based lending known as structural adjustment. Prior to this time, much of the lending of the World Bank had been project or program based. So, for example, in the 1960s the World Bank funded a number of infrastructure projects in Africa (such as dams), slowly transitioning to more programmatic funding related to basic needs in the 1970s (such as rural health care projects). The policy-based lending that began in the 1980s was somewhat different than traditional project or program-based lending in that loans were held out as a carrot for countries that agreed to undertake a series of free market–oriented policy reforms. The basic aim of structural adjustment reform was to balance state budgets and, as either a cause or effect of the first, promote economic growth. According to the World Bank, the basic way to achieve such an end was to cut government expenditures and raise revenues (but not in way that would encumber economic growth).

Policy reforms under the structural adjustment rubric in Africa include the privatization of inefficient state-run enterprises (which, if they are inefficient enough, may be a drain on the state treasury), a reduction of staffing and programming in state agencies in order to cut costs, the devaluation of national currencies in situations where they are deemed overvalued, and a redoubling of efforts in the export sector to foster the generation of foreign exchange. Overvalued currencies are seen as problematic because they may make a country's exports artificially expensive (and may reduce export potential) and its imports artificially inexpensive (and thereby encourage the overconsumption of imported products).

YES

Fudzai Pamacheche and
Baboucarr Koma

Privatization in Sub-Saharan Africa—An Essential Route to Poverty Alleviation

Introduction

Africa continues to face difficult socio-economic conditions and a number of challenges, among them, low level of human development, low levels of productivity, poor infrastructure and investment climate will need to be addressed if the continent is to effectively integrate into the global economy, and achieve accelerated economic growth and sustainable development, and reduce poverty as desired. Achievement of poverty reduction targets requires sound macroeconomic policies, increased national savings, mobilization of resources for productive investment and employment creation as well as rationalizing expenditures so that priority areas such as infrastructure and education are given adequate attention.

Africa recorded a Gross Domestic Product (GDP) growth rate of 5.7% in 2006, compared to 5.3% in 2005 and 5.2% in 2004. This improvement is a reflection of the continent's efforts to enhance macroeconomic management. Increased export prices for commodities such as crude oil and minerals also contributed to a better balance of payments position as seen by the increase in the continent's current account surplus from 2.3% of GDP in 2005 to 3.6% in 2006. In terms of debt position, Africa's total debt stock decreased considerably from 35.9% of GDP in 2005 to 26.2% in 2006. This could be explained by the debt cancellation for 14 African countries under the HIPC and G8 commitments. However, due to high interest rates, total debt service obligations have only changed marginally, from 4.2% to 4.1% over the same period. This debt burden, for some countries, continues to constrain spending on public investment, thereby retarding growth and employment creation. On investment flows, Africa's share in world investment increased from 0.6% in 2000 to 3.3% in 2005.

Governments across the continent have developed Poverty Reduction Strategy Papers (PRSP), which attempt to comprehensively address the problem of poverty, through close collaboration with development partners. These governments have seen the urgent need for appropriate policy frameworks

From *African Integration Review,* vol. 1, no. 2, 2007, pp. 1–22. Copyright © 2007 by African-union.org. Reprinted by permission of African Union Commission.

and regional integration as well as the implementation of political and economic reforms that reinvigorate the economy. As part of this drive for reform, privatization of public enterprises has become a key focus of many African governments.

This paper attempts to convey the message that, through privatization, sub-Saharan African countries can achieve an expanded and more dynamic private sector, more efficient and effective infrastructure provision and increased investment, both domestic and foreign. These positive developments emanating from privatization will subsequently lead to the achievement of poverty alleviation goals, given their direct impact on economic growth and subsequently, job creation in these countries. For such growth to have a direct impact on poverty though, governments must take deliberate targeted measures to address social concerns. For example, many water privatizations in Africa failed to achieve the desired result in several countries (e.g. Ghana). Instead of price reduction and improved services, the poor ended up having to pay more for water, which worsened the poverty situation. Poverty alleviation strategies should include privatization as one of the key elements to avoid this kind of situation. This way, the latter can yield results in favour of the poor.

Another important message is that cross-border linkages between firms in different parts of the continent emanating from the acquisition of stakes in privatized firms in different countries can be a powerful mechanism to enhance the ongoing integration process on the continent. The increased financial dealings and closer ties between companies are important elements of integration in Africa.

A major constraint faced in trying to address privatization issues in Africa is the lack of adequate data on the quantitative impact of privatization. Several studies have been conducted but most of these cover developing countries in general with little specific emphasis on Africa. Despite this shortcoming, however, an attempt has been made to conduct a thorough analysis based on the limited information available. There appears to be very little research in identifying the direct impact of privatization on poverty in Africa and this is a crucial gap that needs to be filled through further research.

I. What Is Privatization

Privatization can be defined in several ways depending on the form it takes. The World Bank defines privatization as *"a transaction or transactions utilizing one or more of the methods resulting in either the sale to private parties of a controlling interest in the share capital of a public enterprise or of a substantial part of its assets,"* or *"the transfer to private parties of operational control of a public enterprise or a substantial part of its assets."* An examination of the different forms of privatization makes it easier to comprehend the above definitions. It is worthy of mention that the particular form adopted depends on the circumstances and the objectives of the privatization.

Forms of Privatization

Privatization can take many different forms, all of which entail some form of private sector participation in product or service delivery. The most common forms are:

1. *Management contract:* Here, responsibility for the provision of services that were hitherto provided by a state-owned firm is passed on to a private provider. Ownership, however, remains with the state and all required capital investments continue to be provided by the state. Usually, a performance contract is signed with outsourced management. Government tends to benefit because management contracts have the tendency to bring market discipline and technical know-how to a state-owned firm and hence, all the efficiency gains that are typical to market-oriented firms are likely to be realized. An example is the Gambia where government issued a management contract to SOGEA to manage Gambia Utilities Corporation.

2. *Lease:* In a lease arrangement, a private firm takes the responsibility of operating and maintaining the assets of a hitherto publicly owned firm. Government retains ownership as well as responsibility for financing capital investments, usually through a special vehicle established for the purpose. Because the new operator has strong incentive to reduce cost and improve efficiency, government will benefit from the efficiency gains that arise as a result as well as dividends where relevant. It is usually of fairly long-term nature, for example 12 years. For example in 1989, the Government of Guinea entered into a lease arrangement for private sector operation of water services in the capital city, Conakry, and sixteen other cities and towns. In the Gambia, a lease contract was issued for the operation of Marine Dockyard.

3. *Concession:* This takes the form of a private firm taking over responsibility for operating and managing the assets of a public enterprise, as in the case of a lease arrangement. However, unlike in the case of a lease, the private firm takes on the further responsibility of financing long-term capital investment of the firm. It also provides incentives for the private operator to minimize cost and increase efficiency. It is usually of very long-term nature, for example 30 years. This form is very common in the utilities sector. Government benefits directly from the improved level of efficiency as well as the reduced burden to undertake long-term capital investments, given that such responsibility is shifted to the private sector operator. In situations where government is faced with resource constraints, this could be an attractive option. Nigeria is an example of a country where concessions were issued for the operation of many ports.

4. *Divestiture:* In divestiture, publicly owned assets are sold to a private sector actor. This means, the management as well as all future capital investment requirements become the responsibility of the new private sector owners. It can be done by auction (where the firm is sold through a bidding process to the highest bidder); private placement (where the shares of a company are sold through direct negotiation); or initial public offering (where shares are sold through a public offering on a local stock exchange). Divestiture can be either partial

(e.g. sale of 49% of shares) or full scale, which entails total transfer to the private sector. Given that public enterprises are usually a drain on government's operating budget, divestiture is desirable because of its potential to stop the negative flow.

An example of a successful divestiture exercise is the sale of 33.3% shares of the Societe nationale des telecommunications (Sonatel) of Senegal in 1997 to a subsidiary of France Telecom which also took over the management of the company.

Depending on the objective of the divestiture exercise, government may implement an *Employee Stock Option Programme.* Here, a predetermined proportion of shares on offer are reserved for the employees of the public enterprise. This has the effect of creating a wide dispersion of equity ownership. It is attractive to governments as a result of its effects on public opinion and the support for the privatization programme. Its drawback is that it can lead to a situation where no one party has sufficient stake in the enterprise to be in a position to influence the board and to press for effective management for the achievement of profit-oriented results. In the example of Senegal's Sonatel above, at the time of the original sale, 10% of the shares were sold to Sonatel employees at a highly discounted rate, easing some of the workers' earlier anxieties about privatization.

5. *Joint venture:* This can take the form of a partnership between an existing public enterprise and a private investor (public-private partnership). It can be an acceptable solution in a situation where full-scale privatization faces much resistance. Risks are generally shared and the struggle for control can be an impediment to success. However, economies of scale and access to new technologies and management expertise can be positive outcomes of joint ventures. An example is Debswana Diamonds, a joint venture between the government of Botswana and De Beers.

6. *Asset sale:* This is the process of selling off the assets of an enterprise, probably following the cessation of operations. It is usually less complex than other forms of privatization. However, the possibility of a number of assets being left over without a buyer is there. Given its nature, this method tends to create much negative public perception. Governments generally tend to be accused of selling off public assets at give away prices in this form of privatization. Such sales are usually inevitable and are usually effected by auction.

II. Privatization Trends

In 2004–2005, sub-Saharan Africa generated US$975 million from 23 transactions, representing just 1% of total global privatization transaction value of developing countries. This is down from 3% in 2000–2003. Nigeria alone accounted for 70%, the bulk of which is due to a 25-year port concession. Tanzania and Ghana together accounted for another 12%, dominated by telecommunications. The rest is accounted for by Sudan, Kenya, Rwanda, South Africa and Madagascar, mainly through telecommunications transactions.

An analysis of sector trends for developing countries in general shows that three sectors accounted for about 90% of total transaction value, namely infrastructure (telecoms, electricity and natural gas, transport, water), finance (banks, insurance, other financial services), and energy (exploration and production of oil and gas, other hydrocarbons). Between 2000–2003 and 2004–2005, infrastructure's share of total value declined from 55% to 42%, the financial sector's share grew from 19% to 30% while the energy sector remained at about 15%.

During the period 2000–2005, sub-Saharan Africa raised US$11 billion in privatization proceeds, representing 3% of the global total for developing countries. A total of 960 transactions were conducted in 37 countries of the region, and this represents the third highest number of transactions, behind Europe and Central Asia and Latin America. However, 70% of the transactions were low-value firms in competitive sectors. Four countries, namely South Africa, Nigeria, Ghana, and Zambia were the largest contributors. South Africa alone has accounted for up to 40% of the region's proceeds in recent years.

III. The Effects of Privatization

In 1998, Peter Young of The Adam Smith Institute, in an article entitled "The Lessons of Privatization" summarized the findings of an earlier study conducted by the Institute with the objective of identifying "The Impact of Privatization in Post-Communist and Developing Countries." The article summarized the findings of the study which reviewed the impact of privatization in these countries by pulling together synthesis of existing research. Conclusions were drawn about the overall impact of privatization, the success of different approaches to privatization, the importance of the policy environment, among others. Below is an overview of the key findings elaborated in the article.

Improving Enterprise Performance
In almost all cases studied, company performance improved after privatization. A World Bank study discovered a performance improvement in eight out of nine developing country cases studied. A sample of sixty company cases studied revealed a substantially improved performance in 75% of the cases. Generally, company profitability surged in a majority of cases and privatization removed existing constraints on new investment and access to capital. Also, through output growth outpacing the growth of labour and other inputs, privatization has the effect of raising productivity and efficiency. This is the situation in a number of cases such as Togo, where performance was observed to have dramatically improved following privatization. In situations like this, enterprises were able to adapt their production to meet real demand.

However, cases in other countries such as Mali did not yield similar results. The lack of improvement in efficiency and productivity was explained by the poor handling of the privatization process itself. Companies were sold to buyers who lacked the ability to run such enterprises or the ability to pay the purchase price, payable in installments. Another source of difficulty in some countries was continued government interference in the aftermath of privatization.

Reducing Government Debt

A number of governments have been able to raise huge sums of money from privatization transactions. These financial resources have enabled the governments to sustain macroeconomic stability and repay huge portions of government debts. As a result of privatization, many governments have also reduced the need for huge subsidies to public enterprises with the consequent impact of strengthening their fiscal positions. In some countries, privatization of unprofitable public enterprises did not only lead to the receipt of large amounts of cash from transactions but also increased government revenues through the cancellation of income tax concessions to those enterprises.

In situations where public enterprises had ceased to operate prior to privatization, or were performing poorly, governments benefited fiscally by removing such enterprises from their books. As a result of the sale, the liabilities that existed in the name of these enterprises at the time of privatization were written off and potential future losses were prevented from growing larger.

Consumers Benefit

The majority of cases studied show that consumers benefit from privatization. This is as a result of lower prices emanating from the efficiency improvements following privatization. For example, privatized energy firms were able to reduce prices sharply as a result of their ability to limit the amount of stolen or unbilled electricity. Also, because investment constraints were removed, privatized firms were in a better position to avail their products to the public. Evidence suggests that privatized firms seek more aggressively to improve quality and introduce new products to satisfy the consumer.

Employing Workers

Evidence suggests that privatization is in the interest of employees, although there are a few exceptions to this. Such benefits take three forms: (a) employment levels tended to increase after privatization, (b) remuneration packages tended to improve after privatization and (c) many employees bought shares at discounted prices in the privatized firms and these benefited when share prices eventually rose. In cases where employees lost their jobs as a result of privatization, such employees tended to receive generous severance packages. Severance and retirement incentives buy labour support and allow privatization and its benefits to happen and, where unemployment insurance systems are not in place, mitigate the social impact of layoffs. It should be noted that in some cases, the reduction in the level of employment took place prior to privatization and as such, could be attributed to the need for greater efficiency, and not just privatization. In cases where shut down enterprises were re-opened by private investors, employees benefited directly.

1. Strengthening Capital Markets and Broader Ownership of Capital

Depending on the type of privatization method used by countries, privatization has led to the strengthening of capital markets and the widening of ownership of capital. Many developing countries have devised schemes for selling shares to employees and these yielded immense benefits. In many cases, shares would be reserved for citizens of the country, with the objective of widening local ownership of the capital.

In the privatization of Senegal's Sonatel, for example, two-thirds of the total shares were reserved for Senegalese nationals and institutions. All of these shares were quickly bought up. In Nigeria, flotation of shares on the local stock markets was seen as an effective tool for greater local ownership of privatized firms. Regional quotas achieved an equitable geographic distribution of shares. Usually, part of a large enterprise is sold to an external strategic investor while a certain percentage of shares is floated through the stock market. The same process was used in the case of Kenya Airways as well as Ashanti Goldfields in Ghana. This exercise is constrained by the fact that many of Africa's stock markets are weak and most people are too poor to afford such shares. However, where such flotation takes place, it tends to have a huge impact on capitalization of the stock exchange.

More Competition

Privatization encourages competition and hence leads to all the benefits associated with it such as improved customer service and reduced prices. In practice, privatization is accompanied by competition and in some cases, privatized firms are given a period of protection while competition is introduced afterwards. Privatization has given impetus to market reforms in many countries. To have an impact, it is important to coordinate the activities of the bodies responsible for privatization and those responsible for competition.

Increasing Investment

Through privatization, many countries have been able to attract significant amounts of foreign investment. This is the case in many Latin American countries. In some African countries, however, privatization accounts for a minimal share of foreign investment due to restrictions placed on such investments.

An indirect benefit is that privatization signals the level of a government's commitment to freer markets and as such, encourages greater greenfield investment and other forms of investment not directly related to privatization. A World Bank study suggests that privatization has a huge impact on investment decisions and further states that an extra 38 cents in new investments is generated for every dollar of privatization revenue. It further documents that financial and infrastructure privatizations have the greatest impact on foreign direct investment.

The findings of the study by Adam Smith Institute support privatization efforts and emphasize the need to pursue privatization more rigorously in the years ahead.

IV. Hostility to Privatization

Why is there so much hostility to privatization in spite of all the potential benefits? Privatization is generally viewed with much skepticism across Africa by all segments of society. African intellectuals and officials have the tendency to view the public sector as the promoter and defender of indigenous interests and to believe that privatization will empower and enrich foreigners at the expense of indigenous people.

African trade unions and workers' representatives equally despise privatization, citing the possibility of job losses or worsening terms of service. A case in point was the protest marches organized by the Zambian Congress of Trade Unions calling on government to rescind its decision to privatize the Zambian National Commercial Bank. In Senegal, the government announced in 1997 its plans to sell majority of shares in Société nationale sénégalaise d'électricité (Senelec), the state electricity company. The company's unions strongly denounced the move and launched a series of strikes and go slow actions which contributed to severe power blackouts country-wide. As a result of such resistance, many governments are now paying more attention to job concerns. Usually, in the process of sale, retention of existing staff is either an explicit criterion or a major consideration. In Burkina Faso, for example, the government received four offers for its sugar complex in Banfora in 1998. It sold the complex to the bidder that offered the lowest price but pledged to maintain the entire workforce while making significant new investments.

Many African politicians and public officials derive huge material and prestige benefits from public enterprises, in the form of loans, gifts, housing, board memberships, future jobs for themselves, procurement kick-backs and so on. All these may no longer be within reach following privatization, hence the resistance.

Domestic private sector also usually have cozy supply relationships with public enterprises. These tend to be threatened by privatization, given the more aggressive, quality-conscious, cost-cutting tendencies of private owners.

What all the above mean is that, in African countries, several powerful groups exist that have material reasons to delay, dilute or sabotage public enterprise reform in general and privatization in particular. They put forward their case by pointing to perceived economic, financial and social shortcomings. In the case of Zambia, a country which ran a privatization programme described by the World Bank in 1998 as the most successful in Africa, many Zambians perceived privatizations as very negative, hence putting pressure on the government to rethink its policy. The case against privatization was based on the following perceptions:

- The programme had been imposed and micromanaged by international financial institutions, without sufficient attention to requisite policy and regulatory frameworks and without adequate involvement of Zambian citizens;
- It would result in the closure of many firms previously run by Zambians;
- It would exacerbate the problem of unemployment;
- It would increase the incidence of corruption; and
- It would benefit the rich, foreigners and those politically well connected.

Do these arguments have merit? In 2001, the Zambian Privatization Agency commissioned a study to assess the impact of privatization. The study showed that 235 of the 254 firms privatized over the period 1991–2001 had continued in operation; 57% of buyers were Zambians and an additional 13% were joint ventures between Zambians and foreigners; post-privatization capital expenditures in the non-mining sector totaled $400 million; 19 firms closed following privatization but 7 of them resumed operations after being resold and efforts were underway to resell an additional 5 at the time of the study; employment declined in the privatized non-mining sector from 28,000 at the time of privatization to 20,000 in 2001 but the workforce expanded in several firms.

Is the situation the same in other parts of Africa? A study in 1998 reviewed 81 privatizations in electricity, agriculture, agro-business and service sectors in Cote d'Ivoire. Their findings were that firms performed better after privatization, they performed better than they would have had they not been privatized, and privatization contributed positively to economic welfare.

Another study in 2001 reviewed 212 privatizations in Ghana. The study showed that privatization had the effect of easing pressure on balance of payments, increasing efficiency, stimulating local capital markets, enhancing the inflow of FDI, creating quality gains for consumers and increasing employment and remuneration post-privatization.

A study in 1998 assessed the impact of privatization in Mozambique and Tanzania and revealed positive changes in operating and financial performance of the divested firms: three-quarters of firms which had ceased operations before divestiture was contemplated, resumed productive operations following privatization, and investment, production, sales and value-added increased significantly following privatization.

Based on the findings of the above studies across Africa, it is evident that privatization of public enterprises yields positive benefits. However, in terms of specific sectors, the limited evidence available suggests that firms producing tradeables do much better in private hands than in state control. The process of privatizing such firms is relatively less complex.

On the other hand, successful privatization of utilities and other infrastructure are much more complex. Privatization of water companies has faced problems in several countries in Africa, including Guinea and Ghana. To achieve success in the privatization of utility and infrastructure firms, there is strong need to create and reinforce the institutional mechanisms that guide and regulate market operations. Some key steps to be undertaken include the enactment of policy framework for and building of implementation capacity in regulatory and competition promotion agencies; capacity building of officials in charge of divestiture and regulation; strengthening of governance and judicial systems; reduction of political interference on enforcement agencies and taking adequate steps to render contracts enforceable.

V. The Potential Effects of Privatization on Poverty

Poverty eradication has become an overriding goal for countries across sub-Saharan Africa. Based on the potential impacts of well-managed privatizations, significant inroads can be made into the fight against poverty. This is possible due to the following factors:

1. Greater Operational Efficiency

In this age of globalization, organizations and nations outperform each other when they are able to provide goods and services more efficiently. Such efficiency implies greater competitiveness than rivals. Given the potential efficiency impacts of privatization, it can be safely argued that sub-Saharan countries stand to become globally more competitive in the world market.

President Olusegun Obasanjo of Nigeria, in a statement delivered on the occasion of the inauguration of the National Council on Privatization in July 1999 said *"it is conservatively estimated that the nation may have lost about 800 million US dollars due to unreliable power supply by NEPA (the national electricity company), and another 440 million US dollars through inadequate and inefficient fuel distribution. And figures like these do not even tell the whole story. They cannot, for example, capture the scope of human suffering and even loss of lives caused by shortage of petroleum products. That is not to mention the frustration and debilitation of the informal sector where business centers, repair workshops, hairdressing salons, etc depend on steady supply of electricity to function."*

The implementation of the Nigerian privatization programme led to positive results in many areas. A study shows that a number of firms (for example Okomu oil, Aba textiles, Flour mils and Niycom) recorded improvements in output in the post privatization years. Such evidence is confirmation that President Obasanjo was right in resorting to privatization to improve enterprise performance.

To demonstrate that privatization can increase the level of a nation's competitiveness, take the example of an electricity company that is able to produce more efficiently following privatization. This means, it is able to reduce its tariff and as a result, processing or manufacturing companies that depend heavily on electricity are able to produce at much lower cost and consequently, are able to reduce the final prices of their products for export markets. According to the law of demand and supply, such lower prices result in higher demand, all other things remaining equal. Since export levels are expected to increase as a result of such efficiency, we would expect higher growth rates, higher levels of employment and subsequently, a positive impact on poverty.

2. Reduced Government Spending on Public Enterprises

As already mentioned, privatization implies that governments no longer spend public finances on supporting poor performing public enterprises. Such savings, coupled with the cash inflows in respect of privatization proceeds provide opportunities for governments to spend adequately towards education and health as well as other development and job-creating investment areas. All other things remaining equal, sub-Saharan African countries are able to reduce poverty as a result of the privatization-related savings in public expenditures.

3. Greater Inflows of Foreign Direct Investment

As has already been described, privatized firms benefit from massive investments by the new owners in the aftermath of privatization. In addition to this, the result of the "signaling effect" of a government's privatization policy is to create more confidence in the economy. This is likely to lead to higher inflows of foreign direct investment. As these investments come, jobs are created, the economy grows and the higher incomes mean reduced levels of poverty in the country. For example, in Ghana, privatization has attracted substantial foreign direct investment. The World Bank's Global Development Finance puts net FDI for 1995 and 1996 at $233 and $230 million, respectively.

4. Strengthened Capital Markets

In many countries, the implementation of privatization programmes led to the strengthening of capital markets as well as increased financial sector activity. The natural consequence of such strengthening is the greater availability of financing for major projects such as infrastructure. Given that poor infrastructure is one of the greatest challenges sub-Saharan Africa faces today, the creation of avenues for developing these through such capital market strengthening will surely have a more or less direct impact on creating jobs, increasing growth and reducing poverty.

For example in 1994, the government of Ghana offered shares in Ashanti Goldfields Corporation for sale. It offered 30% of its 55% stake in the Corporation on the Ghana and London Stock Exchanges. This deal transformed the Ghana Stock Exchange overnight in that the corporation's $1.8 billion accounted for 90% of total capitalization, making it Africa's third biggest stock market.

5. Technology and Skills Transfer

Through privatization, especially in the case of sale to foreign investors, hitherto publicly owned firms benefit from the much needed technologies and skills of more competitive entities abroad. Many countries such as Nigeria, embrace the "core investor" concept whereby at least 51% of the shares are sold to a core investor who must meet certain minimum requirements, among them, the ability to bring in advanced technological and managerial know-how. This leads to better performance in the form of greater productivity which eventually means more job creation, higher individual incomes and reduced poverty.

VI. Lessons of Privatization

What all of the above mean is that the effective implementation of privatization programmes in sub-Saharan Africa could be key to increasing the region's competitiveness, increased growth, higher income levels and hence, reduced poverty. Therefore, the ongoing efforts to fight poverty and to meet the other Millennium Development Goals across the region, must consider the privatization option as a complementary activity to other efforts. This requires an integrated approach, with all the various efforts inter-linked to reach the final goal of poverty alleviation.

Efforts to reduce problems encountered in the process of privatization and increase the benefits have led many to focus on some key issues. These include:

- Greater attention to social and political concerns
- Encouraging prospective buyers to outline future investment plans

- Linking privatization programmes with broader development and private-sector promotion strategies
- Broadening company ownership to include employees and the general public
- Ensuring better follow-up and monitoring

It is also important to note that privatization must not be done in an ill-prepared and hasty manner. Many a time, donors have exerted pressure on African governments to privatize without adequate assessment of the information needs, constraints, resources and time required to carry out effective transactions. They often put too much emphasis on numerical targets, hence creating a situation where the quantity of public enterprises privatized has greater priority over the quality with which a privatization is conducted. In December 1997, the IMF suspended lending to Niger under an enhanced structural adjustment facility (ESAF) when the government failed to implement the privatization of 12 enterprises. Lending resumed the next year following the government's promise to push ahead with the privatization.

What do these lessons mean for privatization programmes? Given that the wrong approach to privatization can lead to overall performance levels worse than pre-privatization, the issue of overall design and implementation of privatization programmes is vital. Hence, It is important that critical success factors are identified and an ideal programme be designed to achieve the potential positive benefits of privatization, particularly within the context of poverty alleviation. These *Critical success factors* include:

1. Commitment (Political Will)
Given the amount of resistance privatization tends to face, there has to be great commitment/political will if the programme is to succeed in any country. Through such commitment, political leaders will be prepared to defend their position on the need to privatize, irrespective of the amount of criticism they face from the different stakeholders. Nigeria, for example, has placed its privatization programme under the office of the vice president, which in some way is an indication of the importance the government attaches to the programme.

2. Clear Objectives
Privatization can be done for different reasons. The benefits that could be attained in the short term could be quite different from the long term potential benefits. This means governments must be very clear on the objectives of privatization in order to be able to put in place the necessary conditions for their successful achievement.

3. Solid Institutional and Regulatory Framework
Many privatizations fail to achieve their objectives, not because they were improperly executed, but because the institutional and regulatory framework is lacking. Privatizing a hitherto public monopoly could change the status of a firm to a private monopoly. Without the introduction of a regulatory

framework to enhance competition, consumers could be in a worse off situation than prior to the privatization. As such, putting in place an appropriate institutional and regulatory framework is vital to the success of any privatization, more so for areas such as telecommunication and air transport.

Mrs. Isatou Njie Saidy, Vice President of The Gambia, in her keynote address delivered at a validation workshop on private sector participation and regulatory framework for The Gambia in March 2004 said *"Government understands and appreciates the virtues of a good regulatory framework for Utilities. An effective regulatory framework, which is pro-competition, is essential as it serves as a signal and assurance to prospective and existing investors and firms about Government's commitment to encourage market entry into a monopolistic market, by providing a level playing field. It deters both public and private monopolies from abusing their dominance at the detriment of new firms and eventually consumer welfare. Apart from its deterrent role, the regulatory framework mandates services providers to provide acceptable levels and quality of service through a performance incentive regime. It also gives them an incentive to deploy innovative and cost effective technologies to meet their license obligations."*

The above summarises the reasons why an effective regulatory framework is vital to the success of privatization, particularly in the case of utilities.

4. Transparency
Many privatizations face difficulties due to the lack of transparency in the whole process. Many see such transactions as a way of making money for a few government and other powerful officials and hence, making the process more transparent ensures that it is more or less acceptable to the general public. This implies that a framework must be put in place with a view to ensuring that the principles of good governance and accountability guide the entire process. The decision to pay more attention to garnering information on stakeholders' views and getting them to participate in policy formulation may slow down the process. However, it is worth the while in the end, given that the chances of success are much higher, following such a participatory approach.

5. Stakeholder Consultation
Stakeholders of privatization include trade unions, employees, consumers, managers and employees of public enterprises, government ministers and so on. Each of these groups has a different interest in public enterprises and their privatization is, therefore, of concern to all. Given the influence of these different groups and their capability to disrupt any proposed privatization, it is of paramount importance that they are adequately consulted prior to and during the entire process of implementing privatization transactions.

6. Social Safety Net Issues
Usually, privatizations will involve some form of retrenchment or layoff of employees. It is therefore important to put in place adequate safety net measures to reduce the potential impact of any negative social impacts. Labour fears can be overcome by a variety of measures and incentives such as

outplacement assistance, transitional training and educational programmes, earmarked unemployment benefits. These can help raise the level of acceptability of privatizations.

Given the importance of each of the above issues in ensuring the success of any privatization, it is necessary to seriously consider them at the planning stage of any privatization programme and adequate measures put in place to address them.

Conclusion

As has already been argued, privatization has a potentially high impact on poverty alleviation. In light of this potential link, privatization should be adequately integrated as a core part of any poverty alleviation strategy. This has implications for the design of the privatization programme. Whatever the objective of the programme, it is important to take note of the concerns of the poor and, where necessary, adequate measures be put in place to guarantee that in the end, the benefits of privatization will reach the poor. Only then can privatization be seen to have a human face.

In addition to its potential impact on poverty alleviation, it must be noted that privatization can also have a considerable impact on the ongoing efforts towards regional integration on the continent. Given the strong possibilities of cross-border linkages and capital movements emanating from privatization transactions, there is a tendency to enhance the integration process in Africa. The acquisition of shares in a Ghanaian firm by a company based in Kenya is likely to bring about closer working relationships between these two companies, leading to sharing of expertise and technology and greater level of financial dealings. Consequently, the more of such transactions there are, the faster the pace of the integration process, particularly in respect of capital movements. Hence, to facilitate such interactions at the desired pace, issues of financial sector development are vital. Basic prerequisites are strong and effective financial systems as well as open and transparent financial sectors.

Finally, it is worthy of mention that there is much room for research on privatization in Africa, particularly with respect to understanding its link to poverty alleviation. Given the efforts across the continent to eradicate poverty and the potential impact privatization can have on the economies of African countries and on poverty alleviation, as we tried to communicate in this paper, focused studies aimed at quantifying the impacts of privatization in Africa will be very much beneficial in guiding African governments and policy makers as they work towards improving the well being of their people.

The Global Economic Context

Africa illustrates, perhaps better than elsewhere, that globalization is very much a policy-driven process. While in other parts of the world it may be credible to view globalization as driven by technology and the "invisible hand" of the market, in Africa most of the features of globalization and the forces associated with it have been shaped by Bretton Woods Institutions (BWIs) and Africa's adhesion to a number of conventions such as the World Trade Organization (WTO) which have insisted on opening up markets. African governments have voluntarily, or under duress, reshaped domestic policies to make their economies more open. The issue therefore is not whether Africa is being globalized but under what conditions the process is taking place and why, despite such relatively high levels of integration into the world economy, growth has faltered.

The word that often comes to mind whenever globalization and Africa are mentioned together is "marginalization." The threat of marginalization has hung over Africa's head like the sword of Damocles and has been used in minatory fashion to prod Africans to adopt appropriate policies. In most writing globalization is portrayed as a train which African nations must choose to board or be left behind. As Stanley Fischer, the Deputy Director of the IMF, and associates put it, "globalization is proceeding apace and sub-Saharan Africa (SSA) must decide whether to open up and compete, or lag behind." *The Economist,* commenting on the fact that per capita income gap between the USA and Africa has widened, states that "it would be odd to blame globalization for holding Africa back. Africa has been left out of the global economy, partly because its governments used to prefer it that way."

Globalization, from the developmental perspective, will be judged by its effects on economic development and the eradication of poverty. Indeed, in developing countries the litmus test for any international order remains whether it facilitates economic development which entails both economic growth and structural transformation. I shall argue that in the case of Africa this promise has yet to be realized. The policies designed to "integrate" Africa into the global economy have thus far failed because they have completely sidestepped the developmental needs of the continent and the strategic questions on the form of integration appropriate to addressing these needs. They consequently have not led to higher rates of growth and, their labeling notwithstanding, have not induced structural transformation. Indeed the combined effect of internal political disarray, the weakening of domestic capacities, deflationary policies and slow world economic growth have placed African economies on a "low equilibrium growth path" against which the anaemic Gross Domestic Product (GDP) growth rates of 3–4 per cent appear as "successful" performance. I will illustrate this point by looking at two channels through which the benefits of globalization are suppose to be transmitted to developing countries: trade and investment.

The [selection] is divided into three sections. The first section deals with what globalization and the accompanying adjustment policies promised, what has been delivered and what has happened to African economies during the era of globalization. The second deals critically with some of the explanations for Africa's failure. And the last part advances an alternative explanation of the failure with respect to both trade and access to foreign finance.

The Promise of Globalization and Achievements
The Promise of Trade

Expanded opportunities for trade and the gains deriving from trade are probably the most enticing arguments for embracing globalization. The Structural Adjustment Programme's (SAP's) promise was that through liberalization African economies would become more competitive. As one World Bank economist, Alexander Yeats, asserts, "If Africa is to reverse its unfavourable export trends, it must quickly adopt trade and structural adjustment policies that enhance its international competitiveness and allow African exporters to capitalize on opportunities in foreign markets." Trade liberalization would not only increase the 'traditional exports' of individual countries but would also enable them to diversify their exports to include manufactured goods assigned to them by the law of comparative advantage as spelled out and enforced by "market forces." Not only would trade offer outlets for goods from economies with limited markets but, perhaps more critically, it would also permit the importation of goods that make up an important part of investment goods (especially plant and equipment) in which technology is usually embodied.

By the end of the 1990s, and after far-reaching reforms in trade policy, little had changed. The few gains registered tended to be of a one-off character, often reflecting switches from domestic to foreign markets without much increase in overall output. Indeed some increases in exports of manufactured goods even occurred as the manufacturing sector contracted. . . .

Furthermore recent changes in Africa's exports indicate that no general increase had occurred in the number of industries in which most of the African countries have a "revealed" comparative advantage. Indeed, after decades of reforms, the most striking trend, on that has given credence to the notion of "marginalization of Africa," is the decline in the African share of global non-oil exports which is now less than one-half what it was in the early 1980s representing "a staggering annual income loss of US$68,000 million—or 21 per cent of regional CDP."

The Promise of Additional Resources

A persuasive promise made by BWIs was that adhesion to its policies would not only raise domestic investment through increased domestic saving but would relax the savings and foreign exchange constraints by allowing countries to attain higher levels of investment than would be supported by domestic savings and their own foreign exchange earnings. One central feature of adjustment policies had been financial liberalization. . . . The major thesis has been that "financial repression" (which includes control of interest rates and credit rationing by the state) has discouraged saving and led to inefficient allocation of the "loanable funds." The suggested solution then is that liberalization of markets would lead to positive real interest rates which would encourage savings. The "loanable funds" thus generated would then be efficiently distributed among projects with the highest returns through the mediation of competitive financial institutions. Significantly, in this view saving precedes investment and growth. After years of adjustment there is little discernible change in the levels of savings and investment.

Perhaps even more attractive was the promise that financial liberalization would lead to increased capital inflows and stem capital flight. Indeed, most African governments' acceptance of IMF policies has been based on precisely the claimed "catalytic effect" of agreements with IMF on the inflow of foreign capital. Governments were willing to enter the Faustain bargain of reduced national sovereignty in return for increased financial flows. Even when governments were sceptical of the developmental validity of BWIs policies, the belief that the stamp of approval of these institutions would attract foreign capital tended to dilute the scepticism.

To the surprise of the advocates of these policies and to the chagrin of African policy-makers, the response of private capital to Africa's diligent adoption of SAPs has, in the words of World Bank, "been disappointing." The market "sentiments" do not appear to have been sufficiently persuaded that the policies imposed by the BWIs have improved the attractiveness to investors. The much-touted catalytic effect of IMF conditionality has yet to assert itself. The scepticism of private investors about the BWIs stamp of approval is understandable in light of the history of non-graduation by any African country. Indeed, there is the distinct danger that, since economies under BWIs' intensive care never seem to recover, IMF presence may merely signal trouble. The BWIs seem to be unaware of the extent to which their comings and goings are a source of uncertainty among businessmen and evidence of a malaise. This said, there is, nevertheless, a trickle of foreign investment into Africa but this has not been enough to increase Africa's share in global FDI flows. The rise in foreign direct investment in the latter part of the 1990s is cited as evidence that globalization and SAPs are working. This celebration is premature. There are a number of significant features of the financial flows to Africa that should be cause for concern over their developmental impact and sustainability.

First, there is the high country concentration of investment with much of the investment going to South Africa. Secondly, there is the sectoral concentration on mining. Little of this has gone into the manufacturing industry. As for investment in mining, it is not drawn to African countries by macro-economic policy changes, as is often suggested, but by the prospects of better world prices, changes in attitudes towards national ownership and sector-specific incentives. Third, there is the problem of the type of investment. The unintended consequence of the policies has been the attraction of the least desirable form of foreign capital. Most of the new investment has taken the form of the highly speculative portfolio investment attracted by "pull factors" that have been of a transitory nature—extremely high real domestic interest rates on Treasury Bills caused by the need to finance the budget deficit—and temporary booms in export prices which attract large export pre-financing loans. It has also been driven by acquisitions facilitated by the increased pace of privatization to buy up existing plants that are being sold usually under 'fire sales' conditions. Such investments now account for approximately 14 per cent of Foreign Direct Investment (FDI) flows into Africa. Little has been driven by plans to set up new productive enterprises. Some of the new investment is for expansion of existing capacities, especially in industries enjoying natural monopolies (e.g., beverages, cement, furniture). Such expansion may have been stimulated by the spurt of growth that caused much euphoria and that is now fading away. It is widely recognized that direct investment is preferable to portfolio investment, and foreign investment in green field projects is preferable to acquisitions. The predominance of these types of capital inflows should be cause for concern. However, in their desperate efforts to attract foreign investment, African governments have simply ceased dealing with these risks or suggesting that they may have a preference for one type of foreign investment over others.

Finally, such investment is likely to taper off within a short span of time, as already seems to the case in a number of African countries. Thus, for Ghana, hailed as a "success story" by the BWIs, FDI which peaked in the mid-1980s at over US$200 million annually due mainly to privatization was rapidly reversed to produce a negative outflow. It should be noted, in passing, that rates of return of direct investments have generally been much higher in Africa than anywhere in other developing regions. This, however, has not made Africa a favourite among investors, largely because of considerations of the intangible "risk factor," nurtured by the large dose of ignorance about individual African countries. There is considerable evidence to show that Africa is systematically rated as more risky than is warranted by the underlying economic characteristics.

Capital Flight

Not only is Africa still severely rationed in financial markets, but during much of the globalization there is evidence that Africa is probably a net exporter of capital. Paul Collier and associates have suggested that in 1990, 40 per cent of privately held wealth was invested outside Africa and that in relation to the workforce, capital flight from Africa has been much higher than in other developing country groups. In a recent more systematic attempt to measure the extent of capital flight, James Boyce and Léonce Ndikumana show that for the period 1970–96 capital flight from sub-Saharan Africa was US$193 billion, and with imputed interests the amount goes up to US$285 billion. These figures should be compared with the combined debt of these countries which stood at US$178 billion in 1996. . . .

So far financial liberalization has not done much to turn the tide. In a World Bank study of the effects of financial liberalization in nine African countries, Devajaran et al conclude that the effects of liberalization on capital flight are "very small. . . ."

All this indicates that financial liberalization in itself may not be the panacea for reducing capital flight. Effective policy measures to reduce capital flight in the African context may need much deeper and more fundamental changes in the economic and political systems. One policy implication of both the reluctance of foreign capital to come to Africa and the huge amounts of wealth held outside Africa has been the calls for policies intended not so much to attract foreign capital but Africa's own private capital. While this is a valid option, the political economy of such attraction and the specific direct policy measures called for are rarely spelled out.

The Failed Promise of Growth

A comparison between Africa's economic performance during the period over which globalization is often said to have taken hold—the last two decades of the 20th century—and earlier periods shows clearly that thus far globalization has not produced rates of growth higher than those of the 1960s and 1970s. Per capita income growth was negative over the two decades, a serious indictment to those who have steered policies over the decades. This slower rate of growth is not peculiar to Africa as is suggested by some of the "Afro-pessimist" literature. During the period of globalization economic growth rates have fallen across the board for all groups of countries. The poorest group went from a per capita GDP growth rate of 1.9 per cent annually in 1960–80, to a decline of 0.5 per cent per year (1980–2000). For the middle group (which includes mostly poor countries), there was a sharp decline from an annual per capita growth rate of 3.6 per cent to just less than 1 per cent. Over a 20-year period, this represents the difference between doubling income per person, versus increasing it by just 21 per cent. The other groups also showed substantial declines in growth rates. The global decline in growth is largely due to deflationary bias in orthodox stabilization programmes imposed by International Financial Institutions (IFIs).

Explaining the Poor Performance: Has Africa Adjusted?

The poor performance of Africa with respect to the channels through which the positive effects of globalization would be gained (i.e., increased access to markets and finance) is now widely accepted. There are, however, disagreements over the cause of the failure. The BWIs have adhered to two explanations.

The first one is simply that African countries have rather incomprehensibly persisted with their doomed "dirigiste" ways and refused to swallow the bitter but necessary pills of adjustment. Inadequate implementation of reforms and recidivism are some of the most common themes running through the literature on African economic policy. In the 1994 report the World Bank's view was that adjustment was 'incomplete' not because of any faults in the design of the programmes but because of lack of implementation.

The second explanation was that insufficient time had elapsed to reap the gains of adjustment and, therefore, of globalization. Coming from the BWIs, this is a strange position. It was these very institutions that, in dismissing the structuralist argument on the inelasticity of the response of developing countries to economic stimuli, claimed that liberalization would elicit immediate and substantial responses and bring about "accelerated development" (the promise of the Berg Report). Indeed, in the early years, the World Bank was so certain about the response to its policies that it measured economic success by simply looking at the policy stance and assuming that this axiomatically led to growth.

Today, there is recognition that the axiomatic mapping of policies into performance was naïve and misleading. There are admissions, albeit grudging, to having underestimated the external constraints on policy and the vulnerability of African economies to them. Equally the responsiveness of the economies and the private sector has been overestimated, while the wrong sequence of policies eroded state capacities and responsibilities ("policy ownership"). However it is still insisted that the passage of time will do its job and the posture recommended to African countries has been to sit tight and wait for the outpouring of gains. There is no recognition that the accumulated effects of past policy errors may have made the implementation of "market friendly" policies in their pristine form more difficult.

By the second half of the 1990s neither of these arguments carried any weight. African countries have made far more adjustments than any other global region. Indeed the BWIs themselves began to point proudly to the success of their programmes, suggesting that enough time had transpired and a large number of African countries had perservered in their adjustment to begin to reap the fruits of the adjustment process. IMF officials talked about a "turning point" and that the positive per capita growth rates of 1995–97 (4.1 per cent) "reflected better policies in many African countries rather than favourable exogenous developments."

President of the World Bank James Wolfensohn, for example, reported in his 1997 address to the Board of Governors that there was progress in sub-Saharan Africa, "with new leadership and better economic policies." Michel Camdessus, the then Managing Director of the International Monetary Fund, at the 1996 annual meeting of the World Bank and the IMF, said, "Africa, for which so many seem to have lost hope, appears to be stirring and on the move." The two vice presidents for Africa at the World Bank, Callisto Madavo and Jean-Louis Sarbib, wrote an article, appropriately titled "Africa on the Move: Attracting Private Capital to a Changing Continent," which gave reasons for this new "cautious optimism." The then Deputy Managing Director of the International Monetary

Fund, Alassane Quattara, would say the following about the good performance: "A key underlying contribution has come from progress made in macroeconomic stabilization and the introduction of sweeping structural reforms." The major World Bank report on Africa of 2000 stated "many countries have made major gains in macroeconomic stabilization, particularly since 1994" and there had been a turn around because of "ongoing structural adjustment throughout the region which has opened markets and has a major impact on productivity, exports, and investment." Even the Economic Commission for Africa (ECA), a strident critic of SAP in the past, joined the chorus.

And so by the end of the millennium, African countries had been largely adjusted. There can be no doubt that there has been a sea-change in the African policy landscape. Africa is very heavily involved in "globalization" and is very much part of the global order and much of the policy-making during the last two decades has been designed deliberately to increase Africa's participation in the global economy. In any case, more devaluations, lowering of tariffs and privatization of marketing were imposed in Africa than anywhere else. By the mid-1980s, with the exception of the franc zone countries, most SSA countries had adopted flexible exchange rate policies and there had been major real exchange rate depressions. Major reforms in marketing, including the abolition of marketing boards, had been introduced. Arguments that African countries had refused or been slow to adjust or that enough time had not transpired became less credible, especially in light to the celebratory and self-congratulatory remarks by the BWIs themselves.

However, by 1997 the growth rates had begun to falter. By 1999, in its report on global prospects and the developing countries, the World Bank made a downward revision of the 1999 growth rate "despite continued improvements in political and economic fundamentals." The report blamed the poor performance on terms of trade and the Asian crisis. In a sense we had been there before. "Success stories" have been told many times before and countries have fretted and strutted on this "success" stage only to be heard of no more. . . .

Rather than abandon the deflationary policies, supporters of adjustment have simply reframed the question to read: "Why is it that when the recommended policies are put into place (often under the guidance of—and pressure from—the International Monetary Fund and the World Bank) the hoped for results do not materialize quickly." The answer was: lack of "good Governance" and of "good Institutions." These assertions conceal a clear loss of certainty and a growing sense of intellectual disarray. This is apparent in the World Bank study, *Can Africa Claim the 21st Century?* Unlike earlier approaches, the report speaks in a much more subdued and less optimistic tone, based more on faith than analysis. There is an admission, albeit grudging, that policies of the past have not worked. The new agenda is much more eclectic and more a reflection of confusion and loss of faith than the discovery of a coherent, comprehensive policy framework. The additional set of reforms is nebulous, eclectic and largely of a more political and institutional character—good governance, participation of and consultation with civil society, democracy etc. Increasingly the World Bank's new solutions suggest that there is little to be done by way of reform on the economic front. . . .

Trade, Low Growth and Absence of Structural Change

The slow growth discussed above has also had an impact on the growth of exports and diversification by weakening the investment-export nexus crucial to the process. Here again, the orthodox view has been that increased trade or openness measured in various ways is a determinant of growth. Consequently, the major policies with respect to trade have involved trade liberalization and adjustments in exchange rates largely through devaluation.

The failure on the trade front is linked to the failure in the structural transformation of African economies so that they could produce new sets of commodities in a competitive and flexible way. Globalization in Africa has been associated with industrial stagnation and even de-industrialization. African economies were the quintessential 'late latecomers' in the process of industrialization. I have argued elsewhere that although the writing on African economies is based on the assumption that Africa had pursued import substitution for too long, the phase of import substitution was in fact extremely short—in most countries it was less than a decade. SAPs have called for policies that have prematurely exposed African industries to global competition and thus induced widespread processes of de-industrialization. African economies have somehow been out of sync with developments in other parts of the world. When most economies embarked on import substitution industrialization, financed by either borrowing or debt default, much of Africa was under colonial rule, which permitted neither protection of domestic markets nor running of deficits. And even later when much of industrialization was financed through Eurodollar loans, Africans were generally reluctant borrowers so that eventually much of their borrowing in the 1980s was not for industrialization but to finance balance-of-payments problems.

Every case of successful penetration of international markets has been preceded by a phase when import substitution industrialization was pursued. Such a phase is necessary not simply for the "infant industry" arguments that have been stated ad infinitum, but also because they provide an institutional capacity for handling an entirely new set of economic activities. The phase is also necessary for sorting out some of the coordinating failures that need to be addressed before venturing into global markets. A "revisionist" view argues three main points. First, that substantial growth was achieved during the phase of import substitution industrialization; second, that even successful "export oriented economies" had to pass through this phase and maintain many features of the import substitution (IS) phase; and finally that important social gains were made.

The IS phase did lead to the initial phases of industrialization. Significantly, UNIDO notes that African countries were increasingly gaining comparative advantage in labour intensive branches, as indicated by revealed comparative advantage (RCA) but then notes:

> It is particularly alarming to note that the rank correlation of industrial branches by productivity growth over 1980–95 and RCA value in 1995 is very low. Productivity has fallen in furniture, leather, footwear, clothing, textiles, and food manufacturing. An export oriented development strategy cannot directly stimulate Total Factor Productivity (TFP). Policy

must focus on increasing technological progress within the export industries—many of which have seen very rapid progress in the application of the most modern technologies (informatics, biotechnological, etc) to their production and distribution system.

Given the conviction that import substitution in Africa was bad and had gone on for too long, there was no attempt to see how existing industries could be the basis for new initiative for export. The policy was simply to discard existing capacity on the wrong assumption that it was the specific micro-economic policies used to encourage the establishment of these industries that accounted for failure at the macro-level. The task should have been to extend and not reverse such gains by dismantling existing industrial capacity. The rates of growth of manufacturing value added (MVA) have fallen continuously from the levels in the 1970s. UNIDO estimated that MVA in sub-Saharan Africa was actually contracting at an annual average rate of 1.0 percent during 1990–97. UNIDO shows that for Africa as a whole in ten industrial branches in 38 countries labour productivity declined to an index value of 93 in 1995 (1990 = 100). Increases in productivity were registered only in tobacco, beverages and structural clay products. In many cases, an increase in productivity has been caused by a fall in employment growth (UNIDO).

The decline in total factor productivity of the economy as a whole is attributed to de-industrialization which it defines as "synonymous with productivity growth deceleration." Output per head in sub-Saharan manufacturing fell from US$7924 in 1990 to US$6762 in 1996. The structural consequence is that the share of manufacturing in GDP has fallen in two-thirds of the countries. The number of countries falling below the median has increased from 19 during 1985–90 to 31 during the 1991–98 period. While admitting such poor performance in manufacturing industry during the era of structural adjustment, supporters of SAP argue that such decline in industrialization is a temporary and welcome process of weeding out inefficient industrialization and also that insufficient time for adjustment has passed to ensure benefits from globalization through the establishment of new industries. Considering that this argument has been repeatedly deployed since 1985, Africa may have to wait for a long time before the gains from globalization materialize.

Students of historical structural changes of economies inform us that structural change is both cause and effect of economic growth. As Moshe Syrquin observes, a significant share of the measured rate of aggregate total factor productivity is owing to resource shifts from sectors with low productivity to sectors with high productivity. We have learnt from the "new trade theories" and studies on technological development how countries run the risk of being "locked" in a permanently slow-growth trajectory if they follow the dictates of static comparative advantage.

To move away from such a path, governments have introduced policies that generate externalities for a wide range of other industries and thus place the economy on more growth-inducing engagement with the rest of the world. For years, UNCTAD economists have pointed to the importance of growth for trade expansion. They have argued that it is the absence of growth, or more specifically an investment-export nexus that accounts for the failure of many countries to expand and diversify their export base. Rapid resource reallocation may not be feasible without high rates of growth and investment. The principle means for effecting export diversification is investment. Many empirical tests of "causation" have been conducted and suggest that there are good theoretical and empirical grounds for taking the reverse causation seriously as the dynamics of high growth lead to even greater human and physical investment and greater knowledge formation; which, by Verdoon's Law, leads to more productivity and therefore greater competitiveness.

Experience of successful export drives clearly shows a strong relationship between rates of structural change and rates of growth in value added in manufacturing the rates of growth of exports. Lessons from countries that have embraced trade liberalization and achieved some degree of success suggest clearly that liberalization should be in conjunction with policies that ensure that relative prices will be favourable to export industries (and not just to non-tradables) and that interest rates will support investment and economic restructuring. Successful export promotion strategies have required a deliberate design of an investment–export nexus. Diversification of exports that is developmental needs to go beyond the multiplication of primary commodities and include industrial products. This requires not merely the redirection of existing industrial output to the external but the expansion of such output and investment in new activities. There is a need to design a system of incentives that favours investments that open up new possibilities or introduce new technologies to the country. In this respect infrastructure and human resource development are important preconditions for the success of pro-export policies.

The instruments used to promote investment have included not only public investment but also the provision of subsidized inputs by public enterprises, direct subsidies through tax incentives including exemptions from duties, industrial policy which, in turn, has meant selective allocation of credit and encouragement of investment by cheapening imported investment goods (often by manipulation of exchange rates in favour of the import of plant and equipment and export sector). While "diversification" has always featured in virtually all adjustment programmes, there has been no clear spelling out how this was to be achieved. In most cases, the need for diversification was overshadowed by the short-term pressures to exploit static "revealed" comparative advantage and reduce public spending. The failure to stimulate new economic activities (especially industrial ones) has meant not only sluggish growth in exports but also failure to diversify.

Under SAP all these instruments have been off limits. Evidence that more successful cases have had some kind of "industrial policy" has been dismissed on the grounds that African countries had neither the type of government nor the political acumen to prevent "capture" of these policies by rent-seekers and patron-client networks. Governments have been left with no instrument for stimulating investment and industrial development directly or for creating an environment for robust demand and profitability in which investment, or complementary public inputs such as infrastructure, research and development, education and training, could thrive. It is this passivity that has led to failure for structural transformation and the establishment of an investment-export nexus that would have led to increase and diversification of exports. . . .

Concluding Remarks

The African policy landscape has changed radically during the last two decades. Liberalization of trade, privatization and reliance on markets have replaced the widespread state controls associated with import substitution. One would expect by now to see some signs of the "accelerated development" promised by the Berg report in 1981. That adjustment has failed as a prerequisite for development, let alone as a "strategy for accelerated development," is now widely accepted. These failures can, in turn, be traced to the displacement of developmental strategic thinking by "an obsession" with stabilization—a point underscored by low levels of investment and institutional sclerosis. The key "fundamentals" that policy has sought to establish relate to these financial concerns rather than to development. The singular concentration on "opening" up the economy has undermined post-independence efforts to create, albeit lamely, internally coherent and articulated economies and an industrial structure that would be the basis for eventual diversification of Africa's export base. The excessive emphasis on servicing the external sector has diverted scarce resources and political capacities away from managing the more fundamental basis for economic development. Even the issue of "poverty" has received little attention, except perhaps when it has seemed politically expedient to be seen to be doing something to mitigate the negative effect of adjustment. SAP, owing to its deflationary bias, has placed African economies on such a low-growth trajectory, which has then conditioned the levels and types of Africa's participation in the global economy.

Over the last two decades, Africans have been faced with not merely a set of pragmatic measures but a full-blown ideological position about the role of the state, about nationalism and about equity, against which many neo-liberals, including Elliot Berg, had ranted for years. It is this ideological character of the proposals that has made them impervious to empirical evidence including those generated by the World Bank itself and it is this that has made policy dialogue virtually impossible. The insistence of "true believers" on the basic and commonsense message they carry has made dialogue impossible. The assumption that those on the other side are merely driven by self-interest and ignorance that might be remediable by "capacity building" has merely complicated matters further. Things have not been made easier by the supplicant position of African governments and their obvious failures to manage their national affairs well. These policies were presented as finite processes which would permit countries to restore growth. With this time perspective in mind, countries were persuaded to put aside long-term strategic considerations while they sorted out some short-term problems. The finite process had lasted two decades.

There are obvious gains to participation in increased exchange with the rest of the world. The bone of contention is: what specific measures should individual countries adopt in order to reap the benefits of increases exchange with other nations? With perhaps a few cases, developing countries have always sought to gain from international trade. Attempts to diversify the export base have been a key aspect of policy since independence. Import substitution was not a strategy for autarky as is often alleged but a phase in eventual export diversification. However, for years the integration of developing countries into a highly unequal economic order was considered problematic, characterized as it is by unfavourable terms of trade for primary commodities, control of major markets by gigantic conglomerates, protectionism in the markets in the developed countries together with "dumping" of highly subsidized agricultural products, volatile commodity and financial markets, asymmetries in access to technology etc. From this perspective gains from trade could only be captured by strategizing and dynamizing a country's linking up with the rest of world.

It is ironic that while analysis in the "pre-globalization" period took the impact of external factors on economic growth seriously, the era of globalization has tended to concentrate almost exclusively on internal determinants of economic performance. Today, Africa's dependence on external factors and the interference in the internal affairs of African countries by external factors are most transparent and most humiliating and yet such dependence remains untheorized. Theories that sought to relate Africa's economies to external factors have been discredited, abandoned or, at best, placed on the defensive. The focus now is almost entirely on internal determinants of economic performance—economic policies, governance, rent-seeking and ethnic diversity.

While the attention on internal affairs may have served as a useful corrective to the excessive focus on the external, it also provides a partial view of African economies and can be partly blamed for the pursuit of policies that were blind to Africa's extreme dependence and vulnerability to external conjuncture—a fact that the BWIs have learnt as exogenous factors scuttled their adjustment programmes. Indeed, unwilling to discard its essentially deflationary policies and faced with poor performance among many countries which have been "strong adjusters," the World Bank's explanations have become increasingly more structuralist-deterministic and eclectic. . . .

It is now admitted that many mistakes have been made during the past two decades. When errors are admitted, the consequences of such errors are never spelled out. . . . Policy failures, especially those as comprehensive as those of SAP's can continue to have effects on the performance of the economy long after policy failures are abandoned. It may well be that the accretions of errors that are often perfunctorily admitted have created *maladjusted* economies not capable of gaining much from globalization. Both the measures of "success" used for African economies and the projections for the future suggest that essentially the BWIs have put Africa's development on hold. This clearly suggests the extreme urgency of Africans themselves assuming the task of "bringing development back in" in their respective countries and collectively. To benefit from interacting with the rest of the world, African policy-makers will have to recognize the enormous task of correcting the maladjustment of their economies. They will have to introduce more explicit, more subtle and more daring policies to stimulate growth, trade and export diversification than hitherto.

POSTSCRIPT

Have Free-Market Policies Worked for Africa?

The debate presented here may have as much to do with the authors' conceptualization of "development" as it does with the effectiveness of free-market policies (which are premised on a certain vision of development) in Africa. Although all of the authors are economists, development is a vague term that has been articulated in vastly different manners.

The type of development that free-market policies are designed to facilitate (and the type of development advocated by Pamacheche and Koma in this issue) is export-led, laissez-faire economic growth. The belief (under this conceptualization of development) is that if you get the general policy environment "right" and minimize interference of the state, capitalist actors will make efficient and rational decisions that lead to the generation of wealth that will eventually trickle down to all members of society. Furthermore, as individual nation states exist in a global economic system, it makes sense for them to specialize in the production of goods that they can create relatively cheaply and to trade for those that others can produce more efficiently. Proponents of this view often point to the Asian Tigers (e.g., Taiwan, Singapore, and South Korea) as examples of countries that successfully pursued export-led economic growth.

Other than questioning the effectiveness of free-market policies along the narrow lines of whether or not these promote economic growth, critics of these policies generally raise at least three issues. First, they assert that laissez-faire economic growth often does not help the majority of the population but, rather, may serve to concentrate wealth in the hands of a few powerful individuals or entities. This is particularly problematic in the African context where large economic operators may face little to no competition. Konadu-Agyemang (in an article in a 2000 issue of *The Professional Geographer* titled "The Best of Times and the Worst of Times"), for example, highlighted the very uneven pattern of development promoted by free-market policies in Ghana, one of the World Bank's success stories. Second, critics assert that it is often problematic for African nations to engage in free trade with the rest of the world because they are relegated to the production of primary commodities for which prices are low and declining. Furthermore, use of the Asian Tigers as examples of successful export-led growth is disingenuous (see, e.g., an article by Carmody in a 1998 issue of the *Review of African Political Economy* titled "Constructing Alternatives to Structural Adjustment in Africa"). This, according to the critics, is because these states actively protected infant industries and provided subsidies to promote the development of their export sectors (at least during the early years of developing this aspect of their economies), rather than pursuing a minimalist role for the government in the economy. Finally, many of the critics would argue that an entirely different vision of development needs to be emphasized, a vision that prioritizes the fulfillment of basic human needs (e.g., adequate food, clean water, and health care) and the development of human capital (via improved education). This is the perspective taken by, for example, Macleans Geo-Jaja and Garth Mangum in a 2001 article in the *Journal of Black Studies* titled "Structural Adjustment as an Inadvertent Enemy of Human Development."

A final point worth noting is that Pamacheche and Koma are writing on one dimension of free-market policy reform (privatization), whereas Mkandawire is addressing a broader range of free-market policies. Although others have written in praise of privatization efforts (e.g., Yaw A. Debrah and Oliver K. Toroitich, "The Making of an African Success Story: The Privatization of Kenya Airways," *Thunderbird International Business Review,* 2005), Mkandawire's critique of free-market policy reform in general is complemented by those who are more narrowly focused on the problematic aspects of privatization efforts in Africa (e.g., Laurence Becker and N'guessan Yoboué, "Rice Producer–Processor Networks in Côte d'Ivoire," *Geographical Review,* 2009).

ISSUE 6

Do Cell Phones and the Internet Foster "Leapfrog" Development in Africa?

YES: Joseph O. Okpaku, Sr., from "Leapfrogging into the Information Economy: Harnessing Information and Communications Technologies in Botswana, Mauritania, and Tanzania," in M. Louise Fox and Robert Liebenthal, eds., *Attacking Africa's Poverty: Experience from the Ground* (World Bank, 2006)

NO: Pádraig Carmody, from "A New Socio-Economy in Africa? Thintegration and the Mobile Phone Revolution," *The Institute for International Integration Studies Discussion Papers* (2009)

ISSUE SUMMARY

YES: Joseph O. Okpaku, president and CEO of the Telecom Africa International Corporation, argues that cell phones and the Internet have fundamentally changed the lives of people and national economies in Africa by delivering needed services more efficiently. He argues that these technologies can foster sustainable economies, build on efforts to reduce poverty, and allow individuals and institutions to prosper through increased access to information.

NO: Pádraig Carmody, a senior lecturer in geography at Trinity College Dublin, questions the transformational capacity of information and communication technology (ICT) in Africa. Although he admits that ICTs can sometimes enhance welfare, their use is embedded in existing relations of social support, resource extraction, and conflict and therefore may reinforce existing power dynamics. Since Africa is still primarily a user (rather than producer or creator) of ICTs, the use of these technologies does not fundamentally alter the continent's dependent position.

The September 26, 2009 issue of *The Economist* magazine featured a special report on telecoms in emerging markets, with a particular focus on Africa. The magazine triumphantly declared, "Once the toys of rich yuppies, mobile phones have evolved in a few short years to become tools of economic empowerment for the world's poorest people. These phones compensate for inadequate infrastructure, such as bad roads and slow postal services, allowing information to move more freely, making markets more efficient and unleashing entrepreneurship." The well-known economist Jeffrey Sachs argues that the cell phone is "the single most transformative tool for development."

As the quotes above suggest, information and communication technology (ICT) is increasingly seen as the key to rapid development in lower income regions, especially Africa. The ICT-based development thesis suggests that these technologies are improving communication, opening new investment opportunities, incorporating the African diaspora in development, and integrating the continent into the global economy. Yet the validity of the ICT-based development thesis in Africa is hotly contested by a number of scholars and analysts. This group argues that ICTs have a tendency to reinforce existing power relations. They further note that there are a number of real obstacles to wider use of these technologies, such as lack of reliable electric power and relevant content, as well as gross social inequalities in access to ICT.

Key to understanding this issue is the notion of "leapfrog" technology or development. The concept of "leapfrogging" refers to a new technology or set of technologies that allows a country or region to either (1) bypass the (hitherto) conventional, slower, and more expensive path of advancement or (2) avoid the mistakes of those countries that developed earlier and take a cleaner or safer path. Cell phone technology is often considered to be a classic example of the first case as it allows countries to bypass installing the costly infrastructure required for landline telephones. An example of the second case might be a country that avoids dirty, coal-fired energy production in favor of cleaner wind and solar power.

Because of their enormous practical value, cell phones and the Internet have emerged as Africa's most popular ICTs. According to Internet World Stats, the top 10 countries in Africa in terms of cell phone penetration for 2009 were Seychelles (1.01 cell phone prescriptions per inhabitant), Gabon (0.96), South Africa (0.92), Botswana (0.89), and Mauritius (0.81). In terms of percentage of the population with internet access, the top five African countries in 2009 were Morocco (32.9%), Tunisia (26.7%), Egypt (15.9%), Zimbabwe (12.5%), and Algeria (12%). Although the penetration of cell phones and the Internet is still quite low relative to the rest of the world, the rate of growth is phenomenal (40% per annum on average for Internet access).

In this issue, Joseph O. Okpaku, president and CEO of the Telecom Africa International Corporation, argues that cell phones and the Internet have fundamentally changed the lives of people and national economies in Africa by delivering needed services more efficiently. He argues that these technologies can foster sustainable economies, build on efforts to reduce poverty, and allow individuals and institutions to prosper through increased access to information. Pádraig Carmody, a senior lecturer in geography at Trinity College Dublin, questions the transformational capacity of ICT in Africa. Although he admits that ICTs can sometimes enhance welfare, their use is embedded in existing relations of social support, resource extraction, and conflict and therefore may reinforce existing power dynamics. Since Africa is still primarily a user (rather than producer or creator) of ICTs, the use of these technologies does not fundamentally alter the continent's dependent position.

YES

Joseph O. Okpaku, Sr.

Leapfrogging into the Information Economy: Harnessing Information and Communications Technologies in Botswana, Mauritania, and Tanzania

In perhaps one of the most dramatic developments since the Industrial Revolution, information and communications technologies (ICTs) have fundamentally changed the lives of people and national economies all over the world. In Africa these technologies offer an important means of delivering much-needed services more efficiently, allowing the continent to reap more value from limited resources. ICT can help build sustainable economies, scale up efforts to reduce poverty, and increase the ability of individuals and institutions to promote their own development through vastly increased access to information.

Africa was late in entering the ICT arena, and many African countries are handicapped by inadequate infrastructure and access. Nevertheless, many countries—including Botswana, Mauritania, and Tanzania, which together represent a microcosm of Africa as a whole—have embarked on embracing ICT as one of the engines to drive their development, especially to combat poverty.

Although the reform process began fairly recently, the framework was laid down through telecommunications reform, as the antecedent of ICT reform. With World Bank assistance in formulating, implementing, and building capacity for policy reform, Mauritania has dramatically expanded the sector by liberalizing and partially privatizing its incumbent public telephone and telegraph. In the process, Mauritania has also attracted foreign direct investment for ICT at a time when the global trend would predict otherwise. Tanzania has built on a long-term strategic effort at developing the sector, which has resulted in the proliferation of access across the country, including through several thousand public access points, as well as in rural access. For its part, Botswana aims to use ICT to build a new sector, that of becoming a regional hub in financial services.

Sector reform has greatly accelerated the development of ICT in the study countries and advanced the use of ICT for development. Liberalization has brought about competition, which in turn has greatly increased access, especially in mobile telephony. Privatization has introduced the dynamics of commercial enterprise with its expected attendant increase in efficiency. The establishment of formal policy, legal, and regulatory frameworks through regulatory authorities has brought much-needed structure to the sector, with much clearer rules, processes, and regulations and the corresponding increase in transparency. Challenges remain, including high costs, the threat of technological obsolescence, and the need to provide human resource capacity building to support the sector. The promise and achievements already made, however, outstrip the threats.

Adoption of Information and Communications Technologies in Africa

Although still modest by comparison with Western economies, an explosion in the development of ICT infrastructure has occurred in Africa over the past decade. The number of fixed telephone lines rose from 12.5 million in 1995 to more than 21 million in 2001 and 25 million in 2003. In addition, the quality of the infrastructure improved tremendously, with many networks switching from purely analog technology to a hybrid of analog and digital systems and a few countries having fully digital networks.

The change in mobile telephone service has been even more dramatic. In 1995 very few African countries had mobile telephone services. Today every country on the continent has at least one mobile operator, and by 2001 the number of mobile subscribers had overtaken the number of fixed-line subscribers. Sub-Saharan Africa had an estimated 18 million mobile telephone subscribers in September 2003, while the continent as a whole had an estimated 51 million subscribers by the end of the year.

This quantum leap in voice telecommunications infrastructure and services has been matched by a parallel development in Internet use. In 1995 only a handful of African countries had Internet service. By the end of 2003 there were about 4.5 million Internet subscribers and some 52,000 Internet hosts in Sub-Saharan Africa (excluding South Africa), and 12.4 million subscribers and 350,000 Internet hosts in Africa as a whole. Today Internet service is available in every country in Africa, especially in cities and towns, and a large number of Africans, mostly students and young adults, have free email addresses. As of the end of 2003 there were an estimated 4.2 million personal computers in Sub-Saharan Africa (excluding South Africa) and an estimated 10.4 million personal computers in all of Africa.

Despite this rapid growth, ownership of telephone and personal computers remains limited. This has been partly mitigated by the explosion in public access facilities such as telecenters, Internet kiosks, and cybercafes, which have mushroomed across the continent. Mauritania alone has about 3,500 public kiosks, and Tanzania has more than 1,000 cybercafes.

The new technology is being used in innovative ways. Wireless technologies are being used to improve "last-mile access," the stretch between the nearest switch and the subscriber's terminal equipment, in countries with inadequate copper wire or optical fiber cable networks. Global positioning technology is being used to communicate with nomads in the deserts of Mauritania.

Advances in technology have increased access, but they have not reduced prices: Africans continue to pay more than $1,000 per access line, regardless of the technology used. These high prices are preventing African countries from taking full advantage of ICT.

The explosion in ICT infrastructure and services in Africa over the past decade reflects several factors:

- The global reach of mass media, which has created local demand for world knowledge and information.
- The increase in public awareness, including greater knowledge of the right to information.
- The focus on self-actualization, as the internecine conflicts once endemic in Africa are increasingly giving way to development priorities.
- The expansion of the African diaspora (and the desire for families to stay in touch).
- The increasing maturity of the knowledge-based economy and its consequences for economic and social empowerment.
- The shrinking job market, which has compelled young people to become entrepreneurs and exposed them to the promise of ICT.

The impact of these developments in ICT in Africa, in terms of both ICT development (increased infrastructure and access) and ICT for development (adoption of ICT applications), has been to advance the process of development itself, in terms of ICT for development. The result of this duality of sector transformation has been itself dually vast. On the one hand, it has facilitated the delivery of services such as education, health, better governance (on the part of both the leadership and the governed), enterprise and business development, as well as their overall contribution to socioeconomic well-being (especially poverty reduction), political stability, and self-actualization. On the other hand, the transformation has increased demand for more and better services, faithful to the adage that once people have tasted honey, they crave more. This, in turn, has been good for governments, as it has provided a flexible and potentially inexpensive means of establishing an interactive dialogue with the people, the basis of democratic governance.

Reform of the Telecommunications Sector in Botswana, Mauritania, and Tanzania

The development of ICT policy in the three countries is fairly recent and still in its nascent stage for different reasons in each country. Botswana does not have a national ICT policy; in November 2003 it awarded a contract to a Canadian consulting firm to draft a policy. The draft will be reviewed through a series of dialogues with various stakeholders in the country. The

National Information and Communications Technologies Policy for Tanzania was published only in March 2003. The Act of Parliament establishing the Tanzania Communications Regulatory Authority (TCRA), which is responsible for implementing and administering the policy, was promulgated on May 23, 2003. The TCRA has since taken over from the Tanzania Communications Commission. Mauritania published its National Strategy for the Development of New Technologies (2002–06) in November 2001, and the Plan of Action six months later. Reform of the ICT sector was preceded by reform of the narrower telecommunications sector through telecommunications policies (as distinct from broader ICT policies).

As a result of the serious commitment of the governments to ICT development in these countries and the growing awareness of the benefits of ICT (and demand for them) by the public, the sector has seen substantial growth in all three countries. For example, the number of telephone subscribers in all three countries has skyrocketed in the past six years, with most of the growth coming from mobile lines. The increase in teledensity (the percentage of subscribers in the population) has been unprecedented. Between 1998 and 2003 teledensity rose from 7.5 to 34.0 in Botswana, from 0.6 to 12.1 in Mauritania, and from 0.5 to 3.0 in Tanzania. . . .

How Has ICT Improved People's Lives?

Reforms, although still in their early stages in Botswana, Mauritania, and Tanzania, are already yielding dividends—for individuals, for the private sector, and for government:

- Access to communications facilities has improved, making communicating and accessing information much easier for many people.
- Increased access to the Internet has allowed more and more people to communicate electronically and to take advantage of the educational, informational, and entertainment content of the Internet.
- The private sector has benefited from improved communication with buyers and suppliers; expanded market access, including access to global markets; higher productivity and profitability (in both agriculture and urban trade); and greater access to information—about markets, competitors, and best practices—from around the world.
- A new economic sector has emerged, spawning entrepreneurship and job creation at Internet service providers, telecenters, kiosks, and cybercafes.
- Financial and administrative efficiency have increased.
- The ability to serve remote populations through online applications—such as distance education, e-health, e-commerce, and e-government—has increased.
- Where governments have created Web sites (as they have in Tanzania and to some extent Mauritania), access to public information has increased.
- The greater efficiency, cost effectiveness, and transparency that ICT makes possible have enabled governments to deliver more and better services.

- Increased knowledge and information obtained through the Internet have improved the quality of public dialogue, which creates an enabling environment for national integration and the establishment and entrenchment of peace and tolerance.
- The increased demand for ICT services as their value has become broadcast is resulting in the emergence of small and medium-size ICT enterprises, such as telecenters, kiosks, and cybercafes, which create much-needed employment for young people.
- The growing adoption of ICT in business and government, spreading increasingly from the national level to the state, provincial, and local government levels, creates a demand for staff with ICT skills. An increasing number of young men and women seek daytime or evening extramural courses to acquire ICT skills to meet these demands.

In all three countries reform has increased competition and expanded access under the guidance of credible regulatory authorities. Competition has promoted the creation of appropriate institutional frameworks for regulation in the utilities sector, benefiting other sectors as well. In Mauritania reform has also improved the investment environment, as reflected in the investment of $28 million in 2000 for a mobile license and $48 million for partial privatization of the incumbent operator, Mauritel.

Building an Information-Based Economy: Key Components of Success

Several factors are critical to building an information-based economy—and to putting ICT to work to reduce poverty. Commitment to developing ICT must be strong. Champions of reform must be willing to spearhead, implement, and support reform. Institutional changes must be made, and partnerships must be forged—among development agencies, donors, nongovernmental organizations, and returning nationals with skills in ICT and entrepreneurship.

Commitment and Ownership

Development partners have played a critical role in promoting the need for sector reform and facilitating a paradigm shift in the way policymakers view ICT. Such paradigm shifts would be short-lived and unsustainable, however, without indigenous commitment. Recognition of the need for reform and the commitment to undertake it must come from within the countries themselves, and the promise of reform must be consistent with a country's strategic vision. The countries and the people must take ownership of the reform process in order to derive maximum value from the effort and make it sustainable for the long term.

African policymakers are committed to increasing access to and use of ICT for a variety of reasons. In recent years African leaders have tried to build their countries' economies and stem the tide of poverty. As the source of economic strength has shifted from material resources to knowledge and intellectual capacity, they have come to realize the importance of acquiring, creating, and mobilizing their countries' own knowledge and expertise as critical tools to drive a new development thrust, especially in scaling up poverty reduction efforts. Several factors have played critical roles in pushing African leaders to move ahead more quickly to increase access to ICT:

- More than ever, Africa's leaders at the beginning of the new millennium share a common desire to turn over a new leaf. As part of this effort, they want to join forces to develop ICT capacity across the continent.
- Policymakers recognize the potential of ICT to overcome the limitations of distance, terrain, and limited human and financial resources to deliver such services as health, education, and recently banking through Internet-based online applications.
- Policymakers recognize the empowering impact of global Internet access, especially for young people, and its ability to enable citizens to discover and share information that can improve their lives in a variety of ways.
- Governments are more aware of the compelling need to deliver good, accountable, responsible, and responsive governance to the people. They also recognize the ability of the people to gauge the performance of their governments as well as to express their dissatisfaction through access to information tools.
- Policymakers understand that, especially where financial resources are very limited, good governance and economic development require greater efficiency. ICT has the potential to improve governance, allowing governments to get more bang for their development buck.
- The rapid transformation of the world into an information- and knowledge-based environment makes the benefits of ICT tangible and palpable.
- The success stories from Asia, especially China, India, and Malaysia, have proved very compelling.

Champions of Reform: Committed Leaders, Dedicated Bureaucrats, Able Regulators—and an Informed and Eager Public

By themselves, motivation, appropriate policies, and privatization of the sector are insufficient to have an impact on development. A champion of reform is needed, someone from the public or private sector who is powerful enough to promote major changes and has sufficient authority and control over financial resources to direct them in support of needed change.

As champions cannot make things happen by themselves, loyal, enlightened, and committed deputies are needed, especially at the ministerial and principal secretary level, and dedicated bureaucrats committed to change are needed throughout the civil service. In addition, a knowledgeable, enlightened, fair, and transparent regulatory authority is needed to lead the development of the ICT sector by encouraging and interacting flexibly with the private sector and creating an environment in which competition is open and fair. Most important of all is an informed public that is enthusiastic about the empowering potential of ICT and willing and able to experiment with new technologies without worrying about jeopardizing enduring value in tradition.

In Botswana the main driver of transformation has been Cuthbert M. Lekaukau, executive chairman of the Botswana Regulatory Authority. Mr. Lekaukau has been active in all of the relevant forums with potential impact on ICT in Africa, including the International Telecommunication Union (ITU) and the Telecommunications Regulatory Authority of Southern Africa (TRASA), the regional grouping of telecommunications authorities in the 16 member states of the Southern African Development Community. He is backed by a team of young, smart men and women who have solid backgrounds in the field and are actively engaged in the regional, continental, and global exchange of ideas in the sector.

In Mauritania the key driving force has been President Maaouya Ould Sid Ahmed Taya. According to Fatimetou Mint Mohamed-Saleck, the secretary of state for new technologies, "The president believes in new technologies as a solution for development, for crossing barriers to development, for creating a short-cut to development." The president's commitment is matched by that of the secretary of state. The director for new technologies for information and communications, B. A. Housseynou Hamady, and the president of the Mauritania Regulatory Authority, Moustapha Ould Cheikh Mouhamedou, play important roles in implementing reforms. The Ministry of New Technologies has accomplished a lot since it was created by presidential decree in 2000, crafting an ICT strategy and a plan of action and implementing changes to the sector.

In Tanzania championship of ICT comes from the University of Dar es Salaam, which is also taking the lead in expanding access to distance education. This is a direct result of the country's modern political history and the legacy of founding head of state and pre-eminent visionary President Julien Nyerere, whose philosophy and commitment centered on education, knowledge, and information. At independence young politicians, academics and intellectuals, the media, and budding entrepreneurs in most African countries belonged to the same social group. Except during a rebellion by the armed forces, which was quickly put down, Tanzania has never experienced a separation between the intellectual class and the political leadership. Academics and intellectuals have continued to play an important role in Tanzania's development process. Together with the stability that Tanzania has enjoyed since independence, their participation has enabled Tanzania to rely heavily on internal expertise to craft its development strategies.

Tanzania's ICT policy was drafted by a broad-based task force chaired by Matthew Luhanga, vice chancellor of the University of Dar es Salaam, following an extended process of dialogue with all stakeholders, including a special session with Parliament. The fact that the minister of transport and communications is an ICT specialist and former professor at the University of Dar es Salaam, and that many university graduates serve in Parliament, has helped spur ICT development in Tanzania. More people with academic training in ICT are probably working in the policy, regulatory, and operational sectors of the economy in Tanzania than in most countries in the world. This cadre of well-trained people has enabled Tanzania to develop and implement a sound ICT strategy.

Leaders in all three countries understand the scope and long-term nature of developing ICT as a tool for reducing poverty. As Dr. Maua Daftari, the deputy minister of transport and communications of Tanzania, noted, "We in Tanzania are trying, but we have a long way to go. We have to provide connectivity for our rural populations."

Institutional Innovation

The most important institutional innovation governments can adopt is establishment of an e-government platform, which serves not only to enhance administration and the delivery of public services but also as an example to nongovernmental institutions of the value of ICT. Tanzania has an ambitious and comprehensive e-government platform. Mauritania recently awarded a contract for the design of an e-government platform. In contrast, Botswana appears reluctant to create a robust e-government platform, limiting its ICT involvement to interconnecting government ministries.

A good indicator of the importance government places on the use of ICT for improved delivery of administrative and public services is the location of the office responsible for ICT. In Mauritania the Ministry of New Technologies is placed within the Office of the Prime Minister. In Tanzania it is located in the Presidency.

Working with Partners

Donors and development partners are playing a key role in all three countries, supporting the development of policy, strategies, and rules for implementing sector reform, such as licensing and privatization tender rules and procedures; helping implement policy; and developing legal and regulatory frameworks and instruments to manage the sector. Many of these interventions have failed to promote indigenous enterprises or to draw on local players to create the ICT sector. The absence of African experts as consultants in strategic situations that require the long-term commitment that only comes with ownership and responsibility is a flaw that development partners need to address.

Even with donor assistance, government cannot do everything. Public–private partnerships are needed, as well as partnerships between government and civil society. In Tanzania NGOs are playing a major role in the development and deployment of ICT for development.

In all three countries foreign nationals living abroad have helped drive the sector. As a result of the cooling off of the high-tech industry in the United States and Europe, many ICT-savvy Africans have returned home and become key players in the ICT sector, especially in Tanzania.

Can ICT Reduce Poverty?

The relationship between ICT and poverty reduction is complex. ICT is not likely to have an effect on the poorest of the poor in the short term, as it affects economies in a structural and systemic way. By simplifying the way things can be done—and often making possible things that might otherwise have been impossible or unprofitable—ICT acts as a driver of economic change. Only in the wake of this transformation does employment—and the education and skills training it demands—begin to respond. Investment in ICT for poverty reduction must therefore be crafted as a long-term strategy in order not to create false hopes of early returns.

Thus, the promise of ICT for reducing poverty should be viewed as a multistep process. Early on ICT can increase the government's ability to deliver basic social services and create two-way communications between leadership and the people. ICT can make government more responsive and more transparent—and therefore more likely to address poverty reduction.

Later ICT can improve education, enabling "the poor to understand their own circumstances" and improve health outcomes (through telemedicine, for example). In this trajectory the long-term strategic impact on poverty reduction becomes self-evident and self-propelling.

In addition to these indirect effects, ICT also has a direct impact on development and poverty reduction by creating jobs. The bustling activity on the main street of Nouakchott, Mauritania has an effect on poverty reduction in urban areas. The challenge will be using ICT to reduce poverty in rural areas, where few people would be able to afford even inexpensive ICT services.

Lessons and Recommendations

Two primary lessons emerge from the brief experience with reform in Botswana, Mauritania, and Tanzania:

- *Sector reform greatly accelerated the development of ICT and advanced the use of ICT for development.* Liberalization brought competition, which has greatly increased access, especially in mobile telephony. Privatization introduced the dynamics of commercial enterprise, with its emphasis on efficiency. The establishment of formal policy, legal, and regulatory frameworks through regulatory authorities has brought much-needed structure to the sector, with much clearer rules, processes, and regulations and a corresponding increase in transparency.
- *ICT is a potent tool for development—and one that is within the reach of all African countries.* To take full advantage of its promise, policymakers need to develop a strategic vision that recognizes the importance of knowledge and information, and affordable public access to both, as key elements of national development. It also requires a recognition that public access to shared knowledge is not a threat to government but rather a facilitator of open and meaningful dialogue between leaders and the people they govern, to the benefit of all concerned. Political leaders need to have the courage to make the bold decisions required to adopt ICT—especially when doing so threatens established power and authority systems and interests. And they need to lead by example, by developing e-government.

Several recommendations also emerge from experience with ICT development in Botswana, Mauritania, and Tanzania and elsewhere in Africa:

- ICT development should be planned, not ad hoc, based on strategies that reflect local conditions and resource availability.
- ICT should be widely adopted, so that economies of scale can be enjoyed and a paradigm shift in the way of doing business produced. Governments can do much to spur this process by increasing e-government capacity at the national, state, provincial, and local government levels.
- Resources should be allocated to areas in which ICT will have the most immediate impact, especially on income and job creation (and therefore poverty reduction) in order to build the momentum necessary to sustain the process until longer-term impacts can be felt. In the meantime, ICT should be used to increase productivity and efficiency in the productive sectors of the economy.
- ICT is an expensive undertaking that does not pay off in the short term. Expectations must therefore be long term in order to avoid disillusionment and retrenchment before the benefits of the technology are realized.
- ICT cannot simply be thrust on a country with the expectation that people will adopt it. Adopting new technology requires a new mindset, which may require public education. One approach would be to include ICT education as part of the basic school curriculum.
- ICT is prone to rapid change—and to rapid obsolescence. African countries should avoid adopting obsolete technology, which industrial countries often dump on them.
- Pilot-testing projects is a good way of avoiding large investments in ICT projects that may not prove cost effective. Pilot projects must perfect each step of the transformation in order to prevent costly mistakes.
- ICT is not sustainable if it is not approached as an industry. Efforts should be made to increase ICT capacity in local content development, software development, technology and applications adaptation, and outsourcing in order to entrench ICT as a productive sector of the economy. Acquiring ICT capacity without developing the collateral scientific and technological expertise or creating production facilities greatly limits the benefits of ICT. Some form of industrialization, possibly undertaken as a collaborative effort by several African countries, is necessary.
- Governments in the region need to support research and development in ICT.

Future Challenges and Opportunities

Each of the three countries studied has its own vision for its future and the future of its people. Each is taking different steps to achieve its vision.

Botswana, one of the richest economies in Africa, is trying to diversify its economic base away from a heavy reliance on mining and agriculture. It hopes to develop into a Hong Kong–style regional center for financial services and outsourcing—sectors that require world-class ICT infrastructure and human resource capacity. Botswana is seeking to build a science and technology sector to support this strategy.

Mauritania discovered offshore oil reserves in 2001 and is expecting the oil industry to come onstream in 2005. The discovery of oil has the potential to help Mauritania overcome poverty, but policymakers are well aware of the trauma that oil discovery has caused other African countries. To avoid the same fate, Mauritania will need to establish new institutional arrangements. ICT can help. The oil industry depends heavily on global information and communications infrastructure and systems—and trained people to service them. Mauritanian officials are eager to implement a program for rapid human resource capacity building in all aspects of ICT support services.

Tanzania is facing new challenges with the impending end of the exclusivity period for its incumbent fixed-network operator, and needs to develop innovative strategies to develop its information and communications infrastructure to meet the needs of all Tanzanians, especially its very large rural population. The country's strategy is to focus on building expert capacity at all levels and using this capacity to chart a new course for the country. The academic expertise at the University of Dar es Salaam, as well as key experts in the public and private sectors, are helping to develop a new strategy to build on the successes of earlier reforms in the telecommunications sector.

Pádraig Carmody

A New Socio-Economy in Africa? Thintegration and the Mobile Phone Revolution

Introduction

. . . Writing in the late 1990s Manuel Castells characterized Sub-Saharan Africa as a "black hole of informational capitalism." At that time there were more telephone landlines in Manhattan or Tokyo than in all of Sub-Saharan Africa. For him this "technological dependency and technological underdevelopment, in a period of accelerated technological change in the rest of the world [made] it literally impossible for Africa to compete internationally either in manufacturing or in advanced services." However, "gloomy predictions of the impending Fourth World of structurally irrelevant 'black holes of informational capitalism' did not anticipate the privatization of the telecommunications industry across much of the African continent," and its consequences.

From 2000 to 2007 Africa was the fastest growing mobile phone market in the world as the number of subscribers rose from 10 to 250 million, with a 66% growth rate in 2005 alone. This aggregate figure disguised even faster growth in some markets. Nigeria had a compound annual growth rate in its mobile market of almost 150% during 2002–4. In South Africa by the end of 2007 mobile penetration reached approximately 84%, and it is projected there will be 560 million mobile phone subscribers on the continent by 2012. Will this development fundamentally alter the nature of globalization in Africa, or does the new landscape of mobile telephony simply represent an overlay on existing economic structures—a form of thin integration? . . .

The arrival of the information technology revolution in Africa is one aspect of globalization on the continent. What is its significance? Does the rapid spread of ICTs in Africa represent a transformative moment in Africa's economic history or are previous power relations merely being partially modified, but reinscribed by the information revolution which has swept over the continent in recent years? What are the different nodes in the global value chains of mobile phones and where is Africa inserted into these? What are the power relations in the networks and webs created by ICT usage and do these alter patterns of extractive globalization which have characterized Africa's relations with other parts of the world for the last several centuries? This paper seeks to interrogate these questions drawing insights from global value and commodity chain analysis and critical ICT studies to assess the transformational potential of mobile phone technology on the continent.

The Information Revolution and the Mobile Value Chain

The potential of information and communication technology for development (ICT4D) is often related to the increased importance of the global knowledge economy or what Peter Evans has called "bit driven growth." For some the revolutionary aspect of ICTs is that they decouple information from their physical repository, allowing for widescale non-rival knowledge diffusion that can contribute to innovative capacity. Some go further and argue that the internet, for example, represents not just a new form of communication but is instead a new form of societal organization.

Increasingly it is argued that as a matter of urgency Africa must compete in the global information economy. For example, the Africa Competitiveness Report argues that ICTs are vital to success in today's globalized economy. Likewise, Obijiofor argues there is a "strong link" between the adoption of new technologies and the development of countries and communities. Consequently in mainstream accounts ICT is often presented as an unambiguous positive flow of globalization: a harbinger of integration into the global economy of the "borderless" world. However, there is a particular vision and ontology which undergrids such conceptualizations which neglects the importance of the differential geography of research and development, production, raw material extraction, and the cultural adoption and adaptation of information and communication technologies.

ICTs can be broken down into distinctive value chains which contain pecuniary and non-pecuniary elements. Different places are integrated into the global mobile phone industry in very different ways. Much of the research and design takes place in the rich world, whereas China concentrates on assembly, while also developing design capability. Africa supplies the precious metal coltan, necessary for many ICTs to function, and serves as a fast growing market for mobiles. There is then a global, but mobile, division of labor in the industry comprised of "hard" networks, involving flows of physical commodities, and "soft" networks of social interaction and information exchange. Elements of the latter are prerequisites to the former. . . .

Ya'u argues that apart from a few assembly plants and some efforts at software production, Africa imports all of its ICT needs. Very little of the research and development which goes into the making of mobile phones takes place in Africa, and what research does appear to take place is based in South Africa and around functionality, rather than innovation *per se*.

Demand for mobile phones in Africa continues to grow strongly, as despite the global economic slowdown Sub-Saharan Africa's economy is estimated to grow at almost 6% for 2008 and 2009. This fast growth, combined with the even more rapid growth in demand for mobile phones has prompted some companies to set up assembly operations on the continent. For example, the Malaysian company M-mobiles is setting up a mobile phone assembly plant in Mozambique which will assemble between 50,000 and 70,000 mobiles a month and is building another plant in Lusaka. On a smaller scale, Link Technologies from China has set up a plant in Rwanda to assemble 200 mobiles a day. After being redesigned by a Chinese company these mobiles can be programed in the national language, Kinyarwanda.

The South Korean company LG is also planning to set up a plant in Kenya and the introduction of television broadcasts to mobiles by Black Star TV has spurred an investment to assemble mobile phones which can receive the service by a Korean manufacturer in Ghana. The Chinese company ZTE is setting up a new plant in Ethiopia given a recent shortage of mobile phones. In part this may have been a *quid pro quo* after ZTE was awarded the contract to expand Ethiopia's mobile phone network. There have also been closures however, paradoxically associated with "corporate social responsibility."

In Nigeria ZTE closed its mobile assembly plant because it could not compete on price with other companies. According to Malakata in most African countries Nokia and Motorola phones are priced from US $40 while the cheapest ZTE phone is $100. As part of the Emerging Market Handset Initiative of the GSM (global system mobile) Association to bring cheap mobiles to the developing world, Motorola now sells "ultra-low cost" handsets for as little as US $21 in Kenya for example. In this way major Western corporations are using Corporate Social Responsibility to outcompete rivals. Even at these reduced prices, however, new mobile phone sellers face competition from second-hand mobile phones imported from Europe and other rich countries and "semi-legal" Chinese producers. . . .

Africa is integrated then as a consumer, and in some places assembler, of mobile phones. However it is connected through, and to, this technology in other ways too.

Mobile Africa and the Coltan Connection

In addition to the "information revolution," the mobile phone value chain has also been associated with other revolutions, violence and forced migrations in Africa. The war in the Democratic Republic of Congo (DRC) from 1998–2003 and the ongoing conflict in the east of that country is partly a resource war over control of the precious metal coltan, which serves as an important component of mobile phones and other "new" ICTs. Coltan is an abbreviation for colombite-tantalite from which the precious metals Colombium and Tantalum are extracted. Tantalum is twice as dense as steel and can capture and release an electrical charge, which makes it vital for capacitators in portable miniaturised electronic equipment such as mobile phones. Eighty percent of known tantalite reserves are in the DRC.

Two days after the new government of Laurent Kabila in the DRC moved to nationalize the main coltan mining company in 1998 the rebellion to overthrow him began, with the support of the directors of the company which was being expropriated. The war in Congo brought in numerous African armies and was partly fuelled by coltan. In 2000 prices for coltan spiked ten-fold, largely as a result of the launch of the Sony Playstation 2 consol and new mobile handsets.

Much of the coltan in Eastern Congo is mined in two world heritage sites: Kahuzi-Biega National Park and Okapi Wildlife Reserve. Outside of these, unregulated coltan mining destabilised hillsides, leading to landslides and damaging future agricultural potential. Half of the land that was seized for unplanned artisanal coltan mining cannot now be used for agriculture. The "resource pull effect" has also been in evidence. According to a coltan miner in the DRC:

> We think that agricultural activities are a good thing, but we cannot see ourselves taking them up again in the short term because we earn much more money from coltan. However we are thinking of investing coltan money in agriculture and cattle once peace returns.

As the British MP Oona King noted "Kids in Congo are being sent down into mines to die so that kids in Europe and America can kill imaginary aliens in their living rooms" or text each other. . . .

When the price of coltan fell dramatically in 2001 rebels in Eastern Congo were forced to look for other sources of revenue and the war appeared to end in 2003. However, recent problems with the Australian coltan supply chain, which accounts for 41% of global production, has again led to rapidly rising prices. The spot price for tantalum ore rose approximately 30% from 2007–8 and is implicated in the current return to large-scale conflict in the Eastern DRC.

While a Conflict Coltan and Cassiterite Act was introduced in the U.S. Congress to prohibit the importation of these minerals from Congo if any rebel groups would benefit from their sale, this may simply lead to geographical substitution effects as coltan mined in Congo is rerouted to other markets. In any event, according to a British journalist, 80% of Congo's coltan is sent to Australia for processing. The growth in demand for mobile phones in Africa then may be implicated in the resumption of large-scale conflict in the Congo.

Diamonds have also been associated with conflict in Africa, and Africa undoubtedly supplies many of the diamonds for encrusted mobile phones, sold to the super-wealthy in the rich countries. One vendor of these notes that "buying an Athem luxury phone is also an alternative way of investment in diamonds. The phone by itself might not be of considerable value but the diamonds encrusted on the phones are" . . . , although this vendor notes that their diamonds are certified "conflict free." Even if they are conflict free however the conditions under which artisanal and formal miners work are often extremely dangerous and highly exploitative, paying poverty wages. We now turn to the extent and usage of mobile phones in Africa and whether they can help reduce, rather than reproduce poverty.

Mobile Phone Penetration, Usage and Social Impacts in Africa

The number of phone subscriptions per 100 people constitutes the subscription "penetration rate." It is also possible to measure the penetration rate based on the number of phones per head of population, with some rich countries recording rates in excess of 100%. By the end of 2007 there were 280.7 million cell phone subscribers in Africa, with roughly one mobile phone for every three people on the continent. In terms of access, one estimate suggests that 97% of people in Tanzania have access to a mobile phone; that is they live under the "footprint" of a mobile phone. This is not necessarily to suggest that they use one on a regular basis, but rather that they could access one if they had to through mobile phone kiosks, for example. In Ethiopia SIM (subscriber identity module) cards can be rented.

There is, however, a highly uneven geography to mobile phone usage and penetration with subscription rates ranging from over 70% in Reunion to under 1% in Burundi, roughly mirroring the distribution of wealth on the continent. However there are significant differences in penetration rates which cannot be accounted for by per capita income. For example, Morocco has a penetration rate twice as high as Namibia, with only half of the gross national income (GNI) per capita and despite the fact that Namibia was the first country in Africa to have built a digital network. World Bank researchers attribute this to better collaboration between the state and the private sector in Morocco and different regulatory environments. However, they neglect the fact that distribution of income may also play a very important role. Namibia has the world's highest level of income inequality, with a Gini coefficient of 70.7, whereas Morocco's is "only" 39.8.

Given the capital cost implied in buying a mobile, there are more mobile phone subscribers than there are phones in Africa. This is because sometimes people buy SIM cards which they then use in other people's phones. In Botswana over 60% of phone owners share phones with their family members, 44% with friends and 20% with neighbours, but only 2% of people charge for this "service." This suggests that in addition to instrumental functions, mobile phones are also used and shared on the basis of non-pecuniary utility or *ubuntu.*

Multi-country studies across Africa have shown that mobile phones are used primarily to maintain social networks, although they are also used to maintain "weak links" to business associates. According to Slater and Kwami, mobiles are used to manage local embedded reciprocities. Rather than being used to connect to the "global economy" the majority of calls in Ghana, for example, are "used to maintain family relations." Adoption may also represent part of a defensive livelihood strategy given widespread poverty and the importance of extended family networks.

Given the relatively high cost per unit of time, people on low incomes in Africa also use mobile phones differently than people in high income countries, by using "beeping" or "flashing," for example. This is where someone calls someone else but hangs up before the call is answered to avoid call charges. Often it signifies that someone should call back, but the number of rings may also serve as a type of code, such as two rings meaning "pick me up." "Flashing involves a clear and much discussed economic rationality, designed to win the fierce battle to keep a mobile in permanent operation. But this battle itself indicates the great importance attached to staying connected by mobile, and this importance we would argue is tied to the costs of maintaining, managing and expanding already existing social networks."

Respondents to one survey noted "mobile phones bring poverty" as a result of the high costs of ownership and some people substitute mobile phone usage for consumption of food and clothing. In South Africa respondents to a survey spent an average of 10–15% of their income on mobile phones. Despite the relatively high cost, the disadvantages of not being in the network would seem to outweigh the costs, however. For some having a mobile phone may be the least bad option. This logic reflects the importance of the extended family in livelihood strategies in adverse economic circumstances.

Donner found in his survey that there was no salient cause effect between people having mobiles and their families being more prosperous, suggesting the importance of other reasons for adoption. . . .

Mobiles make people feel more important and "connected." For yet others they are a livelihood and relationship enhancement and management tool. In a sense they instantiate the contradiction between hierarchization and the desire to offset it. The very lack of socio-economic opportunities up the occupational ladder or through emigration for much of Sub-Saharan Africa's population is offset by the (social) "mobility" offered by cell phones. It represents a form of high-tech connection to the global information society and domestic social peers, which may serve to legitimate unequal globalization. The mobile phone then achieves its "value" through not only its functional utility, but as signifier of inclusion and "development" for populations who are excluded. In Burkina Faso "using this technology is locally perceived merely as a tool for keeping up with global trends, rather than reducing poverty."

While some argue ICTs connect elites in Africa more closely to their counterparts in the rich world, they may be of dubious use to poverty reduction. ICTs are implicated in further and new forms of social stratification between the "information rich" and poor. Rather than leading to spatial and social homogenization new ICTs create geographies and social topologies of "enablement and constraint." That is to say ICTs allow for certain space transcending activities to be undertaken, while being influenced by the agent's position in physical space. ICTs are then implicated in new bordering practices, where participants in the information (IT) revolution are included within the "virtual world" or information economy and society, while others are excluded. Indeed some go so far as to argue that they are helping to constitute new class. . . .

Even for those with access to mobiles, necessity is the mother of invention and Africa has been associated with some of the most innovative uses of mobile phones. . . .

Kenyans are repurposing phones to take the place of other infrastructure they lack, ranging from MP3 players to credit cards. The Kenya Agricultural

Commodity Exchange sends farmers up-to-date commodity information via text message. Tradenet takes this idea further and connects sellers and buyers in 380 markets around the continent by mobile. The Kenyan company, Safaricom, which is 35% owned by Vodafone became the first company in the world to provide a money transfer service by mobile. Mobile banking or m-banking, where accounts can be transacted on over a mobile phone and purchases made, has also become popular in a number of African countries. M-learning, where education is delivered via mobile, is also an emerging area.

Economic and Business Impacts of Mobile Phones

Large-scale claims for the transformative economic impacts of mobile phones are sometimes made. . . .

There is evidence that small and medium sized businesses in Africa have taken up mobile phones with enthusiasm and vigor. Donner in his survey of 31 micro and small enterprises found that there were two perspectives on mobile phone adoption. One saw it as a device for pursuing instrumental business goals and functions, whereas others saw mobiles as satisfying intrinsic emotional needs. Small businesses may also adopt mobile phones without seeing their utility, for fear of the disadvantage that not having them may entail. Consequently they may be used in both "offensive" and "defensive" business strategies.

Mobile phones offer advantages to businesses, particularly as they are a vector of what are termed network externalities. The value of a network increases the more participants there are. They also allow small businesses to access new customers and allow for economies of time, by substituting for time consuming trips for example. There is danger, however, for small and medium-sized enterprises that use of ICT by larger firms exposes them to greater competitive pressure, with which they may not be able to cope.

Mobile phones also offer numerous economic advantages to workers and the self-employed in Africa. For example, casual laborers can leave their phone numbers with potential employers rather than having to wait and see if a job materialises. These "economies of time" may enable them to engage in other economic activities in the meantime and thereby supplement their income. Fisherpeople can call ahead to local markets to see where they will get the best price when they land their catch. However, in some cases market information is not sufficient to redress the balance of power between small farmers and traders. In other instances, traders can use their positions as suppliers of credit to get farmers to sell their produce only to them.

In reference to the adoption of mobile phones by micro-entrepreneurs in Africa it has been noted that "even if the majority of microenterprises are not sources of phenomenal growth, any gains in productivity, profitability, and even basic stability are of the utmost importance to the livelihoods of the households involved." However is there a fallacy of composition here? Might the use of ICTs merely enable some businesses to capture business from other micro and small-scale enterprises (MSSE's) rather than grow the economy *per se*? . . .

Duncombe's research suggests that ICT applications may only bring marginal poverty reduction, but that it may be effective if they are used to build a broader range of social and political assets. That is that microenterprises need to build social capital and trust in local networks more than they need to access new information through ICTs.

Cell phone kiosks, repair shops and unlocking or decoding services also offer small business opportunities and mobile phone companies are often some of the biggest corporate tax payers. Although the amount of ICT capital invested by small and medium-sized enterprises in East Africa seem not to be associated with higher productivity. Chowdury and Wolf attempt to explain this counter-intuitive result by arguing that "a certain threshold of ICT investment may be needed to make it effective, and that this threshold might not have been reached in SMEs in the case of East Africa." Bollou and Ngwenyama found that the adoption of ICT was associated with a once off increase in productivity, followed by falling rates of growth as economies of scale have not been realised in (West) Africa. . . .

From the above discussion it is clear that mobile phone usage by itself is not fundamentally altering Africa's dependent position in the global economy, as much of the research and development for this and other more production-enhancing technologies continues to take place in the rich world. Indeed there is recognition perhaps from somewhat surprising quarters that, cumulatively, technology is widening the rich-poor gap globally. According to the International Monetary Fund, for example, "the main factor driving the recent increase in inequality across countries has been technological progress . . . supporting the view that new technology, in both advanced and developing countries, increases the premium on skills and substitutes for relatively low-skill inputs."

While neoliberal economists increasingly acknowledge the geographically uneven impact of technological development and diffusion, in some cases this assumes almost farcical proportions, with William Easterly claiming that "seventy-five percent of Africa's current income lag relative to Europe can be statistically explained by the technology lag in 1500." This technological determinism is also evidenced in studies of mobile phone usage in Africa. For example, one study argues that "for the average country, with a mobile penetration rate of 7.84 phones per 100 population in 2002, the coefficient of 0.075 on the transformed mobile penetration variable implies that a doubling of mobile penetration would lead to a 10 percent rise in output, *holding all else constant*." However, all else cannot be held constant as the information revolution is only one dimension of increasing "time-space compression" associated with globalization.

Conclusion: Globalization, Mobiles and African Society: M-perialism?

. . . According to Kransberg's first law, technology is neither inherently good or bad, it depends on the uses to which it is put. For example, in the war in Somalia in the early 1990s clansmen used cellphones that the American military were unable to tap. One group which routinely abducts child soldiers,

The Lord's Resistance Army, which has been fighting a guerrilla war in Uganda used a recent ceasefire to reorganize itself and purchased 150 satellite phones with money meant to aid communication during the peace talks.

For exponents of the ICT revolution poor places can "catch-up" with richer ones through the adoption of these technologies; that is to close the digital divide. In this schema a spatial difference is presented as a temporal one as globalization and increased interconnectedness will allow for catch up. However, how equal are the terms of digital integration and to what extent is the digital merely a reflection of other more deep-seated economic divides? According to neoliberal theorists if we can all achieve copresence in the same marketplaces, where real or virtual, the result is a positive sum game. ICTs then in the "borderless world" help eliminate market failures and erase uneven economic geographies. However, this negates the many other factors which contribute to uneven development; particularly political pressures for asymmetric integration, and cumulative economic advantages.

New fibre optic cables such as Africa One and SAT3 reduce costs of bandwidth for external internet and phone connection these are for the most part owned by European and American companies. In 2002 it was estimated that African internet service providers were paying US $1 billion a year for connectivity to these bandwidth providers. . . .

SAT3 was launched in Senegal in 2002, but even with this new fibre optic cable the total available fiber bandwidth for the whole country is less than 1.2 gigabits a second; one tenth that of a university such as Harvard or Chicago in the United States.

The question remains about whether this represents a more favourable integration of Africa into the global economy or whether it is merely reinscribing the continent's dependent insertion into the global division of labor. This is important because this project is "expected to access 90% of Africa's existing sub-Saharan telephone market in which 72% of the sub-Saharan population lives."

What Watts terms "appropriationism" is highly evident in relation to mobile phones in titles of reports such as "The Next Four Billion" by the World Resources Institute and the International Finance Corporation of the World Bank. Phone companies such as Nokia even hire anthropologists to study how people use their mobile phones in the developing world. This is perhaps not surprising, given the relative market saturation in the developed world. By 2006 68% of mobile phone subscribers were already in the developing world.

According to Ya'u "globalisation is not only enabled by ICTs but that the level of connectivity of a country determines to a large degree the possibility of its benefiting from the globalisation process." For him in Africa this development represents a new form of imperialism represented by knowledge dependence. This knowledge dependence takes many forms.

As Clare Mercer argues that "the idea that the Internet [or mobile phones] should be used as a developmental tool across Africa is only plausible if it is taken as axiomatic that Africa is devoid of the information and knowledge necessary for development." For example, according to World Bank researchers mobile phones can also help the rural poor overcome "ignorance of income-earning or market opportunities" and may help in establishing new small businesses in sectors such as prostitution!

Africa's integration into the global information economy is characterised by a missing top and middle. The continent provides raw materials, associated with conflict, and consumes mobiles phone and engages in some limited low-tech assembly operations. The high-value added activities in the chain take place elsewhere. These activities and the use of mobile phone then represent a form of "thintegration" into the global economy. It is not that Africa is excluded from the process of globalization; indeed it is integral to it as a supplier of raw materials. Use of mobile phones, by itself does not change this context.

Are there then pro-poor modes of technical integration into the global economy? Can, as the former Secretary General of the UN put it, digital bridges be built? Perhaps they can, but the key question is where they are developed and built, what kind of traffic crosses them and what direction the flow is. This is structured by previous social relations and rounds of economic incorporation and marginalization in Africa.

POSTSCRIPT

Do Cell Phones and the Internet Foster "Leapfrog" Development in Africa?

In addition to the standard definitions of leapfrogging discussed in the introduction to this issue, there is another take on this term from the economics literature, which may shed further light on this debate. The concept of leapfrogging was originally used in theories about economic growth and industrial innovation, with specific attention to competition between firms. The economist Joseph Schumpeter hypothesized that companies holding monopolies based on existing technologies have less incentive to innovate and that they eventually lose their technological leadership role when technological innovations are adopted by emerging firms that are more willing to take the risks. At some point, these technological innovations become the new standard, and the newcomers "leapfrog" ahead of the former leaders. The situation in Africa is interesting *vis-á-vis* this theory because it is not African firms that are developing ICTs. This is, of course, one of Pádraig Carmody's primary concerns, that Africans are primarily users rather than producers or creators of ICTs. That said, one could argue that ICT innovation is not just about hardware but also about software and actual use. Although Africans have not been at the forefront of hardware innovation, they have been more creative with software and new forms of use.

One of the most interesting examples of African innovation is the way in which cell phones are being used as a delivery mechanism for cash transfers. Kenya's phenomenally successful M-PESA scheme, operated by Safaricom (a local subsidiary of Vodafone), is a service that allows e-money to be transferred by cell phone. Once registered, users can buy electronic funds at any M-PESA agent and send these by SMS to any other cell phone user in Kenya. Electronic funds can be redeemed for cash at any M-PESA agent, exchanged for Safaricom airtime, or used to pay bills. Just two years after its introduction, M-PESA had over 7 million registered users, and 10,000 agents (Katharine Vincent and Tracy Cull, "Cell phones, electronic delivery systems and social cash transfers: Recent evidence and experiences from Africa," *International Social Security Review,* 2011).

A final set of perspectives to consider when examining ICTs in Africa is that offered by Science and Technology Studies (STS). STS examine how social, political, and cultural values affect scientific research and technological innovation, and how these in turn influence society, politics, and culture. Traditionally, STS scholars have explored this synergistic, back-and-forth between technology and society in one region. In the case of CSTs in Africa, it often means examining the impact of distant social, political, and cultural values on the development of a technology that is brought to the African context where it dynamically interacts in a different social setting.

The scholarly literature on ICTs in Africa is increasingly vast and growing quickly. For another example of recent optimistic account of the ICT-based development thesis, see Calestous Juma's *The New Harvest: Agricultural Innovation in Africa* (Oxford University Press, 2011). For a more nuanced view of the developmental potential of ICTs in Africa, see Barney Warf, "Uneven Geographies of the African Internet: Growth, Change, and Implications," *African Geographical Review,* 2010.

ISSUE 7

Is Increasing Chinese Investment Good for African Development?

YES: Barry Sautman and Yan Hairong, from "Friends and Interests: China's Distinctive Links With Africa," *African Studies Review* (2007)

NO: Padraig R. Carmody and Francis Y. Owusu, from "Competing Hegemons? Chinese Versus American Geo-Economic Strategies in Africa," *Political Geography* (2007)

ISSUE SUMMARY

YES: Barry Sautman, associate professor of social science at The Hong Kong University of Science and Technology, and Yan Hairong, the Department of Applied Social Science, Hong Kong Polytechnic University, argue that China's links with Africa represent a distinctive "Chinese model" of foreign investment. They further suggest that this Chinese model represents a lesser evil than assistance offered by the West.

NO: Padraig Carmody, of Trinity College Dublin, and Francis Owusu, of Iowa State University, are less sanguine about Chinese involvement in Africa. Although they also perceive potential benefits from increasing trade with China, they describe how increasing resource flows are strengthening authoritarian states and fuelling conflict.

Chinese trade with Africa has grown by leaps and bounds over the past several years. The value of Chinese trade with Africa rose from $10.6 billion in 2000 to $40 billion in 2005, to just over $70 billion in 2007, and to $115 billion in 2010. Although the figures are not yet in, China likely passed the United States in 2010 to become the continent's leading trading partner. According to the Chinese Ministry of Commerce, trade is projected to continue climbing at a significant rate over the coming decade.

Although increasing trade with China appears to be one of the major reasons why average economic growth rates are increasing in sub-Saharan Africa (in fact, average economic growth rates for sub-Saharan Africa doubled between 2000 and 2005), a number of concepts from the development literature suggest that African countries need to carefully negotiate this situation if they wish to secure lasting development gains.

Dependency theory might be deployed to portray some aspects of China's relationship with Africa. Dependency theorists, like Andre Gunder Frank, wrote about Europe's relationship with its colonies. For Frank, the primary role of the colonies was to provide raw materials to the metropole. Economies were often changed through taxation policies, infrastructure installed, and elites trained for the sole purpose of guaranteeing the smooth export of unprocessed resources. As such, many former colonies were "underdeveloped" as their economies were reshaped to provide such raw materials. The problem is that this type of colonial relationship (which often persists in the form of neocolonialism) produces a situation of dependency wherein cheap unprocessed materials are exported or exchanged for relatively expensive manufactured goods.

Another such idea is that of the "resource curse" phenomenon or "Dutch disease." Although the resource curse idea has several facets, the aspect most relevant to this issue is the notion that the extraction of particularly lucrative resources may lead to the simplification, or concentration, of national economies, with increasing dependence on a single export mineral or fuel. Dutch disease is a slight variation on this same theme and specifically refers to the decline of the manufacturing sector in the face of particularly profitable resource extraction opportunities. This term was first used by *The Economist* magazine in 1977 to describe the decline of the manufacturing sector in the Netherlands after the discovery of natural gas in the 1960s. Although simple trade models suggest that a country should specialize in the export of goods for which it has a comparative advantage, such as oil for an oil-rich country, the reality is often more complicated. If the natural resource begins to run out, or there is a downturn in prices for the good, it will be much more difficult for a country to begin or renew an industrial sector because they will face stiff (if not crushing) competition from other countries that have been investing in this area all along.

In this issue, Barry Sautman, associate professor of social science at The Hong Kong University of Science and Technology, and Yan Hairong, with the Department of Applied Social Science at Hong Kong Polytechnic University, argue that China's links with Africa represent a distinctive "Chinese model" of foreign investment. They further suggest that this Chinese model represents a lesser evil than assistance offered by the West.

Padraig Carmody of Trinity College Dublin and Francis Owusu of Iowa State University are less positive about Chinese involvement in Africa. They describe how African trade flows are reorienting from the "Global North" to the "Global East." They further explore the political and economic impacts of increased geoeconomic competition between the East and the West in Africa. Although they perceive potential benefits from increasing trade with China, they describe how increasing resource flows are strengthening authoritarian states and fuelling conflict.

YES

Barry Sautman and Yan Hairong

Friends and Interests: China's Distinctive Links with Africa

Between countries, there are no friends, only interests.

President Abodoulaye Wade of Senegal,
paraphrasing Lord Palmerston, 2005

Introduction

A remarkable and telling exchange on Chinese policies in Africa occurred in 2006 between the U.S. Council on Foreign Relations (CFR) and the government of the People's Republic of China (PRC). A CFR report on enhancing U.S. influence in Africa had devoted a chapter to China in which it charged that the PRC protects "rogue states" like Zimbabwe and Sudan, deploys its influence to counter Western pressures on African states to improve human rights and governance, and competes unfairly with U.S. firms in contract bids in Africa. These same points have been made by veteran critics of China in the U.S. Congress and by U.S. analysts who see China as a competitor. In response, China's foreign policy elites, which have long regarded the CFR as a "superpower brain-trust" and "invisible government" shaping the U.S. global role, responded by arguing that China has a "strategic partnership with Africa that features political equality and mutual trust, economic win-win cooperation and cultural exchange." The authors affirmed Africa's desire for a more democratic international order and detailed the aid activities of the Forum on China–Africa Cooperation (FOCAC), which convened African and PRC ministers in Beijing in 2000. Addis Ababa in 2003, and Beijing again in 2006. Although the PRC paper eschews the obligation of states to vindicate the rights of oppressed people, and furthermore suggests that China is likely to follow the West in its path of forging bilateral free trade agreements (FTAs) in Africa that bypass WTO (World Trade Organization) regulations, it is firm in its claims that the West ignores African aspirations for a more equitable international distribution of wealth.

While the PRC paper is somewhat defensive, the CFR report, like much Western discourse, actively misrepresents China's role in Africa. The stock notion that China practices neocolonialism in Africa and promotes corruption

From *African Studies Review*, vol. 50, no. 3, March 2007, pp. 77–95. Copyright © 2007 by African Studies Association. Reprinted by permission via Copyright Clearance Center.

is fostered by its portrayal ol China as a country that is uniquely supportive of illiberal regimes, as well as its claims that Chinese activities, such as purchases of illegal African timber, are harmful to the environment. Elsewhere, China has been accused of conducting trade in Africa that is damaging to African antipoverty efforts, although it is rarely acknowledged that Western powers, as well as Taiwan, have long supported authoritarian regimes in Africa. PRC support for Zimbabwe and Sudan is much discussed in the West, but little is said about U.S. support for oil producers such as Gabon, Angola, Chad, and Equatorial Guinea or about its intelligence and other military coopera-tion with Sudan. During his tenure as Uganda's president, Yoweri Museveni, who tried his main opponent for rape and treason and changed the constitu-tion in order to remain in office, was a much-praised U.S. ally. China does purchase illegal African timber, but so does the European Union, and China does not participate at all in the biopiracy in Africa carried on by Western pharmaceutical firms.

Nor can China's trade relations with either Africa or the West be accused of having deleterious consequences for Africans. Only seven sub-Saharan states receive a significant share (5%–14%) of their imports from China. Worldwide, PRC exports compete with African exports almost solely in textiles and cloth-ing, and China, in fact, supplies African firms with most of the cloth they need to compete in their main market, the United States. Some 60 percent of China's exports, moreover, are produced by foreign-owned firms. Inexpen-sive PRC-made household goods brought to Africa by Chinese and Africans do inhibit light industry and may harm the poor as potential producers. Yet, machinery, electronic equipment, and "high- and new-tech products" made up nearly half of China's 2005 exports to Africa, and PRC goods, which are much more affordable than both Western imports and many local prod-ucts, benefit the poor as consumers. African industrialization, moreover, was already severely damaged by Western imports following the imposition in the 1980s and 1990s of World Bank/International Monetary Fund (IMF) Structural Adjustment Programs (SAPs). The CFR report is thus representative of the com-mon Western moral binary in discourses on China's Africa policies. This was exemplified by a German foundation's notice for its panel on "China in Africa" at the NGO forum of the WTO 2005 ministerial meeting, which asked, rhetori-cally, whether "China–Africa trade and investment relations [are] following a pattern of South–South cooperation, guided by development needs of both sides? Or are [they] just replications of the classical North–South model, where Africa's hope of building a manufacturing sector gets another beating? Will the Chinese 'no political strings attached' approach help the African development state regain posture, or is it a recipe for closed-door business with autocrats to get a competitive edge over Western economic interests?"

This article focuses on two sets of China–Africa links that serve to con-trast China's involvement in Africa with what the West has offered through the Washington Consensus (WC) and Post-Washington Consensus (PWC) (a paradigm that adds to WC neoliberalism a discourse of democracy, good gov-ernance, and poverty reduction). These distinctive links stem from China's status as a developing country, its socialist legacies, and its own semicolonial history, as well as from its late entry into Africa in the midst of a decades-long decline in African fortunes associated with WC privatization, liberalization, deregulation, and austerity policies. PRC leaders are usually depicted as having an instrumental approach to dealings with foreigners and are said to have interests, but no friends, abroad. And yet, in comparison with other foreign interests in Africa, China today is often perceived as the lesser evil, a perception that may allow PRC leaders to make good on their claim of including Africans among their "all-weather friends."

China, to be sure, is now a trade-driven industrial power integrated into the world system. Like other nations it practices a realpolitik of aggrandizing national wealth and power, with policies and practices that have been called "neo-liberalism with Chinese characteristics." China increasingly replicates in key ways longstanding developed-state policies in Africa of disadvantageous terms of trade, exploitation of natural resources, oppressive labor regimes, and support for authoritarian rulers. In terms of its Africa policies, the PRC can hardly boast that "China is the best," but neither is China "just like the rest." Although commonalities between PRC and Western approaches are now fundamental, important distinctions also exist: in the distinctive "Chinese model" of foreign investment and infrastructure loans; in the so-called Beijing Consensus (BC), lauded as a model of development that takes seriously developing state aspirations ignored by the West; and in China–Africa aid and migration policies, which are seen by many Africans as serving interests other than those of foreigners and the elite.

The "Chinese Model" of Investment

Before the 1990s, the PRC's Africa policy was purely political: China fostered anticolonial and postcolonial solidarity, and such efforts were repaid through African states' recognition of the PRC. The symbol of China–Africa relations from the 1960s to the 1980s was the Tanzania–Zambia railway (Tazara), built by fifty thousand Chinese laborers. China's practice of supporting developing state initiatives and providing aid that did not enrich elites still resonates with Africans today, even though, since the 1990s, PRC activism on behalf of developing states has waned and much of what it does in Africa is now profit-centered.

Postcolonial Africa is often seen mostly in terms of its problems: as burdened by civil wars, epidemics, and venal regimes that aggravate endemic poverty. These perceptions led to a post–Cold War Afro-pessimism or even Afrophobia and to the downgrading of Africa as a site of interest for policymakers and investors from the developed world. This began to change somewhat in the 1990s, as Western leaders again began to pay attention to the continent, partly because of China's increased presence, which grew by 700 percent during the decade. While many Africans still believe that Africa remains in many respects invisible, especially to the United States, there is no doubt that China, Britain, France, and the U.S. see themselves as competitors in the second largest continent with the fastest growing population: with 900 million people in 2005, less than one-seventh of the world's people, Africa is projected to have nearly a quarter of the global population by 2050, and it has been estimated that Africa's economy may double in a generation.

From 2001 to 2004, Africa's average annual intake from foreign direct investment (FDI) was only $15–18b, despite the continent's providing the world's highest FDI returns, averaging 29 percent in the 1990s and 40 percent by 2005. FDI flows in Africa in 2005 jumped to $29b (of $897b in global FDI flows), but China's FDI stock in Africa was still only $1b of Africa's $96b in FDI stock (two-thirds of it European—half British or French—and one-fifth North American). By late 2006, however, China's investments in Africa were pegged at almost $8b, as pledged investments were actualized. China will soon become one of Africa's top three FDI providers. Since the 2006 Forum on China–Africa Cooperation, that effort has been aided by a $5b China–Africa Development Fund to spur PRC investment. Trade with Africa was a tiny part of the PRC's 2006 $1.76 trillion in world trade, but had grown from $3b in 1995 to $10b in 2000, $40b in 2005, and $55b in 2006, balanced slightly in Africa's favor. There were more than eight hundred Chinese enterprises in Africa in 2006, one hundred of them medium and large state-owned firms. The PRC accounts for only a tiny part of Africa's FDI ($3.6b in 2004 and $6.9b in 2005), but its firms invested $135m and $280m, respectively, in those years. Still, while China is the third largest trader with Africa, after the U.S. and France, its trade was well behind the United States's $91b and represents only one-tenth of Africa's world trade, most of which remains with the E.U. and U.S.

Between the end of the last century and the beginning of this century, Africa's overall share of world trade and global FDI inflows actually declined: in the 1970s Africa received 5 percent of the former and 6 percent of the latter, but in 2005 the figures were 1.5 percent and 3 percent, respectively. Many PRC and African analysts contend, therefore, that increased PRC trade and investment ease Africa's dependence on the West and are mutually beneficial. The U.N. Development Program agrees and underwrites a China–Africa Business Council that promotes PRC investment in Africa.

Africa is the most resource-laden continent, with every primary product required for industry, including (in 2005) 10 million (m) of the globe's 84m barrels per day (bpd) of oil production. Some 85 percent of new oil reserves found from 2001 to 2004 were on west/central African coasts, most of it light, sweet, highly profitable crude. Strong competition for African oil exists because 90 percent of the world's untapped conventional oil reserves are owned by states, and 75 percent of known reserves are in states that exclude or sharply limit outside investment in oil. According to estimates, world demand for oil may reach 115m bpd by 2030. In 2005 the U.S. imported 60 percent of its 20m bpd of oil, 16 percent from Africa. In 2006, however, U.S. imports of oil from Africa equaled or slightly surpassed those from the Middle East, with both at 22 percent of total imports (2.23m bpd). Oil today accounts for more than 70 percent of all U.S. imports from Africa. In 2005, China imported 48 percent of its 7.2m bpd, with 38 percent of its imports from Africa (1.33m bpd). By 2025, its imports should reach 10.7m bpd, 75 percent of consumption. More than 60 percent of the output of Sudan, Africa's third largest oil producer, went to China and supplied 5 percent of PRC oil needs. Angola and Nigeria, the next largest producers, each sent a quarter of their production to China, and in 2006 Angola overtook Saudi Arabia as China's greatest source, supplying 15 percent of PRC oil imports. Overall, however, China consumed less than one-tenth of oil exported from Africa.

Chinese bids for resources fare well because they are packaged with investments and infrastructure loans. China preeminently invests "in long-neglected infrastructure projects and hardly viable industries," and its loans, typically advanced at zero or near-zero interest, are often repaid in natural resources, if they are not canceled entirely. In Angola, China offered $2b in aid for infrastructure projects and secured a former Shell oil block that the largest Indian company had sought. In Nigeria, a promised $7b in investments and rehabilitation of power stations secured for PRC firms oil areas sought by Western multinationals. Chinese companies outbid Brazilian and French firms for a $3b iron ore project in Gabon after pledging to build a rail line, dam, and deepwater port. Its firms had $6.3b in construction contracts in 2005 and now employ many African workers. A Nigerian official has noted that "the Western world is never prepared to transfer technology—but the Chinese do, [and] while China's technology may not be as sophisticated as some Western governments', it is better to have Chinese technology than to have none at all."

The notion of a specifically "Chinese model" of economic growth and foreign relations, in which trade and investment play prominent roles, is common in Africa. African analysts contrast the PRC government's massive investment in infrastructure and support services within China with their own governments' failure to provide these prerequisites for development. They also compare it to Western economic practices, which are seen as exploiting "unequal and disparate exchange" to lock in underdevelopment. The director of the U.K. Centre for Foreign Policy Analysis has observed that "the phenomenal growth rates in China and the fact that hundreds of millions have been lifted out of poverty is an attractive model for Africans, and not just the elderly leadership. Young, intelligent, well-educated Africans are attracted to the Chinese model, even though Beijing is not trying to spread democracy." The president of the African Development Bank has said of the Chinese that "we can learn from them how to organize our trade policy, to move from low to middle income status, to educate our children in skills and areas that pay off in just a couple years."

Many Africans view China's political economy as differing from that of the West in ways that speak to Africa. China's strategy, according to a Nigerian journalist, "is not informed by the Washington Consensus. China has not allowed any [IMF] or World Bank to impose on it some neo-liberal package of reforms. . . ." Their strategy has not been a neo-liberal overdose of deregulation, cutting social expenditure, privatizing everything under the sun and jettisoning the public good. They have not branded subsidy a dirty word. African analysts contend that the PRC government's investment in infrastructure and support services is made possible by China's exemption from "strictures imposed by multilateral and bilateral financiers." Some also note, interestingly, that China has had high growth rates and reduced poverty without adopting Western liberal democracy.

Indeed, suspicion of and cynicism about the West are widespread, and African analysts agree that while China, like the West, wants Africa's oil, the West is more single-minded in its pursuit. While three-fourths of U.S. FDI in Africa has been in oil, 64 percent of PRC FDI in Africa from 1979 to 2000 was in manufacturing and only 28 percent was in resources. According to one African journalist, "the way in which China's demand for oil is framed in Western media—in breathy, suspense-filled undertones— . . . smacks of racist double standards." Even such U.S. allies as Museveni see Africa's virtual "donating" of unprocessed raw materials to the West as allowing a small part of humanity to live well at Africans' expense, and critics contend that Africans need investment that will permit them to sell coffee and not just beans, steel and not just iron ore. They regard China's surging demand for African exports—the PRC's share rose from 1.3 percent in 1995 to 10 percent in 2005—as aiding that effort. Africans also find PRC goods to be cheaper than Western imports, and often cheaper than local goods: a 50 kilo bag of local cement costs $10 in Angola, but imported PRC cement costs $4.

For many Africans, then, there exists a "Chinese model," now often labeled the "Beijing Consensus" (BC), that stands in contrast to FDI/export-led rapid industrial expansion. It is an image of a developing state that does not fully implement WC prescriptions, does not impose onerous conditions on African states' policies, and is more active than the West in promoting industrialism in the global South. According to one South African scholar, China has succeeded in creating "a somewhat idealistic impression of the distant partner or big brother in the East." Whether this positive view of the Beijing Consensus is warranted continues to be debated, although its accuracy may be seen as less important than the fact that it exists and plays a role in how Africans appraise the policies of both China and the Western states.

The "Beijing Consensus" as a Competing Framework

While there actually is no clear "consensus" on the exact components of the "Beijing Consensus," the phrase was coined by Joshua Ramo, a former *Time* magazine foreign affairs editor and Goldman Sachs, China advisor, now managing director of Kissinger Associates. The term describes PRC investments, aid, and trade that are carried on without oversight by Western states and international institutions. While the WC/PWC paradigm has more than a two-decade history in Africa, the BC is now seen as competing with WC/PWC instruments that were set up by the E.U., U.S. and South Africa around the year 2000.

The Cotonou Agreement of 2000, the E.U. framework established with seventy-seven African, Caribbean, and Pacific (ACP) states, is based on free trade (including WTO compliance and subcontinental regionalism), private enterprise, export production, FDI, austerity measures, and conditioned aid. It gives a leading role to the European Commission and individual political, rather than group and socioeconomic, rights. Poverty reduction is seen as a by-product of trade and capital liberalization and FDI secured by compliant labor. Quintennial conferences serve to renegotiate the E.U.–ACP relationship, which also includes bilateral and regional freetrade Economic Partnership Agreements. According to many, these pacts, like those negotiated by the United States, have the effect of weakening solidarity among developing states in the WTO.

The U.S. African Growth and Opportunity Act (AGOA) of 2000 provides that states may receive trade preferences if they marketize, liberalize, privatize, desubsidize, deregulate, and do not undermine U.S. foreign policy interests. Some thirty-seven African countries, many of them authoritarian, have been declared eligible. U.S. and African ministers meet every two years in an AGOA Forum. AGOA trade concessions exceed the U.S. General System of Preferences only slightly, in part because oil and minerals make up more than 80 percent of the value of African exports to the U.S. Only a few countries have gained under AGOA, mainly by exporting agricultural products, such as cut flowers, that are not plentiful in the U.S. Most other African products remain barred by competition from subsidized U.S. agriculture and nontariff health and safety barriers, even though Asian firms produce many of the goods entering the U.S. from Africa. AGOA is also a platform for FTAs between the U.S. and African regional entities. Its appeal for African rulers lies not so much in the benefits that it offers directly, as in closer political ties to the U.S., which result in aid, including military training useful in quelling opposition.

Since 2001, neoliberal principles also have been embodied in the New Partnership for African Development (NEPAD). Based on the idea that integration into the world market is the single antidote to poverty, NEPAD has been endorsed by the African Union and is backed by businesses in South Africa and around the world. China, in fact, voices support for NEPAD and says it implements its principles through FOCAC. But there is no doubt that NEPAD is mostly identified with Western interests. U.S. firms in Africa act as a link between AGOA and NEPAD, and E.U. endorsements of NEPAD link it to the Cotonou Agreement. U.S. leaders praise NEPAD as "extend[ing] democracy and free markets and transparency across the continent," while critics compare it to IMF/World Bank SAPs, claiming that its self-representation as "by Africans for Africans" masks the degree to which it is another mechanism for implementing developed countries' WC/PWC frameworks. In the global South there is a sense of grievance against the policies of neoliberalism embodied in the E.U., U.S. and NEPAD mechanisms, which are seen as promoting an aggravated form of worldwide unequal exchange. At the 2000 FOCAC opening ceremony, Zambia's president stated that developed countries

> are not prepared to discuss the issues of justice and fair play concerning the international trade and commercial sector, which imposes considerable suffering and privation on developing countries. . . . The developing world continues to subsidize consumption of the developed world, through an iniquitous trade system. The existing structure is designed to consign us to perpetual poverty and underdevelopment. . . . It is unrealistic to expect support, relief or respite from those who benefit from the status quo.

China, by comparison, is seen as supporting initiatives by African states to address development problems not solved by neoliberalism's corporate initiatives, and as promoting investment in infrastructure and human capital, rather than just primary products.

PRC aid comes without the strings attached by AGOA and other programs. Even Jim McDermott, the U.S. Congressman known as the "Father of AGOA," has cautioned that "the U.S. cannot rely solely on the private sector to help support Africa's endeavor to develop. Private companies may invest in new manufacturing plants or mineral extracting facilities, but they usually do not drill water in remote villages, or build schools to educate young Africans. Do you know of many venture capitalists who buy malaria or TB drugs for the world's poor to enhance their trade opportunities?" To be sure, China's self-portrayal as Africa's helpmate is often dismissed as propaganda designed to curry favor with African elites. To the most cynical observers the Beijing Consensus reflects little more than China's desire to turn itself into a world leader. Some critics in Western circles hold the view that the BC is simply a more saleable variant of the PWC, or that any "consensus" not enforced by the U.S. is bound to seem attractive. Indeed, while the PRC is seen as offering a new approach, many of the cooperative ventures between China and Africa and its leaders mirror both in form and content the institutions and frameworks of developed countries. Nevertheless, Africans who are disenchanted with Western neoliberalism regard the PRC as a plausible alternative, based on experiences and needs that China shares with Africa.

The most vocal opponents of the BC are U.S. neoconservatives, who dismiss it as "economic growth without the constraints of democratic institutions" or "economic development without political change." Supporters, by contrast, present it as a multifaceted set of policies that encourage constant innovation as a development strategy (instead of one-size-fits-all neoliberal orthodoxy) and use quality-of-life measures, such as promotion of equality and environmental protection (not just GDP), in formulating the strategy. Joshua Ramo asserts that the BC opposes the hierarchy of nations embodied in the WC-related international financial institutions (IFIs) and the WTO. PRC leaders, he says, reject "a US-style power, bristling with arms and intolerant of others' world views" in favor of "power based on the example of their own model, the strength of their economic system, and their rigid defense of . . . national sovereignty."

Ramo cannot be written off as "radical." He is a member of the U.S. Council on Foreign Relations and was declared a "Global Leader of Tomorrow" by the neoliberal World Economic Forum (WEF). He is also affiliated with the Foreign Policy Centre (FPC) in London founded by Tony Blair, and his book on the Beijing Consensus has been praised by the World Bank. Ramo himself claims that the BC is one "model" that exists within the neoliberal paradigm. But he also approvingly quotes an Indian sociologist who has stated that "China's experiment should be the most admired in human history. China has its own path."

China's government denies that it promotes any particular model. Yet, soon after Ramo coined the concept of a Beijing Consensus (and Klaus Schwab, the chair of the World Economic Forum, contrasted it with the WC), a leading PRC economic journal published an article by Ramo, and Chinese television produced a program on his book. His work has been circulated among the top five thousand PRC leaders. China's leading newspaper published an article in which the economists Wu Shuqing (former head of Beijing University and now a Ministry of Education advisor) and Cheng Enfu (head of the Academy of Marxism and a proponent of the "socialist market economy" as a world model) endorsed the "theoretical scientificity and practical superiority" of the BC. Opposing it to the WC, they spoke of its "growing influence in the world, particularly among developing countries." Other PRC works counterpose the BC and WC, and PRC writers praise Africans' supposed move from WC to BC.

To some analysts, the differences between WC and BC amount to an ideological struggle between the U.S. and China, that is, a struggle between a "neo-liberal Anglo-Saxon credo" and an Asian-derived "socially oriented" approach. A U.K. journalist has said that the WC–BC confrontation is "the biggest ideological threat the west has felt since the end of the cold war." Expressing no doubt about which "model" will prevail, he opined that in two decades "the press will be full of articles about 'Asian values' and the 'Beijing Consensus.'" Arif Dirlik, who emphasizes the BC's lack of ideological coherence, has argued, nonetheless, that its appeal may be its acknowledgment of the desirability of a global order "founded, not upon homogenizing universalisms that inevitably lead to hegemonism, but on a simultaneous recognition of commonality and difference." That recognition magnifies China's soft power in Africa, the BC's main testing ground.

Tied and Untied Aid

Worldwide, much aid from developed states is subject to conditions that benefit the donor economically and politically, including its security interests. From the point of view of the recipient countries, such "tied aid" is a particularly inefficient form of development assistance because it does not help them develop their economies by creating new businesses and jobs. Instead, most of the benefits remain in the donor nations. Tied aid is also inefficient because goods and services purchased from the donor often could be available at a lower price from local producers or world markets.

The popular view in the West is that it is generous with Africa, both in terms of aid monies and debt relief. Yet from 1970 to 2002, Africa received $530b in aid and loans and repaid $540b. The G-8 and WC-related IFIs subsequently canceled the debt of only fourteen states, and Africa's debt still stands at about U.S. $300b. An additional $50b in aid was promised in 2005, but more than half was either double-counted or involved money already pledged. Debt relief and refugee-related expenditures in developed countries are also counted by them as part of increased development assistance.

About 80 percent of U.S. grants and contracts to developing countries, moreover, is "tied"—it must be used to buy goods and services from U.S. firms and NGOs. Some 90 percent of Italy's aid benefits Italian companies and exports; 60–65 percent of Canada's aid and much of that of Germany, Japan, and France is tied to purchases from those states. A U.N. study found that such ties cut by 25–40 percent the value of aid to Africans, who are required to buy noncompetitively priced imports. Actual costs of tied direct food aid transfers are 50 percent higher than local food purchases and one-third higher than the costs of third-country food.

Aid from the PRC has differed from other aid programs in a number of ways, including whether the recipient or donor chooses the projects on which aid monies will be spent. China's approach is more commercial than formerly, but it still contrasts with U.S. and U.K.'s insistence on aiding only private enterprise development. For example, the 2005 report of the U.K. Commission on Africa states that Africa should adopt the Private Finance Initiative: all major projects should be built in conjunction with the private sector. China, by contrast, continues to support state-run projects in industry and agriculture.

PRC aid to Africa, while not totally "untied," manages to attenuate any negative consequences to the donor. PRC firms secure many contracts on projects in Africa financed by China's soft loans. One analyst speaks of "indirect conditionalities," an understanding that PRC firms will secure a portion of work financed by PRC loans. The $2b credit line China extended to Angola in 2004, which was used for railroad repair, road building, office construction, a fiber-optic network, and oil exploration, was guaranteed by a contract for the sale of oil from a field that generates 10,000 bpd. The loan, originally at 1.5 percent interest but lowered to 0.25 percent, is to be recouped over seventeen years, including a five-year interest-free period. Its terms reserve for Angolans 30 percent of the value of contracts paid for with its funds. Chinese firms also secure many contracts apart from those financed by PRC aid: in Botswana by 2005 they were winning 80 percent. A study of 505 contracts opened by African states for international bids in 2004 showed that PRC firms won 2.6 percent of the total, but these amounted to 18 percent of the value of all contracts. In 2001, Africa-based contracts were 17 percent of all PRC contracts outside China, but by 2005 they were 28 percent and in 2006, 31 percent. In that year, PRC-won contracts ranked first in new contract value among the totality of contracts secured by foreigners in Africa. In 2006, there were more than six hundred Chinese infrastructure projects in Africa, and financing by China's Exim Bank for African infrastructure had increased to $12.6b, much more than the developed countries' total infrastructure aid to the continent.

PRC winning bids are based on low labor costs and profit margins and quick turnaround. Most Western firms expect 15–25 percent profits rates; most Chinese firms expect less than 10 percent and many accept 3–5 percent. In Ethiopia, some PRC firms make unprofitable bids to get a foot in the door. Lower salaries and profit margins also arise from competition among China's state-owned firms. PRC contractors building Ethiopia's roads seek a 3 percent profit; Western businesses seek 15 percent or more. Efficient, low-cost Chinese construction softens the image of PRC participation in the overall unequal trade and investment relationships with Africa.

Certainly PRC aid to Africa is not politically disinterested. Such aid provides numerous political benefits, such as African support on sovereignty issues and for China's efforts to attain a "market economy status" that will enable it to better resist antidumping actions in the WTO. If China were ever to directly confront Western states in international forums, African states could be counted on as allies. Yet, PRC aid to Africa is not used as a political tool in the same way that it is used by Western political actors. This approach is long-standing: Tanzania's first leader, Julius Nyerere, commenting on the loan for building Tazara, stated that "the Chinese people have not asked us to become communists in order to qualify for this loan. . . . They have never at any point suggested that we should change any of our policies—internal or external." There is no evidence that China attaches political conditions to its aid, except that recipients must maintain full diplomatic relations with the PRC, rather than with Taiwan, as all but five African countries do. In contrast, during the Cold War, the U.S. and U.K. pressured Tanzania to become allied with the West and later to accept IMF/World Bank SAPs. The U.S. is now heavily involved in African politics: for example, through multimillion-dollar programs to support or undermine the governments of Angola, Burundi, Sudan, and Zimbabwe carried out by the self-described "overtly political" Office of Transition Initiatives of the U.S. Agency for International Development. In 2003 the U.S. pressured the three African U.N. Security Council members to endorse the war in Iraq.

In the end, of course, what is most significant about PRC's collaborative projects in Africa is their economic success for both parties. As one analyst explained, "Chinese aid is often dispensed in such a way that corrupt rulers cannot somehow use it to buy Mercedes Benzes. . . . [It] is often in the form of infrastructure, such as a railroad network in Nigeria or roads in Kenya and Rwanda. Or in the form of doctors and nurses to provide health care to people who otherwise would not have access. China provides scholarships for African students to study in its universities and, increasingly, funds to encourage its businessmen to invest in Africa." In 2005, Sierra Leone's ambassador to the PRC, speaking of China's activities in Africa, said, "The Chinese are investing in Africa and are seeing results, while the G-8 countries are putting in huge sums of money and they don't see very much." This difference gives China an advantage in African eyes, even when altruistic motives are discounted.

Draining and Gaining Migrations

Over the past several decades, the centerpiece of PRC aid, infrastructure development, has also facilitated China–Africa migration. This has reached a level unmatched by any Western nation. China has sent sixteen thousand medical personnel to Africa to develop hospitals and clinics, and 240 million patients have been treated by Chinese health care workers. Since the 1960s, ten thousand PRC agrotechnicians have been sent to Africa and have worked on two hundred projects, including setting up farms and agricultural stations and training personnel. In Tanzania, for example, the PRC-built Ubungo Farm Implements Factory turned out 85 percent of the country's hand tools, while the Mbarali Farm produces one-fourth of the rice eaten by Tanzanians. Between 2007 and 2010, China will train ten thousand agricultural technicians for work in Africa. Some 530 PRC teachers have worked in African schools, and many Chinese train African government staff. Many more go to Africa for contract labor service—building railways, roads, telecommunications systems, hospitals, schools, and dams—or to conduct business that makes use of that infrastructure.

Africans go to China to learn how to build the infrastructure themselves or how to work in it as doctors, teachers, and officials. Some sixteen thousand African professionals were trained in China in 2000–2006; fifteen thousand will do so in 2007–9. From 1956 to 1999, 5,582 Africans studied at PRC universities. By late 2004, 17,860 Africans had received PRC scholarships and fifteen thousand had graduated. By 2007, more than twenty thousand Africans had graduated from PRC universities, including several political leaders. China provided fifteen hundred scholarships to Africans in 2005 and two thousand in 2006; by 2010 there will be four thousand a year. Despite difficulties presented by widespread racism, a few Africans remain in the PRC after graduation, some engaging in China–Africa business.

Migration patterns represent another key distinction between China–Africa links and Western connections with the continent. Africans generally perceive Chinese who work in Africa as less privileged and exploitative than Western expatriates. Chinese construction personnel and agricultural advisors live more like their African counterparts, in keeping with the eighth "Principle on External Economic and Technical Assistance" set out by Premier Zhou Enlai during a 1964 trip to Africa: "The experts dispatched by the Chinese government to help construction in the recipient country should enjoy the same living conditions as the experts of the recipient country. The Chinese experts are not allowed to make any special demands and ask for any special amenities." Similarly, the small, mostly short-term African migrations to China are seen as benefiting Africa, while many Africans view the large, permanent migration of African professionals to the West as harming Africa's development.

A surge in Chinese migration to Africa began in the mid-1990s and accelerated in the present decade, soon rendering population estimates obsolete. (See table 1.) An Ohio University (OU) database shows 137,000 Chinese in Africa, the same figure provided for 2001 by Taiwan's government. Its estimates for thirty-four African states are now out-of-date. While some discrepancies may reflect differing conceptions of residence, the magnitude of the differences indicates a very rapid growth in Africa's Chinese communities.

The growth of South Africa's Chinese population is especially significant. The ten thousand "indigenous," or South-African-born, Chinese (SABCs) made up almost the entire Chinese population until 1980, when an immigrant community, 90 percent from Taiwan, began to form. By 1993 there were thirty-six thousand Chinese in South Africa. A decade later, the PRC Embassy in South Africa stated that there were perhaps eighty thousand Chinese residents, while South Africa's ambassador to China has said that there are more than one hundred thousand. Almost all the increase since 1993 has come from Chinese mainland migrants, as the Taiwanese community shrank by half (to 10,000) and the SABC population declined. Estimates in 2004–7 ranged from one hundred thousand to four hundred thousand legal and illegal Chinese residents.

Table 1

Number of Chinese in select African countries, c. 2001 and 2002–7

Country	Ohio U. Database 2001	Estimate for 2003–7
Algeria	2000	8000+ (2006)
Benin	——	4000 (2007)
Angola	500	20,000–30,000 (2006)
Botswana	40	3000–10,000 (2006–7)
Burkina Faso	——	1000 (2007)
Cameroon	50	1000–3000 (2005)
Cape Verde	——	600–1000 (2007)
Congo (Democratic Rep.)	200	500 (2007)
Cote d'Ivoire	200	10,000 (2007)
Egypt	110	6000–10,000 (2007)
Ethiopia	100	3000–4000 (2006)
Ghana	500	6000 (2005)
Guinea	3000	5000 (2007)
Kenya	190	5000 (2007)
Lesotho	1000	5000 (2005)
Liberia	120	600 (2006)
Madagascar	30,000	60,000 (2003)
Malawi	50	2000 (2007)
Mali	——	2000 (2007)
Mozambique	700	1500 (2006)
Namibia	——	5000–40,000 (2006)
Nigeria	2000	100,000 (2007)
Senegal	——	2000 (2007)
South Africa	30,000	100,000–400,000 (2007)
Sudan	45	5000–10,000 (2004–5)
Tanzania	600	3000–20,000 (2006)
Togo	50	3000 (2007)
Uganda	100	5000–10,000 (2007)
Zambia	150	4000–6,000 (2007)
Zimbabwe	300	5300–10,000 (2005–7)

A Belgian diplomat in Tanzania once told a Japanese diplomat in the 1970s that his African employee had queried, "Why are there two kinds of Chinese in Tanzania? One kind wears dirty clothes, looks poor, but works very hard; another kind wears a good suit, rides in a modern car with a camera on his shoulder, and looks like an American." The African worker was conflating as "Chinese" those who had come from the PRC to build Tazara and Japanese businessmen and tourists visiting Tanzania for very different purposes. Only a small number of Chinese in Africa today perform service of the kind rendered by their 1970s compatriots, yet substantial differences remain between the positions of Chinese in Africa and those of citizens of developed countries. According to a South African university official, China provides low-cost technology and its people are willing to work in inhospitable places.

In Africa, the citizens of most developed countries are managers or professionals. Some work for large multinational corporations, others are among the forty thousand NGO-employed expats. They generally command salaries that allow a lifestyle very different from that of most Africans and even most local occupational peers. Larger Chinese communities in Africa do have well-off members, usually businesspeople. But most long-term Chinese residents in Africa are small merchants, with little capital, who sell what one Kenyan has called "down-street merchandise." In Zimbabwe, Zambia, and South Africa, there are also Chinese farmers. And there are also Chinese who study in Africa, mostly in South Africa. In 1999, there were only nineteen students in South African tertiary institutions, but by 2004–5 there were more than three thousand.

Many Chinese work in Africa under labor service contracts. They are paid much less and live more frugally than Western expats doing comparable work. In 1992 Africa employed one hundred thousand developed country expats at a cost of $4b per year, that is, $40,000 per individual or nearly $800 per week. Even now, Chinese wages are not nearly as high. Chinese employees (i.e., managers, engineers, and skilled craftsmen) of one construction firm in Angola receive $500 a month, live two to three to a room, and cook for themselves, while Europeans each rent a house and eat in restaurants. Chinese supervising African workers on construction sites dig along with them. China's largest contract in Africa, worth $650m, is the building of Sudan's Merowe Dam, which in 2003–5 employed eighteen hundred Chinese and sixteen hundred Sudanese. A PRC firm won the bid because it kept expected profit margins and Chinese staff costs low. All project managers, 90 percent of engineers, and 75 percent of technicians are Chinese; locals make up 20 percent of skilled workers and all of the general laborers. Expats earn $220–$600 per week, Sudanese $22–$350. Chinese firms in Africa "reportedly provide good quality projects at a price discount of 25–50 percent compared to other foreign investors," due not only to lower profit margins and access to cheap capital, but also to low wages and living standards for Chinese employees.

While many Chinese now live in Africa, a growing number of Africans live in China. Guangzhou had ten thousand or more Africans in 2006, mainly China–Africa traders and students. Beijing in 2005–6 was said to have at least six hundred Africans, Shanghai five hundred, and Shenzhen one hundred.

Most were students of medicine, engineering, or natural science, and expected to return to Africa in four or five years. That small brain gain contrasts with Africa's brain drain to the West, which originated with SAPs that required deep reductions of state expenditures. Before 1980, Africa had a dozen high-growth countries, averaging 6 percent. One-third of African states had savings rates higher than 25 percent, a level that sustained human resource development. But then SAPs curtailed state funding, including for universities, and costly expats had to replace local intellectuals. Savings rates plummeted to 10 percent, too low for industrialization or adequate education. Now Africa yearly produces eighty-three engineers per one million people; China graduates seven hundred and fifty and developed countries one thousand. Many African engineers emigrate to the West: more are in the United States than in Africa. From 1985 to 1990, sixty thousand African professionals emigrated. In 2000, 3.6 percent of Africans, but 31.4 percent of Africa's emigrants, had a tertiary education. By 2005, three hundred thousand to a half million professionals, including thirty thousand doctoral degree holders, had left and twenty thousand more emigrate each year to the U.S. or Europe. If the half million figure is correct, fully one-third of African professionals had left. On average, each represents a loss of $184,000 to Africa.

Of four hundred thousand African immigrants age 16 and above in the U.S. in 2000, 36 percent were managers or professionals. Many were brought in through the Diversity Visa Program ("Green Card Lottery"), most of whose winners are Africans. By 2005, fifty thousand Africans a year were migrating to the U.S., with perhaps four times that number entering illegally. While such migrants send remittances home, a leading African scientist in the U.S., Philip Emeagwali, estimates that Africans there contribute forty times more to the U.S. than to the African economy. Others stay on after graduation from U.S. universities, which had thirty-four thousand African students in 2000–2001, 6.25 percent of the international students. Most were graduate students, and a survey of Africans who received U.S. Ph.D.s from 1986 to 1996 showed that 37 percent remained there after graduation. The percentages staying were higher in fields that are key to development—engineering (54%), physical sciences (44%), health sciences (44%), and management (67%)—and higher also for two of the three top Ph.D.-receiving peoples, Nigerians (62%) and Ghanaians (61%). Because education levels among entrepreneurs correlate with private enterprise growth, and because human capital is a key determinant of FDI inflows, the brain drain plays a role in continued poverty in Africa. For example, austerity measures and faculty outmigration have collapsed Nigerian tertiary institutions, rendering the country's fifty thousand engineers severely undertrained.

Africa, with 14 percent of world population, has 24 percent of the global burden of disease. The poaching of Africa's human resources is most apparent among medical workers, of which Africa has 1.4 per one thousand people, while North America has 9.9. Of Africa's estimated eight hundred thousand "trained medical staff," twenty-three thousand leave each year for developed countries. The physicians among them cost on average $100,000 to train. Their migration saves a receiving country like Britain $340,000–$430,000 in costs of training a doctor. Ghana, with six doctors for each one hundred thousand people, has lost 30 percent of the physicians it educated to the U.S., U.K., Canada, and Australia, all of which have more than two hundred doctors per one hundred thousand residents. South Africa, Ethiopia, and Uganda have lost 14–19 percent of their doctors. In 2001, Zimbabwe graduated 737 nurses; 437 left for Britain. There are more Ethiopian-trained doctors in Chicago than in Ethiopia and more Beninese doctors in France than in Benin.

Conclusion

In 2005 a PRC official working on WTO affairs, Wu Jiahuang, made a presentation to the United Nations Industrial Development Organization on industrialization, trade, and poverty alleviation through South–South cooperation. He argued that China's high growth rate was fueled by Chinese savings (on average 44 percent of their income) and encouragement of foreign direct investment (half from Hong Kong and Taiwan), which contributed 28 percent of value added to industry in 2004. He said that PRC industrial and trade growth are related, with over half of industrial exports produced by foreign investors. Wu noted that the PRC does not overprotect domestic industry: average PRC tariffs dropped from 43 percent in 1992 to 10 percent in 2005, lower than those of its trading partners. Primary agricultural products and textile tariffs averaged 15.5 percent and 12.9 percent, respectively, while those of China's trading partners averaged 24.5 percent and 17.7 percent. China provides world-class resources and "the cheapest domestic labor," so its businesses can market the world's most competitive products, leading to greater incomes, state revenue, and social welfare. Wu called on the WTO to remove trade-distorting subsidies to farmers in the North to enable farmers in the South to sell their products at a higher price. He explained that Chinese farms are very small, averaging .7 hectares of land, compared to Europeans' 20 ha and U.S. farmers' 200 ha. Wu noted that PRC agricultural tariffs averaged 15.8 percent, compared to 23 percent, in the U.S. and 73 percent in Europe. Meanwhile, state support for China's farmers was only 1.5 percent of their income, while in the U.S. it was 18 percent and in the E.U. it was 33 percent. China and other developing states were thus in the same boat in terms of needing cuts in developed world agricultural subsidies.

Wu's presentation summed up a commonly held perception of PRC practices that relates to distinct China–Africa links: China provides a model for developing states based on rapid industrialization fueled by a high level of investment and concentration on exports and, unlike the West, its low-tariff, low-subsidy regime allows other developing countries to export freely to China and compete with China in world markets. The official thus essentially argued that PRC policymakers are more consistent economic liberals than those of the West and that this greater openness fulfills the common needs of Chinese and citizens of other developing countries. Wu did not explain how China's extraordinary savings rate and its FDI inflow mainly from co-ethnics on its periphery can be duplicated by most developing states. Nor did he recognize that these states are scarcely in a position to take advantage of China's economic liberal policies by competing with PRC producers, either in their domestic market or the world. Still, one point was doubtless convincing: that China, unlike Western states, is not obstructing development in the world's poorer countries. That point, whether it relates to the Beijing Consensus or aid and migration, epitomizes the distinctiveness of the China–Africa link for many Africans.

The practices of Western states associated with past colonialism or present imperialism make PRC practices appear particularly distinctive to Africans. Most prominent among these Western practices are (1) impositions of neoliberal SAPs that have resulted in diminished growth, huge debt, declining incomes, and curtailed social welfare; (2) the use of aid to compel compliance with SAPs and the foreign policies of Western powers; (3) protectionism (despite free-trade rhetoric) in developed states that inhibits African exports; and (4) support for authoritarian leaders (despite talk of democracy and human lights) to secure resources and combat "radicals." In addition, Western disparagement of Africa, through an unremitting negative discourse overlaid with strong implications of African incompetence, remains prevalent. The ideas that on balance colonialism benefited "the natives," and that Africa's troubles have all been postcolonial, are popular among elites of the former colonial powers.

A positive image of China exists despite the prevalence among the Chinese of racist attitudes, which have been experienced both by Africans in China and Africans working alongside Chinese residents in Africa. The PRC government, with its ideology of Social Darwinism (i.e., the richer, the fitter) and characteristic representations of Africa as uniformly poor bears some responsibility for these attitudes. Still, the PRC is careful to identify Africa's problems as the legacy of colonialism. PRC leaders have never termed Africa a "hopeless continent." They would never state, as a U.S. House Sub-committee on Africa member did to a Rwandan human lights activist during the 1994 genocide, that "America has no friends in Africa, only interests, and it has no interests in Rwanda."

PRC leaders, officially at least, celebrate Africa's culture and achievements, and China's sixty-five cultural agreements with forty-six African states have led to hundreds of exchanges. As one scholar has observed, while Africa, to the West, is a "haven for terrorists," the "cradle of HIV/AIDS," and a "source of instability," for China it is a "strategically significant region" and place of opportunity. China, moreover, acknowledges its political indebtedness to Africa for her support of China's entry into the U.N. and continued backing in international forums. That contrasts with Western states' failure to acknowledge their indebtedness to Africa for its contributions to the West's industrialization and cultural development.

Unlike during the Mao era, China today suggests no radical solutions to Africa's predicament. The PRC avails itself of the historically determined disadvantages of Africa in trade, but much of what it sells to Africa is useful to developing manufacturing and providing affordable consumer goods. Some of China's investment in Africa, though apparently directed to non-oil sectors, is nevertheless imbricated with the continent's harsh labor regimes in places like Zambia's Copperbelt. But China is still perceived as different in that it provides some investments of direct benefit beyond elite circles, does not insist that Africa's political economy steer a required course, and contributes to Africa's talent pool rather than draining it.

It is not clear whether the differences outlined here will persist over the long term. Among major powers at any given time, there are always differences in approach to subaltern states. The very process of differentiating super-ordinate and subordinate states and dominant and subaltern peoples tends over time, however, to make the conduct of great powers and their elites more similar than different. In a decade or two we should be able to determine whether that will be the case as well with China in Africa.

Padraig R. Carmody and
Francis Y. Owusu

Competing Hegemons? Chinese Versus American Geo-Economic Strategies in Africa

When elephants fight, the grass gets trampled (African proverb).

Introduction

. . . From tourism in Sierra Leone, to bike factories in Ghana and oil refineries in Sudan, Chinese investment in Africa is rising. Since 2000, Chinese trade with Africa has more than tripled. Whereas China only accounted for 7% of African imports in 2003, imports from Africa grew an astounding 87% in 2004 alone. More than 60% of African timber exports are now destined for East Asia, and 25% of China's oil supplies now come from the Gulf of Guinea. China launched Nigeria's first space satellite, and by the end of 2005, China overtook Britain as Africa's third largest trading partner. . . .

This paper investigates the implications of Chinese and American investment and trade strategies for Africa. It begins by exploring recent Chinese interest and involvement on the continent. It then moves to describe the Chinese geo-economic strategy for the continent, and the advantages it brings to resource competition with the U.S. The economic and political impacts of Chinese investment and trade with Africa are then explored. The paper then examines the impacts of increased American oil investment on the continent and stepped up security assistance, with which this is associated. It concludes by examining progressive alternatives requiring more robust international cooperation.

Resource Colonialist and "Anti-Imperialist"? Chinese Interests and Involvement in Africa

China's desire to become a global economic powerhouse and a counterweight to U.S. hegemony in the international system is now clear. The expansion of the Chinese economy currently accounts for 25% of all global economic growth and by some estimates, at purchasing power parity, the Chinese economy will be as big as the U.S. in 2015. This phenomenal economic growth

From *Political Geography,* June 2007. Copyright © 2007 by Elsevier Science Ltd. Reprinted by permission via Rightslink.

has increased the country's demand for resources, especially oil. In 2003, China surpassed Japan as the world's second largest oil consumer, and it now accounts for 40% of total global growth in oil demand. China's demand for oil increases by 1% for every percentage increase in its gross domestic product, versus 0.4% for the Organization for Economic Cooperation and Development (OECD) countries, whose economies are heavily biased towards services. The search for resources to fuel the country's phenomenal economic growth and the need to find markets for its products requires a global geopolitical strategy and the formation of new alliances. . . .

China's demand for industrial resources is huge, but Africa has the potential to substantially meet this demand, as it is three times larger than China and rich in resources. The annual rate of growth for Chinese consumption of copper is 17%, 15% for zinc and 20% for nickel. China is now the world's largest consumer of copper, the price of which rose from $1319 per tonne in 2001 to $8800 in 2006. It is no wonder that Chinese companies have invested $170m in the copper industry in Zambia, re-opening the Chambishi mine which closed in 1988 and employing 2000 people. While the neoliberal policies promoted by the Western-controlled international financial institutions (IFIs) compounded the continent's economic problems, Chinese investment is partially, and unevenly, reversing their deflationary bias. This comes with a harsh labor regime, however. For example, workers in the Chinese-owned Collum mine in Zambia never get a day off.

China's Geo-Economic Strategy for Africa

"States" are comprised of sets of practices and social relations, rather than unified actors. Nonetheless, developmental state's policies, such as China, are marked by coherence, given the over-riding goal of economic growth and structural transformation. Unlike different branches of the United States government involved in relief, energy procurement or defense, which view Africa largely as a site of humanitarian intervention, resource extraction, and security threat, respectively, the Chinese state appears to "look" at Africa as a strategic-economic space. This geopolitical code reflects the challenges of economic transformation for China, versus international system maintenance for the U.S.

While more focused empirical studies of sectors, firms, countries and regimes are needed in order to shed light on the complexity, specificity and experimental nature of Chinese involvement on the continent, the following emergent elements of China's geo-economic strategy in Africa can be identified: (1) to ensure access to critical natural resources, particularly oil and natural gas, to maintain the country's economic growth, (2) to recycle its massive foreign exchange (forex) reserves into profitable investments overseas, (3) associated with both of these, to facilitate the development of Chinese multinational corporations, (4) to find markets for the products of Chinese industry, (5) to develop African agriculture to provide non-food agriculturals to supply Chinese industry and consumers, and also food products for China's burgeoning cities, and (6) to source knowledge workers in Africa to support

Chinese economic transformation. Different African countries have different resource geographies, from oil to beaches, and macro-economic stability, and they consequently play different roles in the emerging division of labor with China. Nonetheless there is a coherence between the different elements of economic engagement; not driven by a coherent plan, but by structural imperatives. Export-oriented industrialization in China has generated a forex reserve, some of which must be recycled overseas, for example. . . .

Current Chinese aid is concentrated in the productive sectors of physical infrastructure, industry and agriculture. When President Zeming visited Nigeria he concluded deals on Chinese assistance in developing the country's light weapons industry, oil refinery construction, power plants, and possible Chinese rehabilitation of the rail system in deals totalling up to U.S. $7bn. China is also involved in rail, road and fibre optic cable construction in Angola. In many ways, therefore, China's aid and investment in Africa is reminiscent of earlier colonial investments to ensure access to raw materials. . . .

A Benevolent Hegemon? Chinese "Soft Power" in Africa

According to the China's Liang Guixan (2005) Minister Counsellor, the "Chinese model" of development that is currently on offer is based on "sophisticated technology appropriate to African countries' low cost and expertise in poverty alleviation and SMME [small, micro- and medium-sized enterprise] development." It has eschewed outright privatisation and other elements of the "Washington Consensus."

The Chinese . . . African Human Resources Development Fund pays for 10,000 Africans annually to train in Beijing. However, in addition to being an act of goodwill, these knowledge workers are expected to contribute to China's economic transformation.

Chinese and African leaders at a 2003 trade summit agreed to build political and economic ties to counter western dominance and improve the position of poor countries. This is seen to be a way of deflecting U.S. "hegemonism." Moreover, China's own experience of dependence makes it possible for them to make reference to things like structural imbalances in trade relations. This represents a revival of the philosophy of Third Worldism, with China playing the lead roie; but this time with the material resources to back it up.

A Chinese official in Africa argued that "economic rights" are the main priority of developing nations and take precedence over personal, individual rights as conceptualised in the West. Indeed, the view among some senior Chinese officials is that "multi-party politics fuels social turmoil, ethnic conflicts and civil wars." China also sees the human rights discourse as a tool of Western neo-imperialism. This is a particularly attractive philosophy for incumbent African political élites, and is helped by its plausibility. . . .

Because of state involvement, Chinese companies are better positioned to make short-term losses for long-term gains. For instance, the representative of China's state-owned construction company in Ethiopia revealed that he was instructed to bid low on tenders, without regard to profit, and China's largest telecoms manufacturer gifted equipment to Telkom Kenya. However, China is merely "following a very traditional path established by Europe, Japan and the United Stales: offering poor countries comprehensive and exploitative trade deals combined with aid." Chinese companies have other competitive advantages too in that they are often willing to pay bribes and under-the-counter signing bonuses. While this was standard practice in the past, Western companies are now more open to reputational risk and are under pressure to sign up to the Extractive Industries Transparency Initiative, promoted by Tony Blair, and the "Publish What You Pay" campaign supported by George Soros. Corruption scandals involving Western oil companies in Africa nonetheless continue too.

Free Trade Imperialism or South–South Cooperation? The Economic Impacts of China

In some policy circles in the U.S., the rise of China and a new questioning of "free" trade are linked. A study for the U.S. Army War College on Chinese influence in Latin America argues that "in previous decades, dependence on Western capital was a key vehicle for forcing Latin American nations to accept neoliberal economic policies and free trade, limiting the degree to which they could buy social peace with state institutions and government largesse." However, China is now seen to be creating a "neoimperialistic dynamic in the hemisphere" and by displacing domestic manufacturers through imports possibly deepening class disparities and corruption. This represents a shift towards "realist," structuralist, and away from normative neoclassical economics in some U.S. policy circles. In relation to Latin America, Chinese officials have told their counterparts to let manufacturing industry die and concentrate on primary commodity exports. The ability to introduce counter-vailing anti-dumping measures is circumscribed by many countries' desire to grant China "market economy" status under World Trade Organization (WTO) agreements, in order to maintain their access to the Chinese market. However, Latin American countries have also been successful in negotiating voluntary export restraints with China, enabling them to enjoy a continuing trade surplus—again showing China's long-term concern with resource supply, rather than short-term market expansion. The same sorts of competitive displacement pressures are at play in Africa as in Latin America, although with per capita incomes 90% lower, the results are even more devastating there.

Creating industrial employment is central to improved security prospects in Africa, and to democracy promotion. The development of an autonomous civil society, with its own material base in the labor movement and private sector in the "West" have historically held the state accountable, although the exact meaning and content of African "democracy" is context specific and subject to negotiation.

Africa has recently been hit with a Chinese "Textile tsunami." The number of companies registered in Botswana doubled over the last few years; many of them were Chinese (import) trading companies. Out of every 100 t-shirts imported to South Africa, 80 are from China. This has resulted in a "double whammy" as African industry is undermined by Chinese import competition, while the phasing out of the Multi-Fibre Agreement in 2005, meant the value of preferences under the U.S. African Growth and Opportunity Act (AGOA) were undermined for the continent. Textile and clothing exports account for 99.14% of Lesotho's export earnings, however, more than 10 clothing factories closed there in 2005, throwing 10,000 people out of work. South Africa's clothing exports to the U.S. fell from $26m in the first quarter of 2004 to $12m for the first quarter of 2005, with 30,000 people losing their jobs. In the Muslim Kano and Kaduna areas of Nigeria, imports have devastated the local textile and consumer goods industries. The Nigerian textile and clothing workers union estimates 350,000 direct job losses as a result of Chinese import competition and 1.5 million indirectly over the last 5 years. While African trade unionists estimate that 250,000 jobs in the textile and clothing industries have been lost due to Chinese import penetration, roughly the same number as created by AGOA.

The reorientation of trade from manufactures to oil is evidenced by the fact that Angola and Chad surpassed Lesotho in 2004 as exporters to the U.S. The increased emphasis on oil and mineral extraction is resulting in a relative technological downgrading of Africa's economies.

The imposition of temporary restrictions on Chinese textiles entering the U.S. and EU markets is currently providing some respite, as Chinese textile manufacturers are again investing in Africa, in the short-term, as a way around this. For example, all the textile factories which had closed in Lesotho have now been reopened. However under the WTO, the ability to use "safeguard actions" to provide relief from import surges will largely expire in 2008, and completely end in 2013.

The Chinese have noted that "the fundamental reason for the increase in Chinese textile and clothing imports [into Africa] is the high demand for Chinese goods." However critics of Chinese policy in Africa argue that it represents a new neo-colonialism, disguised as South–South cooperation. As Moletsi Mbeki argues, "we sell them raw materials and they sell us manufactured goods with a predictable result—an unfavourable trade balance against South Africa." . . .

China's WTO entry has thereby served to lock in global market access for its export-oriented industrialization. Thus, while there may be some potential for "pro-poor" employment growth in South Africa, in the export-oriented fruit and vegetable subsectors in particular, the growth of China may further exacerbate income inequality there, reinforcing the previous capital intensive growth path. As Neva Seidman Makgetla, an economist for the Congress of South African Trade Unions, argues, "There's no question that for upper classes it's a boon . . . the problem is any lower-class South African's would rather have a job." Only 13% of "black" South Africans are currently employed in the formal sector of the economy compared to 34% in 1970, although there was some employment growth in 2005/6, mostly in agriculture. . . .

The Impacts of Chinese Involvement on African State Restructuring

. . . Sixty percent of Sudanese oil goes to China, and the oil refinery in Sudan was the first one the Chinese built outside China. Debt service payments for it have priority over all others, such as to the World Bank and the IMF. Chinese companies have also built three small-arms factories in Khartoum and sold weapons to the regime. Supplying the Sudanese government with jet fighters allows it to protect the oil fields in the South, where the Chinese have substantial interests. Indeed, during the (North–South) civil war, government troops used the CNPC facility to launch attacks and dislodge southerners in the vicinity of the new oil fields.

A Chinese official at the Ministry of Trade also noted in relation to human rights in Sudan that "we import from every source we can get oil from" or in the words of the deputy foreign minister "business is business." China's new African policy specifically states that it will increase its assistance to African nations with "no political strings attached." Such a position has led to assertions that China has no values, only interests.

In an era of globalized capital markets there are costs to such an approach, however. While the absence of a formal, independent civil society and free press in China frees companies from reputational risk domestically, the flotation of the CNPC on the New York Stock exchange had to be withdrawn and refashioned because of negative publicity over what the proceeds might be used to do in Sudan. After this débacle, China refused to veto Sudan being sent to the International Criminal Court over its actions in Darfur. . . .

The Chinese are also rekindling relations with other highly corrupt former leftist regimes, such as Angola. In Angola, the IMF estimates $1 to 5 billion dollars of oil revenue goes missing every year, while half of Angolan children continue to be malnourished. Yet China's Eximbank provided the country a $2bn line of credit to rebuild its infrastructure as part of an oil deal. China's state-owned oil companies and British Petroleum have a joint venture in the country and Angola is now China's largest supplier of oil. . . .

The Chinese also pay little attention to the environmental impacts of their investments. As explained by an African diplomat "they just come and do it. We don't start to hold meetings about environmental impact assessments and human rights and bad governance and good governance." Western corporations are also substantial polluters, however, with Shell recently fined $1.5bn for pollution of the Niger delta. Thus, Chinese practices in Africa represent "business as usual," rather than a radical break with the past. . . .

Conclusion

. . . America is a somewhat more normative power than China, with a developed civil society exerting pressure for conformity with (civil and political) human rights norms. Nonetheless, it operates on the basis of *realpolitik*. Increasingly close cooperation between Khartoum and Western intelligence agencies in the "war-on-terror" may be blocking war crimes prosecutions in Darfur, for example.

There is already evidence of U.S. interest in cooperating with China in relation to Africa, with Jendayi Frazer recently visiting China. However, if China wishes to promote peace in Africa, it will have to curb its own arms industry and it might also sign up to a code of conduct along the lines of NEPAD and the EU. Enhanced international cooperation will be needed to deliver on the promises of improved governance and economic transformation of NEPAD. . . .

While the revival of economic growth which this has brought to the continent is potentially progressive in its impacts, to date it has largely been confined to enclaves, and upper classes/state élites. Also, while the international community has played a vital role in ending wars in Liberia, Sierra Leone and the Democratic Republic of Congo, among others, oil investment has fuelled local conflict and made many states less accountable to their populations.

A key challenge is to shift African states from authoritarian/patrimonial to developmental mode. Perhaps this is not seen to be in the interests of either the U.S. or China, as these states would then keep their own resources, rather than placing them on the international market. Given the general underdevelopment of the continent, Africa is the only region of the world where net oil output is predicted by the U.S. government to grow substantially faster than consumption—by 91 compared to 35%, respectively, from 2001 to 2025. However, there is a price to be paid by the international community for the current set up. For the moment it is a price they are willing to (let other people) pay.

POSTSCRIPT

Is Increasing Chinese Investment Good for African Development?

$\mathbf{A}$s is demonstrated in the selections for this issue, increasing trade between China and Africa is viewed in very different lights depending on the commentator. The United States has tended to view the situation with increasing alarm as it views African oil resources in particular as crucial given the uncertainty of supplies from the Middle East and its precarious relationship with Venezuela. Others in the human rights community are concerned about China's support for undemocratic regimes (such as those in Zimbabwe and Sudan) in countries where it wishes to maintain or expand access to certain types of resources. In China's defense, it contends that it provides substantial levels of direct foreign investment in those countries where it trades. In the absence of such investment, vital infrastructure might not be maintained or expanded.

African commentators are also clearly split on the issue. Although the oil-rich countries in particular have benefited from increased export revenues and Chinese investment, those countries that possessed a small or medium-sized manufacturing sector have been very hard hit. Although West African markets used to be replete with cheap plastic consumer goods manufactured in Abidjan or Lagos, it is now rare to find anything of this sort that is not made in China. The textile industries in Zimbabwe and South Africa have been similarly devastated by inexpensive Chinese exports.

One aspect of Chinese investment that was not discussed in either of the selections is the involvement of ethnic Chinese in the manufacturing sectors of some African nations. In a 2003 article in *African Affairs,* titled "Close Encounters: Chinese Business Networks as Industrial Catalysts in Sub-Saharan Africa," Deborah Brautigam argues that Chinese nationals have been a catalyst for industrial development in Mauritius. In contrast, their involvement in the Nigerian industrial sector has been more problematic. In her book, *Disabling Globalization* (University of California Press, 2002), Gillian Hart documents the problematic nature of mainland Chinese and Taiwanese investment in South Africa.

If there is a middle road to be taken, it would be for those African countries that are benefiting from trade with China (or with any other global power for that matter) to strategically reinvest the proceeds from this trade in the development of human capital (via increased education expenditures) and in other sectors of the economy that show promise. Although the article does not explicitly deal with the Africa–China relationship, Henry J. Bruton makes an argument along these lines in his 1998 piece in the *Journal of Economic Literature* titled "A Reconsideration of Import Substitution."

For other perspectives on this issue, see Domingos Jardo Muekalia's 2004 article in the *African Security Review* titled "Africa and China's Strategic Partnership" or Chris Alden's 2005 article in *Survival* under the title "China in Africa."

ISSUE 8

Does Foreign Aid Undermine Development in Africa?

YES: Dambisa Moyo, from "Why Foreign Aid Is Hurting Africa," *The Wall Street Journal* (March 2009)

NO: Apoorva Shah, from "Slamming Aid," *Policy Review* (June/July 2009)

ISSUE SUMMARY

YES: Dambisa Moyo, who has worked for both Goldman Sachs and the World Bank, argues that aid to Africa has made the poor poorer and economic growth slower. She further states that aid has left African countries more debt ridden, prone to inflation, vulnerable to currency markets, and unattractive to high-caliber investment.

NO: Apoorva Shah, research fellow at the American Enterprise Institute, does not necessarily disagree with many of Moyo's assertions. His concern is that Moyo is not offering anything new to the debate and that she ignores how new stakeholders are moving beyond the old debates. He discusses how the Millennium Challenge Corporation only works in countries where progress is being made on the consolidation of institutions, accountable governance, and freer peoples. He further finds it difficult to accept some of Moyo's broader generalizations, such as a characterization of aid as "large systematic cash transfers from rich countries to African governments."

What role does foreign aid have to play in African development? The answer to this question may depend on your political philosophy. If you are a political realist (a school of international relations that prioritizes national interest and security, rather than ideals or ethics), you may be more likely to believe that foreign aid is always given to advance the agenda of the donor country or to advance the agendas of international institutions that are "in the employ" of powerful nations. In contrast, if you subscribe to liberal international relations theory (based on a belief that absolute gains can be made through cooperation and interdependence), you may be more apt to believe that international development assistance could actually help some countries.

International development assistance comes in all shapes and sizes. It varies in terms of the provider (e.g., international organizations such as the United Nations, a national government, or a nonprofit organization) as well as in terms of its form (e.g., a loan for a dam, emergency food relief, technical assistance to a government agency, or a fund for a community well). But, as discussed above, there are sharply contrasting views on the value of international development assistance. Let us first examine the arguments against international development assistance followed by an exploration of more positive views.

There is ample evidence to support the view that international development assistance in Africa is problematic. The vast majority of foreign aid provided by countries tends to go to close allies, rather than to those nations most in need of assistance. Some classic critiques of foreign assistance document the inherent self-interest in much foreign assistance. Donors, for example, have often required that goods and services be purchased from suppliers in their own countries. A case in point is the U.S. Food for Peace program, which traditionally purchased surplus food stuffs in the United States for distribution in other countries. Although this program may have been providing needed assistance in food-deficit areas, it also served to support American farmers. Worse yet, emergency relief efforts using American grains sometimes hurt farmers in the target areas (by depressing demand for their crops) or encourage locals to develop tastes for nonlocal grains (such as wheat). The best course of action in most cases would have been to purchase grains locally if they were available.

Given the history of troubled development efforts, one reaction is to write development off as a failed "project." Another reaction is to acknowledge past problems with development, but then to seek to rethink this process. Is there a way to approach development that might be less problematic?

In this issue, Dambisa Moyo, who has worked for both Goldman Sachs and the World Bank, argues that aid to Africa has made the poor poorer and economic growth slower. She further states that aid has left African countries more debt ridden, prone to inflation, vulnerable to currency markets, and unattractive to high-caliber investment. Apoorva Shah, research fellow at the American Enterprise Institute, does not necessarily disagree with many of Moyo's assertions. His concern is that Moyo is not offering anything new to the debate and that she ignores how new stakeholders are moving beyond the old debates. He discusses how the Millennium Challenge Corporation only works in countries where progress is being made on the consolidation of institutions, accountable governance, and freer peoples. He further finds it difficult to accept some of Moyo's broader generalizations, such as a characterization of aid as "large systematic cash transfers from rich countries to African governments."

YES

Dambisa Moyo

Why Foreign Aid Is Hurting Africa

A month ago I visited Kibera, the largest slum in Africa. This suburb of Nairobi, the capital of Kenya, is home to more than one million people, who eke out a living in an area of about one square mile—roughly 75% the size of New York's Central Park. It is a sea of aluminum and cardboard shacks that forgotten families call home. The idea of a slum conjures up an image of children playing amidst piles of garbage; with no running water; and the rank, rife stench of sewage. Kibera does not disappoint.

What is incredibly disappointing is the fact that just a few yards from Kibera stands the headquarters of the United Nations' agency for human settlements, which, with an annual budget of millions of dollars, is mandated to "promote socially and environmentally sustainable towns and cities with the goal of providing adequate shelter for all." Kibera festers in Kenya, a country that has one of the highest ratios of development workers per capita. This is also the country where in 2004, British envoy Sir Edward Clay apologized for underestimating the scale of government corruption and failing to speak out earlier.

Giving alms to Africa remains one of the biggest ideas of our time— millions march for it, governments are judged by it, celebrities proselytize the need for it. Calls for more aid to Africa are growing louder, with advocates pushing for doubling the roughly $50 billion of international assistance that already goes to Africa each year.

Yet, evidence overwhelmingly demonstrates that aid to Africa has made the poor poorer, and the growth slower. The insidious aid culture has left African countries more debt laden, more inflation prone, more vulnerable to the vagaries of the currency markets, and more unattractive to higher-quality investment. It's increased the risk of civil conflict and unrest (the fact that over 60% of sub-Saharan Africa's population is under the age of 24 with few economic prospects is a cause for worry). Aid is an unmitigated political, economic and humanitarian disaster.

Few will deny that there is a clear moral imperative for humanitarian and charity-based aid to step in when necessary, such as during the 2004 tsunami in Asia. Nevertheless, it's worth reminding ourselves what emergency and charity-based aid can and cannot do. Aid-supported scholarships have certainly helped send African girls to school (never mind that they won't be able to find a job in their own countries once they have graduated). This kind of aid can provide band-aid solutions to alleviate immediate suffering, but by its very nature cannot be the platform for long-term sustainable growth.

Whatever its strengths and weaknesses, such charity-based aid is relatively small beer when compared to the sea of money that floods Africa each year in government-to-government aid or aid from large development institutions such as the World Bank.

Over the past 60 years at least $1 trillion of development-related aid has been transferred from rich countries to Africa. Yet, real per-capita income today is lower than it was in the 1970s, and more than 50% of the population—over 350 million people—live on less than a dollar a day, a figure that has nearly doubled in two decades.

Even after the very aggressive debt-relief campaigns in the 1990s, African countries still pay close to $20 billion in debt repayments per annum, a stark reminder that aid is not free. In order to keep the system going, debt is repaid at the expense of African education and health care. Well-meaning calls to cancel debt mean little when the cancellation is met with the fresh infusion of aid, and the vicious cycle starts up once again.

In 2005, just weeks ahead of a G8 conference that had Africa at the top of its agenda, the International Monetary Fund published a report entitled "Aid Will Not Lift Growth in Africa." The report cautioned that governments, donors, and campaigners should be more modest in their claims that increased aid will solve Africa's problems. Despite such comments, no serious efforts have been made to wean Africa off this debilitating drug.

The most obvious criticism of aid is its links to rampant corruption. Aid flows destined to help the average African end up supporting bloated bureaucracies in the form of the poor-country governments and donor-funded non-governmental organizations. In a hearing before the U.S. Senate Committee on Foreign Relations in May 2004, Jeffrey Winters, a professor at Northwestern University, argued that the World Bank had participated in the corruption of roughly $100 billion of its loan funds intended for development.

As recently as 2002, the African Union, an organization of African nations, estimated that corruption was costing the continent $150 billion a year, as international donors were apparently turning a blind eye to the simple fact that aid money was inadvertently fueling graft. With few or no strings attached, it has been all too easy for the funds to be used for anything, save the developmental purpose for which they were intended.

In Zaire—known today as the Democratic Republic of Congo—Irwin Blumenthal (whom the IMF had appointed to a post in the country's central bank) warned in 1978 that the system was so corrupt that there was "no (repeat, no) prospect for Zaire's creditors to get their money back." Still, the IMF soon gave the country the largest loan it had ever given an African nation. According to corruption watchdog agency Transparency International, Mobutu Sese Seko, Zaire's president from 1965 to 1997, is reputed to have stolen at least $5 billion from the country.

It's scarcely better today. A month ago, Malawi's former president Bakili Muluzi was charged with embezzling aid money worth $12 million. Zambia's former president Frederick Chiluba (a development darling during his 1991 to 2001 tenure) remains embroiled in a court case that has revealed millions of dollars frittered away from health, education and infrastructure toward his personal cash dispenser. Yet, the aid keeps on coming.

A nascent economy needs a transparent and accountable government and an efficient civil service to help meet social needs. Its people need jobs and a belief in their country's future. A surfeit of aid has been shown to be unable to help achieve these goals.

A constant stream of "free" money is a perfect way to keep an inefficient or simply bad government in power. As aid flows in, there is nothing more for the government to do—it doesn't need to raise taxes, and as long as it pays the army, it doesn't have to take account of its disgruntled citizens. No matter that its citizens are disenfranchised (as with no taxation there can be no representation). All the government really needs to do is to court and cater to its foreign donors to stay in power.

Stuck in an aid world of no incentives, there is no reason for governments to seek other, better, more transparent ways of raising development finance (such as accessing the bond market, despite how hard that might be). The aid system encourages poor-country governments to pick up the phone and ask the donor agencies for next capital infusion. It is no wonder that across Africa, over 70% of the public purse comes from foreign aid.

In Ethiopia, where aid constitutes more than 90% of the government budget, a mere 2% of the country's population has access to mobile phones. (The African country average is around 30%.) Might it not be preferable for the government to earn money by selling its mobile phone license, thereby generating much-needed development income and also providing its citizens with telephone service that could, in turn, spur economic activity?

Look what has happened in Ghana, a country where after decades of military rule brought about by a coup, a pro-market government has yielded encouraging developments. Farmers and fishermen now use mobile phones to communicate with their agents and customers across the country to find out where prices are most competitive. This translates into numerous opportunities for self-sustainability and income generation—which, with encouragement, could be easily replicated across the continent.

To advance a country's economic prospects, governments need efficient civil service. But civil service is naturally prone to bureaucracy, and there is always the incipient danger of self-serving cronyism and the desire to bind citizens in endless, time-consuming red tape. What aid does is to make that danger a grim reality. This helps to explain why doing business across much of Africa is a nightmare. In Cameroon, it takes a potential investor around 426 days to perform 15 procedures to gain a business license. What entrepreneur wants to spend 119 days filling out forms to start a business in Angola? He's much more likely to consider the U.S. (40 days and 19 procedures) or South Korea (17 days and 10 procedures).

Even what may appear as a benign intervention on the surface can have damning consequences. Say there is a mosquito-net maker in small-town Africa. Say he employs 10 people who together manufacture 500 nets a week. Typically, these 10 employees support upward of 15 relatives each. A Western government–inspired program generously supplies the affected region with 100,000 free mosquito nets. This promptly puts the mosquito net manufacturer out of business, and now his 10 employees can no longer support their 150 dependents. In a couple of years, most of the donated nets will be torn and useless, but now there is no mosquito net maker to go to. They'll have to get more aid. And African governments once again get to abdicate their responsibilities.

In a similar vein has been the approach to food aid, which historically has done little to support African farmers. Under the auspices of the U.S. Food for Peace program, each year millions of dollars are used to buy American-grown food that has to then be shipped across oceans. One wonders how a system of flooding foreign markets with American food, which puts local farmers out of business, actually helps better Africa. A better strategy would be to use aid money to buy food from farmers within the country, and then distribute that food to the local citizens in need.

Then there is the issue of "Dutch disease," a term that describes how large inflows of money can kill off a country's export sector, by driving up home prices and thus making their goods too expensive for export. Aid has the same effect. Large dollar-denominated aid windfalls that envelop fragile developing economies cause the domestic currency to strengthen against foreign currencies. This is catastrophic for jobs in the poor country where people's livelihoods depend on being relatively competitive in the global market.

To fight aid-induced inflation, countries have to issue bonds to soak up the subsequent glut of money swamping the economy. In 2005, for example, Uganda was forced to issue such bonds to mop up excess liquidity to the tune of $700 million. The interest payments alone on this were a staggering $110 million, to be paid annually.

The stigma associated with countries relying on aid should also not be underestimated or ignored. It is the rare investor that wants to risk money in a country that is unable to stand on its own feet and manage its own affairs in a sustainable way.

Africa remains the most unstable continent in the world, beset by civil strife and war. Since 1996, 11 countries have been embroiled in civil wars. According to the Stockholm International Peace Research Institute, in the 1990s, Africa had more wars than the rest of the world combined. Although my country, Zambia, has not had the unfortunate experience of an outright civil war, growing up I experienced first-hand the discomfort of living under curfew (where everyone had to be in their homes between 6 p.m. and 6 a.m., which meant racing from work and school) and faced the fear of the uncertain outcomes of an attempted coup in 1991—sadly, experiences not uncommon to many Africans.

Civil clashes are often motivated by the knowledge that by seizing the seat of power, the victor gains virtually unfettered access to the package of aid that comes with it. In the last few months alone, there have been at least three political upheavals across the continent, in Mauritania, Guinea, and Guinea Bissau (each of which remains reliant on foreign aid). Madagascar's government was just overthrown in a coup this past week. The ongoing political volatility across the continent serves as a reminder that aid-financed efforts to force-feed democracy to economies facing ever-growing poverty and difficult economic prospects remain, at best, precariously vulnerable. Long-term political success can only be achieved once a solid economic trajectory has been established.

Proponents of aid are quick to argue that the $13 billion ($100 billion in today's terms) aid of the post–World War II Marshall Plan helped pull back a broken Europe from the brink of an economic abyss, and that aid could work, and would work, if Africa had a good policy environment.

The aid advocates skirt over the point that the Marshall Plan interventions were short, sharp, and finite, unlike the open-ended commitments which imbue governments with a sense of entitlement rather than encouraging innovation. And aid supporters spend little time addressing the mystery of why a country in good working order would seek aid rather than other, better forms of financing. No country has ever achieved economic success by depending on aid to the degree that many African countries do.

The good news is we know what works; what delivers growth and reduces poverty. We know that economies that rely on open-ended commitments of aid almost universally fail, and those that do not depend on aid succeed. The latter is true for economically successful countries such as China and India, and even closer to home, in South Africa and Botswana. Their strategy of development finance emphasizes the important role of entrepreneurship and markets over a staid aid-system of development that preaches hand-outs.

African countries could start by issuing bonds to raise cash. To be sure, the traditional capital markets of the U.S. and Europe remain challenging. However, African countries could explore opportunities to raise capital in more non-traditional markets such as the Middle East and China (whose foreign exchange reserves are more than $4 trillion). Moreover, the current market malaise provides an opening for African countries to focus on acquiring credit ratings (a prerequisite to accessing the bond markets), and preparing themselves for the time when the capital markets return to some semblance of normalcy.

Governments need to attract more foreign direct investment by creating attractive tax structures and reducing the red tape and complex regulations for businesses. African nations should also focus on increasing trade; China is one promising partner. And Western countries can help by cutting off the cycle of giving something for nothing. It's time for a change.

Slamming Aid

In 1947, following the deadly ravages of World War II, the United States introduced the Marshall Plan, a multibillion dollar effort to support the reconstruction of a withered Europe. At this point, American economic engagement abroad was intended for recovery rather than development, and it was not until 1954, when President Dwight D. Eisenhower amended the Mutual Security Act, that the U.S. created the foreign aid system as we know it today. The Act, which attempted to organize the multitude of foreign assistance programs from the postwar period, also introduced the concepts of development and security assistance—particularly for poor and underdeveloped countries in the world. This was exemplified by the Food for Peace program, which formalized U.S. food aid and, as President Eisenhower said, was intended to "lay the basis for a permanent expansion of our exports of agricultural products with lasting benefits to ourselves and peoples of other lands." Since then, arguments for and against foreign assistance have permeated the foreign aid intellectual establishment.

For example, in 1958, Milton Friedman argued that "despite the intentions of foreign economic aid, its major effect, insofar as it has an effect at all, will be to speed the Communization of the underdeveloped world." RAND economist Charles Wolf Jr. countered in 1961: "[O]ften the effect of aid has been to *reduce* the encroachment by government on the private sector" and aid can indeed be useful for "reasonable" objectives. One year after Friedman's essay, Lord Peter Bauer suggested that foreign aid to India "would be much more likely to retard the rise of general living standards in India than to accelerate it, and to obstruct rather than promote the emergence of a society resistant to totalitarian appeal." And Walt Whitman Rostow, economist and adviser to Presidents Kennedy and Johnson, opined in 1960 that advanced countries have an "important role to play in assisting the laggard societies." So it would go for the next 50 years.

The modern-day extension of this debate has been inherited by New York University economist William Easterly and Columbia University economist Jeffrey Sachs, whose foreign aid feud penetrated the development blogosphere and seemingly drew the line between aid champions and aid critics. Published in 2005, Sachs's *The End of Poverty* claimed that Africa could not achieve the Millennium Development Goals without another big push of aid from the West, which would allow the continent to kick start itself on the path to

From *Policy Review*, June/July 2009, pp. 106–112. Copyright © 2009 by Apoorva Shah. Reprinted by permission of Policy Review/Hoover Institution/Stanford University and Apoorva Shah.

growth. The following year, Easterly published *The White Man's Burden,* which chided foreign aid "planners" for perceiving development as a mere technical challenge that can be resolved through clinical diagnosis and medicine in the form of foreign aid.

Dambisa Moyo, a Harvard- and Oxford-educated Zambian economist and author of *Dead Aid: Why Aid Is Not Working and How There Is a Better Way for Africa,* falls on the side of Friedman, Bauer, and Easterly—and rightly so. The evidence assessing the impact of aid on economic growth (or the lack thereof) is comprehensive and convincing. In 1987, for example, a study by Paul Mosley and two of his colleagues found no statistically significant correlation between aid and economic growth in the developing world. Nearly two decades later, in 2005, the International Monetary Fund (IMF) conducted its own comprehensive assessment of the aid-growth relationship, concluding that there is "little evidence of a robust positive impact of aid on growth."

In her short but strident debut, Moyo takes up the argument that aid is often ineffective and can inhibit growth as if it were new, with eager urgency and passion, chiding the vast "development community" from liberal humanitarians to subsidy-craving farmers and from development economists to international celebrities. The *New York Times* even dubbed Moyo "The AntiBono" for her accusation that "the pop culture of aid has bolstered misconceptions" about development aid, making Africa "the focus of orchestrated world-wide pity." But, of course, the Dead Aid argument is anything but new.

These days, it is difficult even for aid advocates to justify foreign assistance on the basis of its purported contributions to economic progress: Instead they generally suggest that well-directed aid in the appropriate policy environment can accomplish focused objectives such as reducing malaria infection rates, educating children, or feeding the hungry. Bono's ONE Campaign, for example, does not attempt to make the link between aid and growth. And Bono himself is widely known in policy circles to acknowledge some of the critics' misgivings about foreign assistance.

The *Dead Aid* take on foreign assistance is merely another iteration of an irresolvable dispute over matters of principle (or, if you will, ideology) that has existed since the creation of foreign aid itself. The book adds little to this debate, lacking new quantitative data and only lightly referencing previous findings. But where it truly lags behind is in its understanding of a new growing consensus, which recognizes the inevitability of a foreign aid system that is motivated by political interests—both at home and abroad—and compelled by humanitarian concern. While the debate will inevitably continue, the majority of actors in the development community—philanthropists, entrepreneurs, NGO workers, and investors—have begun to move beyond it. These new stakeholders have changed their strategy, working to influence how and for what aid can be best spent.

In the first part of her book, Moyo attempts to summarize seven decades of foreign assistance history in a terse 19-page chapter. She breaks down foreign assistance into a category for each decade, starting with:

> Its birth at Bretton Woods in the 1940s; the era of the Marshall Plan in the 1950s; the decade of industrialization of the 1960s; the shift towards

aid as an answer to poverty in the 1970s; aid as the tool for stabilization and structural adjustment in the 1980s; aid as a buttress of democracy and governance in the 1990s; culminating in the present-day obsession with aid as the only solution to Africa's myriad of problems.

Moyo does not characterize the coming decade in foreign aid, but if existing evidence can be of any indication, it will be the decade beyond traditional development aid. In every sector—public, private, and non-profit—new innovations created over the past two decades have transformed the relationship of the rich world to the developing world.

In the public sector, the United States' Millennium Challenge Corporation (MCC) embodies this new tendency. Understanding that if aid is to work, it can only work in countries where institutions are consolidating, government is accountable, and people are becoming freer, the MCC developed an indicator-based approach to aid in which countries must make progress on a series of global benchmarks in order to receive assistance. Publicly available data on corruption, economic freedom, investment in people, and civil liberties allows the MCC to systematically assess the progress of recipient countries, and it also gives countries aspiring to receive aid a blueprint for reform. Guatemala, for example, adopted the MCC's indicator benchmarks in its reform plan, even though the country was not even eligible for MCC aid. And when these benchmarks are compromised, as in the case of Madagascar's recent military coup, aid is put on hold. Also acknowledging the risks of transferring money directly to government leaders, who with copious and consistent inflows of free wealth lose the incentive to tax their populace and be accountable to them, the MCC instead opens up the bidding for aid contracts to local businesses and firms. As lawmakers and voters demand more accountability, even the slow-to-react public sector has taken steps towards smarter development aid.

The public sector still lags behind the nonprofit world and private sector initiatives in terms of aid innovation, but considering examples like the Millennium Challenge Corporation, Moyo's sweeping caricature of government aid as "large systematic cash transfers from rich countries to African governments" is hard to swallow. The characterization invokes an image of a rich behemoth recklessly dumping money on helpless African states, even though the reality is much more nuanced than that. Nevertheless, the MCC is merely a trickle in the rapidly filling African aid bucket. While President Obama and his allies in Congress make the push for more development aid in their new strategy for "smart power," including by increasing the budget of USAID, the MCC concurrently has lost funding and political support. The fiscal year 2009 Omnibus Appropriation Act cut the organization's funding almost in half, from $1.5 billion to $875 million, limiting the MCC's capacity to expand to more countries and setting back existing projects.

Dead Aid, which has received enormous publicity in the policy hubs of Washington, D.C., New York, and London, could have used its valuable space to make the case to the Obama administration and his counterparts in Europe that smart power must also mean smarter aid. But the book leaves little room for new, reformed government aid programs, recommending instead that

African countries get cash by issuing bonds in the world market. Moyo's background as a Goldman Sachs economist becomes evident in this chapter, as she discusses with expert detail possibilities for African governments to securitize bonds, pool risks, and guarantee credit in order to finance their cash-strapped coffers. Considering the path of the global economy since the book was written, the author's prescription would not only be bitter medicine for African countries, it also appears less plausible (and perhaps dangerous) within the current gridlock of international liquidity.

Nevertheless, development policy no longer resides exclusively in the realm of government. As the Hudson Institute's "Index of Global Philanthropy" reveals, private flows of aid from the developed world to the developing world—excluding remittances—are now twice as high as official flows of development assistance. In the United States, official development assistance makes up only 12 percent of the country's total economic engagement with the third world.

Nonprofit organizations and "social entrepreneurs," both in the West and in Africa, have been the beneficiaries of this uptick in private aid. And as individual donors and foundations give more, they have also become more discerning and demanding in their giving. Take for example Geneva Global, a *for-profit* philanthropic advisory service that helps donors design charitable projects and then monitors these projects using quantitative benchmarks. Such advisories mostly work with private donors, but their methods and techniques could soon be transferred to government-funded aid projects.

Then there is the Skoll Foundation, which invests in social entrepreneurs in the developing world with ideas that can permanently impact public policies. These innovators are often homegrown, dedicated to creating feasible and scalable solutions to local challenges. They operate similarly to business entrepreneurs and hope to replicate their impact at national and international levels. Social entrepreneurs came to prominence with the creation of Ashoka: Innovators for the Public, an organization designed to find and fund "fellows," in the 1980s. But in the past decade they have widened their reputation through innovations such as the PlayPump, a cheap and efficient water pumping and filtration system for African villages, and small start-ups like DotSavvy, a Kenyan internet company that built an e-learning application to train doctors and nurses on HIV/AIDS treatment. Such technologies produce immediate, marketable impact, so they are more attractive to Western donors and investors.

Dead Aid only skims this vast and growing field, mentioning Kiva.org, a California-based online philanthropy that allows American donors to fund microenterprises in Africa. But microfinance is only a drop in the ocean of enterprise solutions to poverty and in a fast-paced, technology-driven sector, one of its more antiquated ideas.

Western donors have learned that private-sector development needs more than liquidity, and they have supported diverse initiatives to bolster the capacities of businesses in the African private sector to market their products, build consumer bases, and refine technical business skills. The "base of the pyramid" (BoP) concept, introduced by professors C.K. Prahalad and Stuart

Hart in 2002, approaches poor Africans in a new paradigm, as discerning producers and consumers who can be integrated into global markets. According to the World Resources Institute, there are four billion people in the world living on less than two dollars a day. Almost 500 million of them live in Africa, and they make more than 70 percent of the continent's total income. In the BoP paradigm, these Africans are not merely poor mendicants in need of humanitarian service—they are also picky buyers and potential links in indigenous African supply chains.

As Vijay Mahajan notes in his book *Africa Rising*, "[Africans] want cell phones, bicycles, computers, automobiles, and education for their children." This not only reflects a new way in which the West can interact with poor Africans (selling them products and services they need and ask for, rather than giving them what the West *thinks* they need), but it also reduces the paternalism that is coupled with traditional philanthropy. When people do business, the first priority is to get things done cheaply and quickly. There is little room for posturing or condescension. Take, for example, Western support in breaching the African "digital divide." Many Western firms, like the U.S. IT giant CISCO, now work with local African firms to lay cables and build telecommunications infrastructure.

In addition to designing new products and services specifically based on African needs, Western donors and entrepreneurs have created mechanisms to support and consult local businesses. Social venture capital organizations like Acumen Fund and Endeavor merge the rigor and profit-driven analysis of investment banks with the social goal of supporting indigenous entrepreneurs in the developing world. It is already a given that African businesses need to trade more and need more open markets to trade with. Social venture capitalists place more effort on improving African products and services to international standards, refining supply chains, and cultivating human resources. Trade in Africa should not be slave to fluctuating commodity prices, and budding entrepreneurs and their firms need technical support and business education in order to move up the value chain. That is where the Western donor's capacity can fill existing gaps.

The model for building trade capacity in the ideal *Dead Aid* world, on the other hand, would follow China's charm offensive in Africa. In the bluntly titled chapter, "The Chinese Are Our Friends," Moyo praises China's opportunistic investment style, noting that "rather than conquer Africa through the barrel of a gun, [China] is using the muscle of money." She acknowledges that the Chinese are in it for commodities such as oil and copper and attempts to present a nuanced assessment of the issue: While China's acquiescence to African human rights violations and support for corrupt African leaders are worrying, Moyo notes that the majority of Africans have favorable views of the Chinese and have benefited from the infusion of foreign direct investment. And African leaders have every right to prefer China's "straightforward approach" to the "endless nitpicking of the IMF." Moyo does not deploy this same nuance, however, in her analysis of Western objections to the Chinese presence. Western concerns—about China turning a blind eye to the genocide in Darfur or its support of Robert Mugabe in Zimbabwe—are merely rooted in the "Western liberal consensus" whose members believe "(often in the most paternalist way) it is their responsibility to look after Africa." By building a false dichotomy between a mutually beneficial Chinese–African relationship and a haughty and paternalistic Western–African relationship, *Dead Aid* leaves little room for the middle ground, which in this case would allow Western investors and donors to support local businesses and build trade capacity while working to secure human rights and improve social conditions.

But is it a total fantasy to envision, as does Moyo, a world without aid? Perhaps not. Imagine a corrupt, underdeveloped, post-conflict nation, suffering from high inflation, minimal savings, and stagnating exports. Led by an authoritarian general who came to power in a recent military coup, this country is so reliant on foreign economic assistance that the aid it receives accounts for more than half of the government's revenue. In fact, the nation's previous leadership, complacent with this assistance, designed policies specifically to maximize the amount of aid it received.

Now imagine that aid donors, finally fed up with the corruption, anti-democratic practices, and failing economic policies of its recipient, decide that within the next decade, they will irreversibly terminate all economic assistance to the country. The recipient nation, realizing it must stand on its own two feet, embarks on a series of swift and radical reforms and structural adjustments in order to restart its economy and make up for the imminent loss of free revenue. A generation later, the country has a thriving democracy, a highly educated populace, and has joined the elite club of OECD industrialized countries.

This donor–recipient tale is not merely an economic daydream. It is the true story of what happened in South Korea, arguably the most successful of all Asian Tigers. Recovering from the Korean War and heavily reliant on the United States for its economic buoyancy, South Korea was an early example of how aid can inhibit a country's progress and how the prospect of losing aid can catalyze its reform and economic growth. As Nicholas Eberstadt writes, "there is intriguing evidence to suggest that South Korea's transition to a regimen of outward-oriented growth was a direct consequence of foreign aid polices: more specifically, of a warning by Washington that it would be *terminating* its programs for Seoul."

If Dambisa Moyo had her way, this would be the path of aid-addicted countries in Africa, whose fate of misery, corruption, and economic stagnation, according to her, has come as a consequence of unquestioned and unrestricted Western aid and not in spite of it. Moyo recommends that Western donors pick up the phone, call African leaders, and "[tell] them that in exactly five years the aid taps would be shut off—permanently." After this, they must no longer expect their ineffective and dysfunctional governments to be bankrolled by copious flows of well-intentioned official development assistance.

But the South Korean miracle and Moyo's dream of an African parallel are long shots rather than blueprints. In today's world, it is not realistic for aid to ever truly go away, simply because its purpose is not always economic growth or poverty reduction and because the political benefits of aid often outweigh its failures. Thus, many development actors have decided to refocus their attention on how aid is disbursed and spent. The above-described innovations of the development community suggest a world *beyond* aid rather than a world without aid.

Dead Aid has tinges of glamour and provocative assertions—the anti-celebrity bit has been embraced by mainstream media, and the proposal to cut off all aid in five years is a brazen challenge to the West from an African-born economist. It is also spot-on in its analysis of aid's side effects and unintended consequences. The West's aid policies will continue to harm Africans, as Moyo convincingly argues, if they cling to the status quo. Yet while *Dead Aid* would make it appear otherwise, much of the foreign aid community—donors, entrepreneurs, nonprofit volunteers—has moved beyond tired arguments such as "trade not aid." None of the new concepts or innovations are foolproof, and in fact many are deeply flawed, but they reflect the growing understanding among this diverse "development community" that while traditional foreign assistance has not worked and is often corrosive to the developing world, it will not go away anytime soon. The most critical question, then, is: What's next?

POSTSCRIPT

Does Foreign Aid Undermine Development in Africa?

As discussed in the introduction, it is not unusual to become disenchanted with foreign assistance and development after reading about a history of failed development projects in the African context. Whether development work is locally funded or supported by outside donors, some have argued that we need to fundamentally rethink the way we approach this work. One theme that has received considerable attention in foreign assistance circles since at least the early 1990s is participatory development. The emphasis on participation came about in reaction to a history of "top-down" development projects that left local people with little to no sense of ownership of the projects undertaken by outsiders in their communities. Besides being problematic in its own right, a lack of participation often inhibited the longer term sustainability of programs (because local people had little interest in maintaining projects they did not initiate or request). In order to foster community participation, some development actors have employed a range of techniques to facilitate community problem identification and group problem solving. Development practitioners and researchers have devised a number of techniques that enable greater community input (a group of which are known as participatory rural appraisal techniques).

The problem, and a latent contradiction, is that communities may identify problems that many development agencies are not prepared to solve (at least in the quick and efficient manner that their funders expect). The result is that communities often identify problems and select programmatic solutions that they think a development agency can deliver on. One of the most acclaimed nongovernmental organization (NGO) initiatives in Africa was the Six-S network in Burkina Faso, Mali, and Senegal. Six-S was founded in 1977 by Bernard Quedrago, a teacher and school inspector from Burkina Faso, and Bernard LeComte, a French national with considerable development experience. Both Quedrago and LeComte were frustrated with traditional development approaches that tended to be top-down and emphasize projects. Six-S (which is a French acronym meaning "making use of the dry season in the Savannah and the Sahel") was an NGO created to support traditional village-based groups, known as NAAMs, in a variety of locally inspired initiatives, from microcredit schemes to soil conservation. One of the great innovations of the Six-S network was its emphasis on flexible funding, an approach that allowed the village-based NAAM groups to propose their own projects that were then selected by a committee of village representatives, rather than a donor. By all accounts, Six-S was amazingly successful over a 15-year period, serving several hundred thousand people organized into 30,000 groups in 150,000 villages (Bernard J. LeComte and Anirudh Krishna, "Six-S: Building upon Traditional Social Organizations in Francophone West Africa," in Anirudh Krishna, Norman Uphoff, and Milton Esman, eds., *Reasons for Hope: Instructive Experiences in Rural Development* [Kumarian Press, 1997]).

Internet References . . .

AFROL News—Agriculture

AFROL News of the Heifer Project International provides information on agriculture from across Africa, including general agricultural trends for the continent and country-specific news.

**http://www.afrol.com/Categories/Economy_Develop/
Agriculture/msindex.htm**

African Conservation Foundation

The African Conservation Foundation Web site allows one to search its databases of African environmental topics by country, organization, type of information, and environmental category.

http://www.africanwebsites.net

African Data Dissemination Service

The U.S. Government's African Data Dissemination Service offers a warning system for possible famine or flood conditions. It also allows one to download data, both map based and tabular, regarding crop use, hydrology, rainfall, and elevation (as well as several other data sets).

http://earlywarning.usgs.gov/adds/

United Nations Environment Programme: Regional Office for Africa

The United Nations Environment Programme: Regional Office for Africa provides information on UN environmental initiatives in Africa, both in restoring the natural environment and in harnessing the environment for human use in an environmentally friendly way.

http://www.unep.org/ROA/

Food and Agriculture Organization of the United Nations

The Food and Agriculture Organization of the United Nations Web site contains news articles and other information about agriculture, food, and environmental issues. It contains information with both African and international emphases.

http://www.fao.org/

Agriculture, Food, and the Environment

*A*s a region equally known for the abundance and destruction of its resources, questions regarding environmental management, agriculture and food production have loomed large in debates concerning Africa. One of the more recent preoccupations is a rising concern about the impact of climate change on African farmers. Real and imagined crises also have fueled discussions over how to prevent famine and augment food production without undermining the integrity of the ecosystem. Debates also rage about the best way to approach wildlife conservation on the continent.

- Is Climate Change a Major Driver of Agricultural Shifts in Africa?
- Is Food Production in Africa Capable of Keeping Up with Population Growth?
- Does African Agriculture Need a Green Revolution?
- Is Community-Based Wildlife Management a Failed Approach?

ISSUE 9

Is Climate Change a Major Driver of Agricultural Shifts in Africa?

YES: **Pradeep Kurukulasuriya et al.**, from "Will African Agriculture Survive Climate Change?" *World Bank Economic Review* (2006)

NO: **Ole Mertz, Cheikh Mbow, Anette Reenberg, and Awa Diouf,** from "Farmers' Perceptions of Climate Change and Agricultural Adaptation Strategies in Rural Sahel," *Environmental Management* (2009)

ISSUE SUMMARY

YES: Pradeep Kurukulasuriya, of the United Nations Development Programme, and colleagues argue that agricultural revenues for dryland crops in Africa will fall under global warming scenarios. They suggest that irrigation is a practical adaptation to climate change in Africa.

NO: Ole Mertz and colleagues, of the Universities of Copenhagen and Dakar, suggest that farmers in Africa's Sahelian region have always faced climate variability at annual and decadal time scales. Although households at their study site in Senegal are well aware of climate change, they attribute most changes in farming practices to economic, political, and social factors, rather than environmental ones.

Although human-induced global climate change continues to be debated in some policy circles, the majority of the mainstream scientific community accepts this as a reality. An international body of climate scientists known as the International Panel on Climate Change (IPCC) meets regularly to review the evidence on the rate, character, causes, and effects of climate change. In 2001, this group concluded that "the balance of evidence suggests that there is a discernable human influence on global climate." Although the IPCC initially predicted smaller increases in global temperatures, these have been slowly revised upward over time. The current policy objective is to try to hold global temperature increases to 2°C (3.8°F). The major driver of anthropogenic climate change is greenhouse gas emissions, of which the most significant is carbon dioxide emissions from fossil fuel combustion.

Just as the sources of carbon emissions driving climate change are distributed unequally around the world (with the industrialized nations being the major producers of these gases), the impacts of climate change are also varied and distributed unequally across the planet. In most instances, increasing temperatures will also be accompanied by changing rainfall patterns. In general, the impact of climate change on Africa will be more significant than in the temperate regions. Through the "downscaling" of climate change models, atmospheric scientists have been able to predict the impacts of climate change on the African continent (and its various subregions). Current predictions are for the Sahara and the semiarid parts of southern Africa to warm by as much as 1.6°C (2.9°F) by 2050. The equatorial countries would see temperature increases of up to 1.2°C (2.2°F) over the same period. Sea-surface temperatures in the open tropical oceans surrounding Africa will rise by less than the global average (i.e., only about 0.6–0.8°C [1.1–1.4°F]), so the coastal regions of the continent will warm more slowly than the continental interior. Rainfall changes projected by most models are relatively modest to significant. In general, rainfall is projected to increase over the continent, with the major exceptions being southern Africa and parts of the Horn of Africa. In these areas, rainfall is projected to decline by 2050 by about 10 percent.

The way in which climate change will impact different African livelihood systems will also vary widely. The number of people in Africa who are involved in agriculture is not insignificant. Approximately 75 percent of the female labor force, and 62 percent of the male labor, is involved in farming in sub-Saharan Africa. This includes over 140 million smallholders and approximately 6 million commercial farm workers. Those engaged in small-scale, rain-fed agriculture are the most likely to be hard-hit by increasing temperatures and changing rainfall patterns. But the vulnerability of these farming households to climate change is conditioned by the particular mixtures of crops they are planting (with some crops being more drought resistant than others), as well as ability to fall back on other strategies when agricultural production declines or fails.

In this issue, Pradeep Kurukulasuriya, of the United Nations Development Programme, and colleagues argue that agricultural revenues for dryland crops in Africa will fall under global warming scenarios. They suggest that irrigation is a practical adaptation to climate change in Africa. In contrast, Ole Mertz and colleagues, of the Universities of Copenhagen and Dakar, suggest that farmers in Africa's Sahelian region have always faced climate variability at annual and decadal time scales. Although households at their study site in Senegal are well aware of climate change, they attribute most changes in farming practices to economic, political, and social factors, rather than environmental ones.

YES

Pradeep Kurukulasuriya et al.

Will African Agriculture Survive Climate Change?

The increasing concern about climate change has led to a rapidly grow-ing body of research on the impacts of climate on the economy. Quantitative estimates of climate impacts have improved dramatically over the last decade. Sub-Saharan Africa is predicted to be particularly hard hit by global warming because it already experiences high temperatures and low (and highly variable) precipitation, the economies are highly dependent on agriculture, and adop-tion of modern technology is low.

Despite the estimated magnitude of the potential impacts on Africa, there have been relatively few economic studies. Most of the quantitative pro-jections are interpolations from empirical studies done elsewhere. A limited number of agronomic studies on Africa have confirmed that warming would have large effects on selected crops, but these studies reflect only a small share of Africa's crops, they fail to capture how farmers might respond to warming, and they do not quantify overall economic impacts. The economic impact on the livestock sector in Africa has gone largely unstudied.

This study uses farm-level data collected across diverse climate zones in 11 African countries to explore how the current climate already affects African farmers, specifically net farm revenues. Total net farm revenue is defined as the sum of incomes from three main farm activities: dryland crops, irrigated crops, and livestock. Irrigated crops rely on at least some irrigated water (from surface flows or ground water). Dryland crops rely only on rainfall that falls on the farm. Livestock in Africa largely depend on grazing on natural lands or pasture.

This information is used to estimate the impacts of changing tem-perature and precipitation on the net revenues of African farmers using the Ricardian method. Net revenues are regressed on climate, soils, and other control variables. Separate regressions are estimated for the three main farm activities to shed light on the climate response of each of these components of farm income. The amount of land that was planted could be accurately measured for the crop regressions used to estimate net revenue per hectare. Since most African farmers rely on common land for livestock grazing, it was not possible to determine how much land was used. The livestock regressions are consequently based on revenue per farm. Although these analyses are

From *World Bank Economic Review,* vol. 20, no. 3, 2006, excerpts pp. 368–386. Copyright © 2006 by The World Bank. Reprinted by permission of Oxford University Press Journals via Rightslink.

therefore different, total farm income is still the sum of the incomes from these three sources. . . .

Data and Empirical Specifications

The study relied on long-term average climate (normals). These long-term data for districts in Africa were gathered from two sources. Satellite data on temperature was measured by a Special Sensor Microwave Imager (SSMI) on U.S. Department of Defence satellites for 1988–2003. The SSMI detects microwaves through clouds and estimates surface temperature. The satellites conduct daily overpasses at 6 a.m. and 6 p.m. across the globe. The precipitation data come from the Africa Rainfall and Temperature Evaluation System created by the Climate Prediction Centre of the U.S. National Oceanic and Atmospheric Administration. It is based on ground station measurements of precipitation for 1977–2000. Thus, the temperature and precipitation data cover slightly different periods. This discrepancy might be a problem for measuring variance or higher moments of the climate distribution, but it should not affect the use of the mean of the distribution.

The 11 countries in this study were selected across the diverse climate zones of Africa and precipitation of each country in the sample. Although Africa is generally hot and dry, there is a great deal of variation across the continent. Egypt and South Africa are much cooler than the rest of the countries in the sample. Similarly, relative to the other countries in the sample, Cameroon is very wet, followed by Kenya, Zambia, Ghana, and Ethiopia; the other countries, especially Egypt, are drier.

Within each country, districts were selected to capture representative farms across diverse agroclimatic conditions. Between 30 and 50 districts were sampled in each country. In each district, surveys were conducted in 2002–04 of randomly selected farms (seven countries were surveyed in the 2002–03 season and four countries were added in 2003–04). Sampling was clustered in villages to reduce the cost of administering the survey. A total of 9,597 surveys were administered. The final number of surveys with usable information on crop production was 9,064. Of these, 7,238 farms had dryland crops, 1,221 had irrigated crops, and 5,062 had livestock. Many farms had both crops and livestock. The total number of farm surveys per country varied from 222 in South Africa to 1,288 in Burkina Faso.

The relative importance of dryland crops and irrigated crops varies considerably. For example, Egypt is entirely dependent on irrigated crops because the climate is too dry to support crops without irrigation. In contrast, most farms in East Africa and the Sahel have very little irrigated crops. Livestock net revenue varies widely across countries, but it is particularly important in relatively dry countries.

Data on hydrology were obtained from a continental scale hydrological model of Africa. Using climate data and local typography, the model estimated the potential monthly long-term stream flow for each district. Potential water flows were used because water can be withdrawn from many places along a watershed. Water flow measures the amount of water coming from other districts and is an important complement to the water generated in each district from precipitation. Water flow is particularly important in Africa, where water is generally scarce. For example, the Nile delta would be completely unsuitable for farming without the water from the Nile River.

Data on the composition, coarseness, and slope of the major soils in each district were obtained from the Food and Agriculture Organization. . . .

The marginal impact of temperature and precipitation on each country is also calculated (figures 1 and 2). The analysis reveals that the impacts of climate change differ across countries. Cooler countries such as Egypt, South Africa, Zambia, and Zimbabwe are likely to suffer livestock losses from warmer temperatures because of the loss of beef cattle (figure 1). Irrigated crops in currently hot regions such as Ethiopia and West Africa will suffer with warming, whereas irrigated crops in the Nile Delta and Kenyan high-lands will gain. However, some effects are fairly universal. Dryland crops in all countries throughout Africa will be damaged by any warming. Figure 2 suggests that the marginal impact of precipitation is mostly beneficial, compared with that of warming, and that livestock and irrigated farms will mostly benefit from rising precipitation and lose from declining precipitation.

[Findings] and Policy Implications

This study examined the net revenues of farmers in 11 African countries and provided quantitative confirmation of what scientists have long suspected. Although African dryland farmers have adapted to local conditions, net revenues would fall with more warming or drying. Dryland crop and livestock

Figure 1

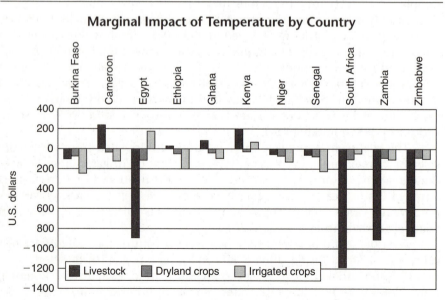

Source: Authors' analysis based on data described in the text.

Figure 2

Marginal Impact of Precipitation by Country

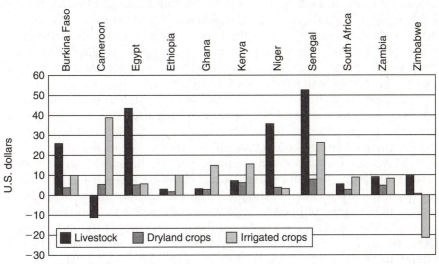

Source: Authors' analysis based on data described in the text.

farmers are especially vulnerable, with temperature elasticities of −1.9 and −5.4, respectively. Irrigated cropland benefits slightly from marginal warming because irrigation mutes climate impacts and because these farms are currently located in relatively cool places in Africa.

With precipitation elasticities of 0.4 for dryland crops and 0.8 for livestock across Africa, net revenues for dryland crops and livestock will increase if precipitation increases with climate change and decrease if precipitation decreases with climate change. Net revenues for irrigated land will follow in the same direction but to a much smaller extent (elasticity of 0.1). Increases in precipitation will have an unambiguously beneficial effect on African farms on average, whereas decreases in precipitation will have a harmful effect. However, country effects and within-country effects can differ.

The revenue effects for dryland crops, irrigated crops, and livestock are assessed independently. When the marginal temperature effects across all three sources of revenue are summed, increases in revenues on irrigated cropland at first offset losses for dryland crops and livestock. As temperatures continue to rise, however, the net effect on African farms becomes steadily more harmful. Total farm revenue decreases as precipitation falls but rises as precipitation increases. Climate scenarios that entail either significant warming or substantial drying will consequently be quite harmful. However, climate scenarios that entail only mild increases in temperature and more rainfall may actually be beneficial. The total impact on African agriculture will depend on the climate scenario.

The analysis reveals that net farm revenue has a quadratic relationship with both temperature and precipitation. The marginal impact of climate change consequently will depend on each farm's initial temperature and precipitation. Farms that are located in hotter and drier areas are at greater risk because they are already in a precarious state for agriculture. Dryland farming throughout Sub-Saharan Africa is vulnerable to warming. Dryland farming in the East, West, and Sahel regions of Africa are especially at risk. In contrast, irrigated crops in places that are relatively cool now, such as the Nile delta and the highlands of Kenya, enjoy marginal gains from warming. Finally, drier locations such as Egypt, Niger, and Senegal get big livestock gains from increased precipitation relative to wetter locations in Africa.

Because Sub-Saharan African economies as a whole depend more heavily on agriculture, total GDP and per capita income is also vulnerable. In contrast, nonagricultural GDP in Northern Africa is more diversified, and so the economies of these countries are less vulnerable to climate change.

This study measures the marginal impact of climate change. It does not predict the future. Simulating likely future climate impacts is a large undertaking. First, one must examine the projections of several climate models to get a sense of the range of plausible climate scenarios. Second, one must project how agriculture is likely to change in the future, both in technology and in land use. For example, the average dryland farmer currently earns $319 a hectare and the average irrigated land farmer earns $1,261 a hectare. The more technologically advanced irrigated farms earn even more. The adoption of technology and capital is very important to the future of agriculture in Africa. Third, one must estimate by how much carbon fertilization is likely to increase crop productivity over time. These gains will reduce the magnitude of the damages in Africa, although it is not clear by how much.

Will Africa survive climate change? The results of this study suggest that Africa will be hit hard by severe climate change scenarios. Some countries are more vulnerable than others, so it is important to focus on the countries that really need help. In fact, in several scenarios, many African farmers gain whereas others lose from climate change. This study also notes that African farmers already practice some forms of climate adaptation. Policymakers may want to pay special attention to these successful adaptation practices. For example, irrigation water (including related inputs) and livestock are already used in some areas to alleviate climate hardships such as droughts and low levels of precipitation.

One adaptation that has moved very slowly in Africa is technology adoption. Africa lags behind the rest of the world in adopting irrigation, capital, and high-yield varieties. Some technologies may help farmers adapt to drier or hotter conditions, such as the development of new soybean varieties in Brazil. However, even climate-neutral technical advances will help farmers increase productivity and counterbalance losses from climate change. Through research and outreach, governments could encourage the development and use of varieties with more tolerance for the hot and dry conditions of many of Africa's agroclimatic zones.

The quantitative results, especially the sizable differences between irrigated and dryland agriculture and livestock in Africa, suggest that promoting irrigation could help alleviate the likely effects of climate change in Africa.

Where water is available, moving from dryland to irrigated agriculture would increase not only average net revenue per hectare but also the resilience of agriculture to climate change. Governments could make public investments in infrastructure and canals for water storage and conveyance, where appropriate and where the public good nature of these investments prevent adequate private sector investment. Investment in successful irrigation in Sub-Saharan Africa ranges between $3,600 and $5,700 a hectare in 2000 prices. This analysis suggests that the difference between dryland and irrigated agriculture runs between $150 and $5,000 a hectare, depending on the country. This range of investment values implies that farmers in some countries could repay irrigation investments within a very reasonable period. Policymakers may want to consider supporting such coping interventions for climate change, where appropriate.

Finally, in addition to encouraging direct adaptations, both local and national governments and international organizations could invest in infrastructure and institutions to ensure a stable environment to enable agriculture to prosper. Such policy interventions may not only achieve the long-term goal of helping vulnerable populations adapt to climate change, but may also increase the likelihood of achieving the more immediate Millennium Development Goals, such as halving hunger, reducing poverty, and improving health.

Ole Mertz, Cheikh Mbow,
Anette Reenberg, and Awa Diouf

 **NO**

Farmers' Perceptions of Climate Change and Agricultural Adaptation Strategies in Rural Sahel

Introduction

There is increasing evidence that climate change will strongly affect the African continent and will be one of the challenging issues for future development, particularly in the drier regions. The challenge is composed of the likely impacts of climate change on ecosystem services, agricultural production, and livelihoods, as well as the limited resilience and high vulnerability characterizing regions dominated by economic poverty, subsistence food production, and a low and highly variable natural production potential. An economic analysis of 9000 farmers in 11 African countries predicted falling farm revenues with current climate scenarios, and using two global circulation models, a study at national level in Mali predicted future economic losses and increased the risk of hunger due to climate change. It seems clear that the combination of high climatic variability, poor infrastructure, economic poverty, and low productivity will constitute important challenges for Africa and the Sahelian countries in particular.

Nevertheless, rural communities in the Sahelian zone of West Africa have always managed their resources and livelihoods in the face of challenging environmental and socio-economic conditions. They have to a large extent been able to develop their livelihood strategies in a way which enables them to constantly cope with and adapt to an erratic climate, severe pest attacks, changing policies at local, national, and global levels, and so on.

The scientific agreement that climate change is happening and will continue well into the future regardless of the effectiveness of mitigation measures has reiterated the need to understand how farmers and pastoralists in the Sahel have coped with climate variability and change in order to guide the strategies for adaptation in the future. However, there are at least two complications in arriving at such an understanding:

- First, while the climate change scenarios for West Africa point to increased temperatures in the Sahel, climate models are in disagreement regarding changes in precipitation, suggesting either increases

From *Environmental Management*, vol. 43, 2009, excerpts, pp. 804–816. Copyright © 2009 by Springer Science and Business Media. Reprinted by permission via Rightslink.

or decreases. Indeed, based on 21 global model outputs, the Fourth IPCC Assessment states that "it is unclear how rainfall in the Sahel, the Guinean Coast and southern Sahara will evolve."

- Second, in all rural communities, climate is only one of many factors influencing their coping and adaptation strategies to environmental changes and this may be even more pronounced in the relatively poor and vulnerable communities of the African drylands. Seemingly marginal changes in subsidies, market conditions, labor supply, seed availability, energy supply, etc., may lead farmers with low economic resilience to radically change their strategies regardless of the climatic parameters.

This article will only address the second problem. The first, we have to leave to climate modelers in the hope that more precise climate scenarios can be established to serve as foundation for realistic impact prediction and adaptation strategies. The objective of this article is closely linked to the general objective of the human dimensions work package of the African Monsoon Multidisciplinary Analysis (AMMA) project that aims to improve the understanding of coping and adaptive capacities of natural-resources dependent societies in the Sahel. We will investigate how farmers are coping and adapting to past and current changes, recognizing that while the concept of coping capacity is more directly related to short-term survival, the concept of adaptive capacity refers to a longer time frame and implies some learning. Using case studies from central Senegal, we will assess how development trends in the Sahel, such as diversification of production, new technologies, migration, off-farm activities, trade, government policies, and land and resource degradation/restoration affect the land use and livelihood strategies of households. We will assess which coping and adaptation strategies are adopted in the face of these changes or trends and to what extent these strategies are a specific response to climate change and variability (CCV). Our specific aim is to attempt a separation of climate causes from other drivers of change in order to inform policy-makers on the relevance of the increased focus on climate adaptation in development efforts.

Adaptation of Agriculture to Climate Change and Variability in Africa

The study of coping and adaptive resource management strategies is not new, particularly not in the Sahelian region, where a poor and vulnerable population has always dealt with a highly fluctuating natural environment. There are diverging opinions on how well rural populations are dealing with their environmental and economic conditions. Recent studies point to dryland populations as the most ecologically, socially, and politically marginalized lagging behind on most economic and health indices and that climate change will be yet another stress factor in a vulnerable system. Others stress the resilience of Sahelian farmers and their proven ability of coping with even the hardest crises, such as the droughts of the 1970s and 1980s and question the notion of persistent crisis in Sahelian households. Moreover, it is argued that the value of local knowledge in climate change studies has received little attention.

Farmers possess valuable indigenous adaptation strategies that include early warming systems and recognize and respond to changes in climate parameters, for example, by maintaining flexible strategies with short and long cycle crop varieties.

The complexity of identifying climate as a direct driver of change in agriculture was alluded to in the introduction. Large differences between regions, villages, and households exist. [For example, it was] found that in one community in Sudan drought was perceived as a key driver for change in gum arabic production, while another community was relatively unaffected and supposedly more resilient. Previous studies have pointed to the complexity of drivers of change in the Sahel as have several studies in southern Africa.

There are, however, examples of good correlations between climate parameters and change. Using agent-based modeling in a vegetable garden system of South Africa, scholars showed that wealthier households benefit more than the poor from weather forecasts and that subsistence farmers are the most vulnerable to short-lived droughts even if average rains are good. Also in South Africa, dry spells caused farmers to reduce cropping efforts and focus on livestock, and in Mali, farmers responded to shorter rainy seasons by using short cycle varieties of sorghum, although long-cycle varieties were still farmed as they give higher yields and have better taste. More long-term adaptation was observed in Burkina Faso and Niger, where shifts in farming location between sandy dunes and more clayey pediplains and piedmonts were related to precipitation patterns, whereas short-term coping with the 1997 drought in Burkina Faso caused farm households to implement a range of food saving strategies, encourage migration, sell livestock, and even resorting to borrowing and mortgaging of the following year's crops. In this case, the ramifications of one year's drought were felt in the following year in terms of lacking seed and labor for cultivation and it sparked interest in drought resistant varieties, but longer term adaptation measures were not assessed.

Many of these studies provide recommendations for actions to improve the living conditions of rural populations, and there have been many efforts aimed at reducing the vulnerability of rural communities to CCV. Various early warning systems, such as FEWS (Famine early warning system), are used in Sahelian countries, and a range of support initiatives, including improvement of access to markets for inputs and products and commodity chains; promotion of investment incentives; revision of food sufficiency assumptions; supporting access to labor markets; opening of land markets; promotion of communication systems (e.g., cell phones); and supporting the most vulnerable with flexible micro-credit have been implemented with varying success. Not all policy instruments have been sensitive to the adaptive resources and constraints of African farming households, which may be partly caused by the difficulty in measuring and finding the right indicators for the adaptive capacity as well as the problems of determining driving forces and cause-effect relationships. Many of the development efforts only indirectly address the problem of CCV, and although policies today more frequently make explicit that they are a response to predicted climate change, it seems that they are relatively similar to past policies favoring economic development. . . .

Results

Perceptions of Climate Change and Variability

The local communities have a very clear memory of the years dominated by extreme climatic conditions and other significant events leading to disturbances of the production. In some cases, the same years are characterized by both drought and excessive rains.

The perception of climate parameters was also assessed at household level. Households generally found that the overall trend was a decrease, but several households nuanced this view. Four mentioned that the "rains have returned" since the early 2000s, three that rains simply vary too much to determine the trend, and one that the quantity of rain is irrelevant as both drought and excessive rains may destroy the crops. Households generally agreed on increased temperatures throughout the year and that cold periods have become shorter and hot periods longer. Finally, wind was almost unanimously pointed to as having become stronger, especially in the dry season.

The statements in the group interviews generally corroborated the household questionnaires as the respondents were very concerned with the increasing dry and wet season wind speeds. However, the nuanced views on rainfall seen in the household interviews were not repeated in the group interviews where there was agreement on a negative trend in rainfall. The perception of climate parameters of male and female groups was similar.

Perceptions of Climate Impacts and Adaptation

After evaluating climate change parameters, households were asked to identify impacts of climate change and variability and their adaptive responses.

Most of the identified impacts are negative and related to effects of strong winds and excessive rain, which is interesting considering that most households and groups of farmers perceived reduced rainfall as the main problem. Problems related to lack of rain are only mentioned eight times as the cause of negative impacts—compared to excessive rain, which is mentioned 14 times. Wind is mentioned 30 times as a cause of the problems. Positive impacts are mentioned by a few respondents and include better possibilities for business (if not hampered by bad weather) as well as recovering vegetation and fauna as a result of returning rains and a perceived decline in the area of cultivation leading to more fallow land and better soils. Migration by young people was identified as both an impact and adaptation measure: the absence of young people is by older people remaining in the household felt as an indirect impact of climate, which is considered to be partly responsible for declining yields and few opportunities within in agriculture thus causing people to leave; but it is, of course, also an adaptation measure by the family to secure income from remittances and thereby counter economic difficulties that may be directly or indirectly caused by climate factors. Migration was mainly perceived in a negative sense as more work was left for the older people—the most frequently mentioned adaptation measure to migration—though the positive aspect of income was acknowledged. Otherwise, the households generally mentioned few adaptation measures—new crops or crop varieties (mostly vegetables); keeping animals in stables; replacing draught horses with cattle, which are cheaper to feed; and using manure were the main measures mentioned to counter perceived climate impacts on agricultural production. While questions focused on climate induced responses, it is not completely clear whether all of these are in fact adaptation to climate change. The new crops are mainly introduced to diversify and secure better income, both of which may be indirectly related to climate.

The six group interviews were initiated by asking respondents to agree on five positive aspects of living in the community and five major challenges. Climate factors were not identified by any of the groups as a main positive or challenging aspect of village life, although it could be an underlying cause of several of the issues mentioned. Health, education, and road access could be considered the main concerns as they were usually mentioned first and are the most frequently mentioned. Only when asked directly about climate issues did the group interviews largely corroborate the impacts identified in the household interviews, and they reiterated that rainfall variability during the rainy season is of major importance. Identified impacts include declining crop yields as a result of strong winds, intensive rainfall events and extended periods of cloud cover (these statements relate to the past 1–5 years). Both heat and cold are mentioned as having detrimental impacts on livestock, which is partly contradictory to the household statements mentioning problems with cold weather to be on the decline. A decline in the population and yields of useful wild plant species was also mentioned as a consequence of more extreme weather and of increased damage from bush fires caused by strong winds in the dry season. Poor health was raised as a problem during periods with dust storms and prolonged rain, and "reduced solidarity" was identified as an indirect impact of adverse climate in the community as everybody is increasingly trying to keep their own households going without considering their neighbors. Finally, household work is directly and indirectly affected: cooking is hazardous during periods with strong winds because of the risk of fire, and migration of young people results in reduced labor availability.

Adaptive measures mentioned in the group interviews were also similar, though a few new activities emerged: credit schemes and support from NGOs were mentioned as important adaptive measures needed within agriculture, and a focus on revitalizing the traditional solidarity measures, especially for aiding each other with family events and during crises, was emphasized. The main barriers identified in the group interviews toward appropriate adaptation to CCV were lack of funds to initiate small businesses (credit access), lack of success in doing business (low income, low benefit), underpaid manual work, high price of basic supplies and food, and theft of livestock.

Observed Changes in Land Use and Livelihood Strategies

In order to assess whether the farmers' perceptions of climate change impacts and adaptation in agriculture are in accordance with changes in household activities, selected parameters are analyzed.

The average cultivated area per household in the three villages in 2006 was 19.8 hectares with a range of 3–120 ha. If one large farm of 120 ha is excluded, the average is 14.8 ha. Sixty four percent of the farms are less than 20 ha. Farmers were asked an open question to assess the main determinants of how they decide on the area cultivated in any given year. Climate factors play a very minor role in that decision, whether it is today or before the drought of the 1980s—rain was the only parameter mentioned and only by two households. Availability of farm equipment and seeds are by far the most important and fertilizer is by close to half of the respondents ranked second. Before the mid 1980s, this tendency was the same, although "food needs" are ranked first by 20% of the farmers indicating that there was a more urgent need to be self-sufficient in food at that point in time.

Another indicator of adaptation is the introduction of new crops. Only three farmers mentioned this as an adaptive measure, but a wide range of new crops have been taken up by farmers. The most important new crop is water melon (*Citrullus lanatus*) and the number of households mentioning this crop as new exceeds the number of households farming it. This may be because they forgot it when mentioning crops in the fields or because they cultivated it in previous years but not in 2006. Cowpea (*Vigna unguiculata*), maize, and sesame (*Sesamum indicum*) are also new to several households. There is a wide range of reasons for adopting different crops, but none of these reasons are directly related to climate factors. A desire for increased income, especially in the "lean season"—the months before the annual grain harvest—is common. Although new to the households interviewed, none of these crops are new to Senegal and they are promoted by crop diversification programs—an objective also specified in the National Adaptation Program of Action.

Data were also collected for changes in livestock, but sample checks revealed that the household data are highly inaccurate, probably because farmers are unwilling to reveal their livestock holdings.

Discussion

Farmers in the Eastern Saloum are strongly aware of the climate and have clear opinions on changes, especially in wind patterns and the intensity of climate events. This is corroborated by studies in other parts of Africa, e.g., in the Limpopo Province of South Africa, where a large majority of respondents in three regions of Limpopo Province related changes in long-term climate patterns to increased variability and unpredictability and identified climate as a "livelihood-affecting risk."

Linking household strategies for agricultural and livelihood change directly to climate parameters is, however, very complex inasmuch as cultural, socio-economic, and environmental drivers of change are intimately linked. The data presented in this article do not show climate as a main driver of change, but the results may of course be limited by the sample size. The qualitative methods of semi-structured group interviews are the most valuable in getting information on the issue, but when comparing with the results of the household survey, relationships pointed to in qualitative group interviews are not easily rediscovered in the quantitative data. The changes are, for the most part, the same, but the perceived causal links are often different. This does not necessarily mean that climate factors are not the underlying cause, but the direct cause is seldom linked to excessive rainfall, heat or strong wind that are frequently mentioned as the most adverse climate conditions. The lessons for these discrepancies in results from household and group interviews are twofold: group interviews appear to be influenced by narratives that may bias responses; on the other hand, household questionnaires may not capture underlying causes or drivers that are more easily explored in semi-structured interviews. Hence, both methods should have built-in checks for these biases and complement each other.

The household interviews showed some nuances in the opinions on changes in rainfall as some claimed that the rainfall had in fact increased. That the group interviews showed a unanimous negative trend could be related to well established local (and international) narratives of declining rainfall that are not easy to go against in a group and which have also been observed elsewhere in the peanut basin of Senegal. On the other hand, climate was not an issue of concern when groups were asked to identify main positive and challenging aspects of village life. Several respondents stated that the main characteristic of the rainfall patterns is unpredictability and the official climate records more or less corroborate the statements related to extreme years.

It is clear that many concerns other than climate are equally or more important for decision-making in agriculture. Availability and affordability of farm equipment, seed, and fertilizers are the main constraints, and these are intricately linked to the political-economic situation, especially agricultural policies and market development, which also appear to overrule climate concerns elsewhere. The complexity of drivers of change and adaptation was shown in the Diourbel area of Senegal for the period between 1954–1989. They identified structural adjustment and "decapitalization" in 1984 as the main drivers of farming system change, including increased fallowing, disposal of equipment, decline in the use of fertilizer, migration, and investment in nonfarm activities. Adaptation to the situation included adoption of new crops, such as cowpea and roselle and more livestock, measures that are similar to what farmers are doing today.

In fact, current Senegalese policies to strengthen the productive capacity of agriculture through credit systems, new crops, and improving soil fertility are now labeled as climate change adaptation policies, although they, to a large extent, address general economic problems that have many different causes and are similar to earlier anti-desertification plans. The implementation of the policies include distribution of agricultural equipment and food during 2004 and 2005, subsidies on new crop types and varieties, and establishing a processing facilities, but the impact of these schemes is not easily discerned in the responses of households in the present study, except for the adoption of new crops by few farmers. Moreover, it is not clear whether these schemes adequately respond to climate concerns. For example, several farmers stated that the current short season ground nut varieties are adapted to drought—an adaptive measure useful in the past during the drought crises—but the recent high rainfall years have caused yield losses as the ground nuts mature when the weather is still very humid.

Many farmer statements indicate concerns related to rainfall, but rarely without also including the economic situation, e.g., "The climate has become harsher. The drought and economic activities degrade the natural vegetation." Wind remains the strongest concern especially for the health of animals, although this may also be masking a generally poor animal health caused by insufficient veterinary services or funds to pay for them. Generally, farmers have a rather fatalistic approach to climate concerns and the typical statement, "weather is a divine phenomenon that we are not in charge of," is also reported from other parts of Africa. Degradation of the vegetation, mentioned by many respondents, was not only linked to climate but also to (mis)management. However, none of the respondents mentioned the recent "invasion" of powerful Marabouts (religious leaders) who moved to the area with hundreds of "talibé" (farm workers) and have proceeded with extensive land clearing to grow ground nuts and millet. This is probably due to the sensitivity of this issue as many people in the region are followers of the Marabouts, again underlining the complexity of arriving at clear understandings of local strategies in a "noisy" reality of social, cultural, economic, and environmental change.

The lessons for policy-making on adaptation to climate change are, despite the complexity of the issue, relatively simple. On the one hand, the uncertainty of future precipitation trends in the Sahel and—as the farmer observed above—good intentions with extension programs that may have adverse effects if weather patterns change call for great caution when policies are prepared. For example, many National Adaptation Programs of Action in the Sahel were developed before the results of the IPCC Fourth Assessment Report and base their recommendations and project proposals on the overall trends towards drying observed from 1950–2000 and on single model scenarios depicting future drying trends. Obviously, this may lead to unfortunate results if the climate develops differently. On the other hand, if policies—like the recent Senegalese agricultural policies—promote general economic development that provide farmers with a range of options rather than a specific climate or drought driven focus on certain crops and water conservation techniques, this will allow greater flexibility for local people to adjust to future changes no matter which direction they take. In short, strengthening general adaptive capacity to cope with CCV will, especially in the agricultural sector, be better than devising narrow climate adaptation solutions with uncertain outcomes.

Conclusion

Farmers in the Sahel are concerned with CCV and climate parameters play, according to statements in group interviews, an important role in decision making. However, it is difficult to rediscover the climate parameters when analyzing livelihood strategies in household interviews, unless the questions are directly focusing on climate issues. Reasons for changes are seldom directly attributed to climate, though these may be one of the underlying causes. Concerns with increasing speed and duration of wind and its destructive characteristics are by far the most noteworthy.

The present study has been too limited in scope to provide firm conclusions on the adaptive capacity of the communities. Although the problems in measuring adaptive capacity, finding the right indicators, and separating climate effects from other impacts have been dealt with by combining household surveys and group interviews and taking an open-ended approach when asking questions, the findings point in different directions as household responses do not always corroborate group responses. Hence, our findings support the view that strong narratives on climate exist in the local communities. Adaptive measures directly linked to climate parameters are found, but they appear less important in shaping the dynamic of rural livelihood strategies than adaptation to economic, political and social factors. The implications for policy-making is to exercise great caution before joining the quest for adaptation solutions until a better understanding of local and regional climate change scenarios, as well as local adaptive strategies and capacities, is obtained. Sensible focus on economic development that allows flexibility for adjusting to various CCV scenarios is likely to be most successful.

POSTSCRIPT

Is Climate Change a Major Driver of Agricultural Shifts in Africa?

A key aspect of this issue concerns how and why farmers adapt their farming systems in the African context. In the short run, most farmers will deal with climate variability through coping strategies. For example, in the event of harvest shortfalls, rural households may collect wild foods, use grain stores, sell livestock, or send family members to the city or another region to work. But if short-term climate variability turns into longer term change, farmers may be forced to permanently change or adapt. Adaptation could range from changing the type of crops one grows to investing in irrigation infrastructure or getting out of farming all together. In this issue, Pradeep Kurukulasuriya and coauthors see irrigation as one of the best ways to deal with changing rainfall patterns. The problem with irrigation infrastructure is that it is quite expensive, making it inaccessible to most Africans. Robert Kates has described how irrigation infrastructure in most poor African countries tends to disproportionately benefit the wealthy ("Cautionary Tales: Adaptation and the Global Poor," *Climatic Change*, 2000).

The other interesting issue touched on in the NO article by Mertz et al. is that many African farmers are not solely trying to cope with climate change, but a variety of economic factors as well. As such, farmers are often placed in very challenging circumstances when they must cope with both climate change and economic change (or globalization). This problem has been described as "double exposure" by Karen O'Brien and Robin Leichenko ("Winners and Losers in the Context of Global Change," *Annals of the Association of American Geographers*, 2003). As discussed in the NO article for this issue, farmers in Senegal were more influenced by economic factors in their cropping decisions than by climatic ones. This is not unusual as economic conditions are often more pressing in the short run. A major problem, however, may ensue if economic conditions are leading African farmers to engage in agricultural practices that are actually more vulnerable to climate change. Such a scenario is not rare as many African farmers have shifted away from diverse food production systems and into cash crops for the market (which are often simpler farming systems that are more susceptible to environmental variability). For a more specific discussion of this problem in Mozambique, see Julie Silva's article titled "Double Exposure in Mozambique's Limpopo River Basin" (*The Geographical Journal*, 2010).

ISSUE 10

Is Food Production in Africa Capable of Keeping Up with Population Growth?

YES: Michael Mortimore and Mary Tiffen, from "Population and Environment in Time Perspective: The Machakos Story," *People and Environment in Africa* (John Wiley & Sons, 1995)

NO: John Murton, from "Population Growth and Poverty in Machakos District, Kenya," *The Geographical Journal* (March 1999)

ISSUE SUMMARY

YES: Michael Mortimore, a geographer, and Mary Tiffen, a historian and socio-economist, both with Drylands Research, investigate population and food production trajectories in Machakos, Kenya. They determine that increasing population density has a positive influence on environmental management and crop production. Furthermore, they found that food production kept up with population growth from 1930 to 1987.

NO: John Murton, with the Foreign and Commonwealth Office of the British government, uses household-level data to show that the changes in Machakos described by Mortimore and Tiffen "have been accompanied by a polarization of land holdings, differential trends in agricultural productivity, and a decline in food self-sufficiency." As such, he argues that the "Machakos experience" of population growth and positive environmental transformation is neither homogenous nor fully unproblematic.

There is a long-standing debate about the ability of agriculture to keep up with population growth in Africa. Those who are concerned that population growth will outstrip agricultural production are often referred to as neo-Malthusians. This perspective is designated as such in deference to the eighteenth-century British clergyman, Thomas Malthus, who posited such a scenario as inevitable in his famous 1798 tract, *Essay on the Principle of Population.* The neo-Malthusians generally have dominated contemporary debates concerning global population growth and food supplies and are led by such figures as Paul Ehrlich (*The Population Bomb,* Simon and Schuster, 1968) and Lester Brown (see numerous World Watch Institute publications). The major contrarian perspective in African studies is the Boserupian view point, named after Ester Boserup (*The Conditions of Agricultural Change,* Aldine Publishing, 1965). She established that increasing population densities may induce farmers to intensify their efforts and thereby produce food at a rate that keeps pace with population increase.

A third perspective, often referred to as the technocratic or cornucopian view, is somewhat akin to the Boserupian population thesis. This view, most commonly associated with the late Julian Simon (*The Great Breakthrough and Its Cause,* University of Michigan Press, 2000), asserts that human ingenuity will resolve resource constraints created by population growth if the free market is allowed to operate and appropriate price signals are transmitted to producers. The major difference between the Boserupian and technocratic viewpoints is that the former tends to emphasize indigenous methods of adaptation whereas the latter stresses the importance of market-led change.

To some extent, both of the perspectives in this issue are sympathetic to the Boserupian view, but they arrive at different conclusions in the final analysis. The case presented by Michael Mortimore, a geographer, and Mary Tiffen, a historian and socio-economist, about the situation in Machakos, Kenya, is probably the most well-referenced piece of analysis supporting the Boserupian thesis in African studies. Mortimore and Tiffen assert that increasing population density has had a positive influence on environmental management and crop production. Furthermore, they find that food production kept up with population growth from 1930 to 1987. The study shows that farmers were able to reverse land degradation, enhance their livestock, invest in their farms, and increase productivity despite increasing population density.

John Murton, with the British Foreign and Commonwealth, agrees with the findings of Mortimore and Tiffen in the aggregate, but then uses household survey data to demonstrate that there is a difference between the trends for wealthier households with access to nonfarm income and poorer households without access to such revenues. It is households with nonfarm income that are able to invest in the inputs needed for increasing agricultural productivity. For those households unable to make such investments, yields are declining. As such, Murton is essentially saying that increasing labor alone will only go so far in the agricultural intensification process, after which point capital expenditure on inputs is critical.

YES

**Michael Mortimore
and Mary Tiffen**

Population and Environment in Time Perspective: The Machakos Story

Introduction: Linkages Between Population Growth and the Environment

The linkages between population growth and environmental degradation are controversial. The view, widely held, that rapid population growth is incompatible with sustainable management of the environment is influenced, knowingly or not, by the neo-Malthusian belief that resources are limited. According to literature prepared for the United Nations Conference on Environment and Development (the Rio Earth Summit), 'the number of people an area can support without compromising its ability to do so in the future is known as its population carrying capacity.'

In agricultural terms, each agro-ecological zone is believed to have a carrying capacity which must not be exceeded if environmental equilibrium is to be maintained. In the words of Mustapha Tolba, formerly head of the United Nations Environment Programme, 'when that number is exceeded, the whole piece of land will quickly degenerate from overgrazing or overuse by human beings. Therefore, population pressure is definitely one of the major causes of desertification and the degradation of the land.' According to such a view, degradation threatens to diminish food production, and thereby the human carrying capacity, in a cumulative downward spiral.

More sophisticated estimations of population supporting capacities take account of technological alternatives to the low-input systems that are found in much of the tropical world. A recent study shows that if technology is varied (or levels of inputs increased), the limits rise accordingly. The critical constraint, for practical purposes, is access to technology. However, poverty inhibits investment, and the poor are said to be incapable of conserving their environment: rather, 'poverty forces them to exploit their limited stocks just to survive, leading to overcropping, overgrazing, and overcutting at unsustainable rates. A vicious circle of human need, environmental damage and more poverty ensues.'

A negative view of the effects of population pressure on the environment, which has been underwritten by several UN organizations, the Rio Earth

Summit, and influential writers carries great weight in the environmental debate. Nevertheless it is not supported by some well-documented situations in Africa.

One of these, Machakos district of Kenya, is the subject of a recent study of resource management by African smallholders. The study covered the period 1930–90, which is long enough to control for rainfall variability, and for changes in the political economy. Profiles of change were constructed for all the major environmental and social variables. The linkages in what the World Bank has called the 'population, agriculture and environment nexus' were systematically investigated. The study shows positive, not negative influences of increasing population density on both environmental conservation and productivity.

Characteristics of Machakos District, Kenya

Machakos lies in south-east Kenya. Its northernmost point is about 50 km from the capital, Nairobi, from which it stretches some 300 km southwards. Since at least the eighteenth century it has been inhabited by agropastoralists known as the Akamba, who also populated the neighbouring Kitui district. Men looked after the livestock and cleared new land, while women cultivated a small plot for food crops. . . .

Setting the Scene for an Ecological Disaster

When British rule was imposed on Kenya, the Akamba people were confined in the Ukamba Reserve by the colonial government's Scheduled Areas (White Highlands) policy. . . . It was bounded by European settlers' farms and ranches on the north and west. To the east and south were uninhabited Crown lands, on which the government allowed only grazing, by permit. Thus encircled, the Akamba grew in numbers, and in livestock, while clearing extra land for shifting cultivation of maize and other crops, and chopping down trees for fuel burning and construction of their homes. Despite their protests, the government refused to relax its policy of containment.

During the period 1930–90, the population of Machakos district grew from 238 000 to 1 393 000, and an annual rate of increase of over 3% was maintained from the 1950s until after 1989. After 1962, the Akamba were allowed to settle on the semi-arid former Crown lands, and also took over some of the Europeans' farms in government schemes. Thus the land available to them effectively doubled. However, the growth of the population reduced the amount of land available to less than a hectare per person by 1989.

This conjunction of rapid population growth with unreliable rainfall, frequent moisture stress, low soil fertility and high erodibility, suggests the likelihood, on the premises outlined above, of population-induced degradation on a grand scale. This was indeed the diagnosis offered in assessments of the reserve in the 1930s. A disastrous series of droughts (in 1929, 1933, 1934, 1935 and 1939) caused major crop failures, losses of livestock, pest outbreaks, the deterioration of vegetal cover and accelerated erosion. In 1937, Colin Maher, the government's soil conservation officer, wrote despairingly:

> The Machakos Reserve is an appalling example of a large area of land which has been subjected to uncoordinated and practically uncontrolled development by natives whose multiplication and the increase of whose stock has been permitted, free from the checks of war and largely from those of disease, under benevolent British rule.
>
> Every phase of misuse of land is vividly and poignantly displayed in this Reserve, the inhabitants of which are rapidly drifting to a state of hopeless and miserable poverty and their land to a parching desert of rocks, stones and sand.

No less than eight official visits, reports and recommendations were commissioned between 1929 and 1939, strongly reflecting an official consensus view centred on overstocking, inappropriate cultivation, and deforestation in a reserve thought already to be overpopulated in relation to its carrying capacity. Did events bear out this gloomy prognosis? . . .

A Farming Revolution

Change in Machakos was multi-faceted. Technical innovation was not restricted to soil and water conservation. . . . An inventory of production technologies in Machakos identified 76 that were either introduced from outside the district or whose use was significantly extended during the period of the study. They included 35 field and horticultural crops, 5 tillage technologies and 6 methods of soil fertility management. The technical options available to farmers were thereby extended, adding flexibility to the farming system. Such flexibility is a great advantage in a risky environment.

Making Money from Farming

In the 1930s, capital was mostly locked up in livestock, and occasional sales provided needed cash. From cultivating maize, beans and pigeon peas for subsistence, farmers have since moved into marketing crops. The most successful of these, until its price fell in the 1980s, was coffee. Some Akamba learned to grow it while employed on European coffee farms, but in 1938 the government expressly forbade 'native' coffee growing in order to protect the European producers' interests. After the overturning of this ban in 1954, strict rules were enforced in the growing, processing and marketing of coffee. African producers successfully achieved high grades. Coffee was an attractive component of rehabilitation programmes, since it was profitable and had to be grown on terraces. Coffee output increased spectacularly in the 'boom' of the later 1970s. It generated investment funds, and supported improved living standards, in the sub-humid zone. This had spill-over effects in the drier areas, through the demand of coffee-growing farmers for agricultural labour, for food and livestock products in which they might no longer be self-sufficient, and for a whole range of consumer goods, housing improvements and services. By

1982–83 over 40% of rural incomes in Machakos district were being generated by non-farm businesses and wages.

In contrast to coffee, cotton, which is recommended for the drier areas, was not a success in the long term. Its price was only rarely high enough to compensate for its tendency to compete with the food crops for labour and capital, and the profit margins were reduced by marketing inefficiencies. Three attempts to promote cotton—in the 1930s, 1960s and 1978–84—ran into the sand. Output limped along, and in 1991 the closed ginnery at Makueni offered silent testimony of failure.

Both coffee and cotton were sold to monopsonist parastatal marketing boards and required government support in extension and supervision, and in supervising officially sponsored co-operatives for input provision, processing and grading. By contrast, expanded growing of a great variety of perennial and annual fruit and vegetables was closely linked with the growth of Kenya's canning industry, the Nairobi and Mombasa retail markets, and exports of fresh vegetables by air. Itinerant Asian buyers, firms operating contract-buying and enterprising Akamba, as individual traders or in formal co-operatives or informal groups, have all played a role. A generally high value per hectare facilitated the skillful exploitation of wet micro-environments, even in the driest areas, and the development of technologies such as micro-irrigation and the cultivation of bananas in pits. Fruit production is attractive to women farmers, as trees do not compete with food crops (for which they often have the main responsibility).

Akamba farmers have adapted rather well to the opportunities provided by the market. No amount of promotion can succeed without incentives. But given these, innovation in both production and marketing aspects is commonplace. Meanwhile, livestock sales, on which they depended for market income in the 1930s, have steadily declined.

Achieving Food Sufficiency

The staple food of the Akamba is white maize. The shortness of the two growing seasons, and the high probability of drought, call for varieties that either resist drought or escape it by maturing quickly. In the 1960s, the government's local research station began a search for drought-escaping varieties that culminated in the release of Katumani Composite B (KCB) maize in 1968.

The new maize was promoted by the extension service, and it was steadily, if unspectacularly, adopted by the farmers. Various surveys suggest that from two-thirds to three-quarters use it, but it is not known how much of the maize area is planted to it, nor what proportion of output it contributes. Of 40 farmers interviewed in five locations in 1990, only a third said they used it exclusively, and another third used it together with local and hybrid varieties.

This was no 'green revolution.' The ambivalent response has, however, an explanation. Given the unpredictable rainfall, farmers need to keep their options open. Their local varieties, though slower to mature, are more resistant to drought, and hybrids do better in wetter sites. KCB is liked because, in combination with other varieties, it strengthens this flexibility rather than undermines it. Some farmers cross-pollinate it with their local varieties, further enhancing their adaptive choice.

With and without KCB, food crop production per person kept up with population growth from 1930 to 1987 although imported foods remained necessary after a series of bad seasons. The district's dependence on imported food in the period 1974–85 was less than in 1942–62 (8 kg per person annually compared with 17), notwithstanding major droughts in both periods. Food output per person in 1984 (after three seasonal droughts) was slightly higher than in 1960–61 (after two seasons with drought, and one with floods).

Faster Tillage

In view of the reputation then enjoyed by the Akamba for resistance to change, it is surprising that the ox-plough, introduced to the district as early as 1910, had spread to about 600 (or 3%) of the district's households by the 1930s. Farmers trained their own cattle, and ploughs were cheap (about equivalent to the price of a cow in 1940); furthermore the technology was being tested and developed on nearby European farms where some Akamba worked. Ownership greatly increased the area a farmer could cultivate, and enabled him to sell maize or cotton. Its adoption called for the cessation of shifting cultivation, and facilitated the adoption of row planting and better weeding, in place of broadcasting seed.

After the Second World War, ex-soldiers who had seen ploughs in India returned with the capital to buy their own. Proceeds from trade and employment outside the district were also invested in ploughs. The government made ox-ploughs the basis of a new farming system imposed on a supervised settlement at Makueni location. The government, and traders, provided some credit. Coffee (after 1954), horticulture and cotton (in some years) generated investment funds. Adoption accelerated in the 1960s and was more or less complete by the 1980s. Surveys found 62% or more of farmers owning a plough, the remainder being too poor, or having fields too small and steep for its use.

The plough proved to be both a durable and a flexible technology. The first ones in use were adapted to opening new land, with teams of six or eight oxen. Farmers later selected a lighter, two-oxen instrument suitable for work on small, terraced, permanent fields. The Victory mouldboard plough, though much criticized on technical grounds, is used everywhere, and for several operations—primary ploughing, seed-bed preparation and inter-row weeding—and attempts to promote its replacement by a more expensive tool-bar have failed. It saves labour, and is also used by women. The 'oxenization' of Akamba agriculture was, in a measure, a triumph of capitalization in a capital-poor, risk-prone and low productive farming system.

Fertilizing the Soil

Shifting cultivation used to rely on long fallows for replenishing the soil. In the 1930s, there was very little systematic manuring. But the fertility of arable land, as measured by yields, was low. The Agricultural Department favoured farmyard manure over inorganic fertilizers. It also, unsuccessfully, promoted composting.

It was not until the 1950s that manuring became widespread, in the northern sub-humid areas. By this time, arable fields were fixed, and cultivated every year. The silent spread of this practice can be judged from the fact that by the 1980s, 9 out of 10 farmers were doing it, in both wetter and drier areas. Now, most arable land is cultivated twice a year—in both rainy seasons—and composting is being adopted by small farmers with few livestock.

By contrast, the use made of inorganic fertilizers is minimal, the bulk of it on coffee. Manure is made in the *boma* (stall or pen) and supplemented with trash and waste. The amount applied depends on how many livestock there are, and how much labour and transport are available when needed. Every farmer knows that, under present technical and economic conditions, sustaining output depends on the use made of boma manure. Few can afford inorganic fertilizers in quantity.

Feeding the Livestock

At the beginning of our period (the 1930s), Akamba women cultivated food crops at home, while their men used to take the livestock away to common grazing lands for several months of the year. However, common grazing land vanished as it was transformed into new farms. After about 1960 settlement on Crown lands could no longer be restrained, and thousands of families moved into them. Each household must now keep its animals within the bounds of the family farm, or obtain permission to use another family's land, often in return for some rent or service. More than 60% of the cattle, sheep and goats are stall-fed or tethered for a part or all of the year. When in the boma, cut fodder and residues are brought to them, which requires additional labour. Fodder grass is grown on terrace banks. These changes are most advanced in the sub-humid zone. A third of the livestock are grazed all the time, mostly in the dry semi-arid zone. The effort required to maintain livestock is making grade or crossbred cattle popular (estimated to number about 9% of the total in 1983 and to have grown rapidly since), whose milk yields and value are superior to those of the native zebu, though their increased health risks call for frequent dipping.

Farming the Trees

From the 1920s, the Forest Department believed that reafforestation was necessary to arrest environmental desiccation, and supply the growing need for domestic fuel and construction timber. For several decades the department struggled, under-resourced, to reserve and replant hilltop forests. In 1984, however, estimates of household fuel requirements put the need for new plantations at 226 000 ha (15 times the area of gazetted forest reserves!). The destruction of surviving natural woodland seemed an imminent possibility.

But sites photographed in 1937 and 1991 showed little sign of woodland degradation. A fuel shortage has failed to develop on the expected scale and the district does not import wood or charcoal in significant quantities. Indeed it exports some. Part of the explanation for this expert miscalculation lay in ignoring the use made of dead wood, farm trash, branch wood from farm trees, and hedge cuttings, for domestic fires. The other part lay in failing to appreciate a major area of innovative practice: the planting, protection and systematic harvesting of trees. Forest policy in the 1980s was shifting towards farm forestry promotion, but in this it was following, not driving, farmers.

Tree densities on farmland in one location, Mbiuni, averaged over 34 per ha (14 when bananas are excluded) by 1982. Furthermore, the smaller the farm, the greater the density. The range of trees planted includes both exotic and indigenous, both fruit and timber species. Akamba women generally manage fruit trees, while the men look after the timber trees. Owners of grazing land manage the regeneration of woody vegetation, which is used for timber, fuel, browse, honey production, edible and medicinal products.

Producing More with Less

These were some of the features of a revolution in farming wrought in unpromising circumstances. What was the driving force behind these changes?

The growth of population had two important outcomes: the subdivision of a man's landholdings among all his sons, according to Akamba custom, and the increasing scarcity of land as former communal grazing and Crown land became new private farms.

As holdings shrank in size, the arable proportion rose, leaving less and less land for grazing while the cultivated area per person stagnated. [T]he percentage of arable land increased from the older settled areas to the new, and from the wetter to the drier areas.

These changes created the imperative for intensification. By intensification we mean the application of increasing amounts of labour and capital per ha to raise crop yields. Crop–livestock integration is intrinsic to this process in dryland farming systems in Africa. It was driven by the needs for draft energy on the farm, for fodder (the stalks of maize and haulms of beans, for example), for manure and milk.

Two changes—to intensive livestock feeding systems, and to permanent manured fields, often under plough cultivation—were pivotal in this transformation, whose outcome was an increasingly efficient system of nutrient cycling through plants, animals and soil. The changes could not have occurred without security of title. Akamba custom had already recognized individuals rights in land, including the right of sale, in the older settled areas by the 1930s. Security has been reinforced by statutory registration of title, a slow legal process which began in Machakos in 1968 and has still not covered all areas.

Equally important were sources of investment capital. This was not only required for terraces and ploughs. To clear and cultivate new land, build hedges or plant trees requires labour which often has to be hired, as well as tools and expertise. The off-farm incomes earned by Akamba men inside and outside the district have contributed for decades to agricultural investment. Such incomes are often high in households with small farms, and there is little evidence that investment per ha falls where farmers are poorly resourced in land.

The outcome of this process of intensification was an increase in the value of output per square kilometre (at constant prices) from 1930 to 1987.

This was calculated by taking output data for the only three available years before 1974 (1930, 1957 and 1961), selecting two later years which were climatically average (1977 and 1987), and converting all the values into maize equivalent at 1957 prices. The year 1957 was an unusually good one and 1961, as already noted, unusually bad, hence the upward trend was interrupted. The trend continued despite the additions of large areas of the more arid types of land in 1962. Output per capita closely reflected this curve.

We conclude that, contrary to the expectations expressed in the 1930s, the Akamba of Machakos have put land degradation into reverse, conserved and improved their trees, invested in their farms, and sustained an improvement in overall productivity. . . .

No Miracle in Machakos?

What happened in Machakos did not contravene the laws of nature, as the Malthusian paradigm would express them, but rather grew logically from a conjunction of increasing population density, market growth and a generally supportive economic environment. The technological changes we have described, in conservation and production, cannot be adequately understood as exogenous, as mere accidents that gave breathing space on a remorseless progression towards irreversible environmental degradation and poverty. Rather, as argued long ago by Ester Boserup and more recently by Julian Simon, they were mothered by necessity. Technological change was an endogenous process, in which multiple sources and channels were employed, involving selection and adaptation by farmers.

Increasing population density is found, then, to have positive effects. The increasing scarcity (value) of land promoted investment, both in conservation and in yield-enhancing improvements. The integration of crop and livestock production improved the efficiency of nutrient cycling, and thereby the sustainability of the farming system.

The Machakos experience offers an alternative to the Malthusian models of the relations between population growth and environmental degradation. Elsewhere in Africa, there are more documented cases of positive associations, though it would be foolish to ignore the differences.

Successful intensification under rising densities has certain preconditions. These are peace and security, for trade and investment, and a marketing and tenure system in which economic benefits are shared by many, rather than monopolized by a few. Degradation may occur, as it did in Machakos, when a change from a long fallowing system is first needed, but when population densities or other conditions are not conducive. Normally, as population grows, so do the opportunities for specialization and trade. To the stick of necessity the market adds the carrot of incentives and resources for investment in new technologies.

In the past, development planners tried to transform farming systems that were seen as inefficient and technically conservative. In fact, they are changing themselves, as studying them in time perspective shows, and there is scope for supporting positive change with appropriate policies. The guiding principle must be to go with the grain of historical change. This means encouraging investment, by encouraging trade and by improving farm-gate prices (for example, by improving roads and by avoiding heavy taxation of agricultural products). There is also a need to protect investment when crises (e.g. famines) threaten to force households to sell their assets. Increasing the technical options available to local resource users in a risky environment is one of the most productive avenues to pursue, by encouraging endogenous experimentation; by increasing information through general education as well as agricultural extension, and by creating new avenues for technological development and transfer.

John Murton  **NO**

Population Growth and Poverty in Machakos District, Kenya

The influential book *More People Less Erosion* investigated population growth and environmental change in Machakos, Kenya. Tiffen *et al.* argued that population growth need not lead to environmental degradation, since if smallholder farmers were allowed to operate freely within competitive markets, responding profitably to new agricultural opportunities, they would logically manage their resources in a sustainable way, with a resultant benefit to society and the economy. Although environmental improvements in Machakos seem beyond dispute, Rocheleau argues that the aggregate District level data used by Tiffen *et al.* mask social and economic differentiation as a result of these changes, and that many farmers in Machakos are experiencing declining welfare amidst the improving environment.

The current paper seeks to address this debate by examining the results of a household-level village study in the old Machakos District of Kenya. The new research found that agricultural intensification has not been a homogeneous experience. Rising living standards have been experienced largely by those families who have access to non-farm (and usually urban-derived) income. This income facilitates security in agricultural crises such as drought, and enables a virtuous cycle of on-farm investment, leading to higher agricultural yields, rising incomes and higher standards of living. In contrast, families without access to such income were found to be experiencing a cycle of declining soil fertility and declining yields per head. These divergent experiences of vicious cycles and virtuous spirals were found side by side within neighbouring households in the study area. In the past it was easy for farmers to obtain non-farm income and so make social and economic progress. However, the combination of population increase, slow economic growth, and structural changes in the Kenyan economy mean that poor families are now enjoying less social mobility than they did previously.

This paper highlights how differential access to non-farm income has driven a polarization of land holdings within the village, with the result that, for poorer families at least, environmental sustainability is no longer proving to be a guarantee of livelihood sustainability under conditions of rapidly rising population.

From *The Geographical Journal*, March 1999. Copyright © 1999 by Royal Geographical Society with Institute of British Geographers. Reprinted by permission of Wiley-Blackwell.

The Study and Study Site

In common with other recent work studying environmental transformation, and work in Kenya investigating agricultural change and differentiation, change in Machakos was studied by means of a longitudinal data set. The results of research into agricultural practices and land holdings conducted in Machakos in 1965 by Frederick Owako were compared to new data from the same area today. Both sets of research were conducted in Ndueni village of Mbooni location, in what is now Makueni District; Makueni District having been created when the old Machakos District was carved in two in 1993.

Owako's research formed part of a PhD thesis entitled "The Machakos Problem," and investigated the ability of local agriculture to cater for the growing number of people in the District. His work compared the state of agriculture in study villages in Mbooni, Iveti, Masii, Kangundo and Nzaui locations. He found uneven distributions of land and livestock in each of the villages, with more skewed distributions in the most densely populated areas. Owako believed that agriculture would not be able to continue to support the growing population of Machakos District, and that a landless class would evolve which supported itself through non-agricultural activities and professions.

The current research sought to investigate temporal processes of change, and was carried out between January and May 1996, involving a survey of households from Ndueni village, followed by rapid rural appraisal work conducted in Iveti, Masii and Kangundo, and archival research in Nairobi. For the purposes of comparison, fieldwork in Ndueni followed the door-to-door sampling procedure adopted by Owako, but utilized improved methods of field measurement and crop and income recording. In this way the current survey was able to conduct questionnaires in 180 households in Ndueni, with only three households refusing to answer the questionnaire. After the questionnaire a stratified sample of 56 households was revisited for more in-depth interviews regarding the results of the first phase of the study. In addition to the data collected by Owako on land and livestock holdings, the present research also elicited information on agricultural and non-farm income, as well as investigating the agricultural labour market.

The study was conducted in Ndueni village, Mbooni location, 35 kilometres from Machakos town (population 80 000) along a dirt road which is often muddy, and occasionally impassable during the rainy seasons. Nairobi is a further 60 kilometres from Machakos, and is the destination of many of the cash crops from the village. Whilst the dominant crops are mixes of maize and beans, export crops such as French beans grown in Mbooni are purchased by export companies and flown out of Nairobi International Airport. Other horticultural produce is trucked down to Nairobi and Mombasa by local traders. Coffee from the area is processed at the Kenya Planters Cooperative Union in Nairobi, as well as at independent millers in adjacent Thika District.

The study village, Ndueni, lies on the side of a highland massif which rises out of the Makueni lowlands. The bottom of the village lies near the valley floor at 1400 metres, whilst the upper parts of Ndueni reach a height of just over 1850 metres. Most of the village is thus quite cool and temperate, and similar to other high-potential areas of Machakos such as Kangundo and Iveti. Rainfall is moderate but unpredictable, with an average of 1250 millimetres a year in a bimodal pattern. The area is of a fairly high agricultural potential, lying between agro-ecological zones 2 and 3, and supporting coffee and rainfed horticulture. Despite this the most widely-grown crops are the maize, beans and peas which form the basis of most people's diets. The area also has many zero grazed dairy cows, and sends milk down to the nearby lowlands on a daily basis.

Change in Ndueni

Much of the change seen in Ndueni mirrors the development of the wider area described by Tiffen *et al*. Because of its high agricultural potential, Mbooni (of which Ndueni is a part) was the first area of Machakos or Makueni to be settled by the people of the Akamba tribe, who make up over 95 per cent of the population of the two districts. However, as land grew more scarce in Mbooni, farmers began to move to other hill areas, and thereafter down on to the less well-watered lowlands of Machakos and Makueni.

Just as in the rest of Machakos and Makueni, Ndueni village has been experiencing rapid population growth. Ndueni was first settled by six families from nearby villages in around 1840, and 161 of the 180 households in the village's survey were traced to these original pioneers. This rapid population growth has forced a decrease in the average size of landholdings in Ndueni, as the original holdings carved out by pioneers have been subdivided amongst their descendants. This decline has driven the adoption of more intensive agricultural practices as families seek to maintain agricultural livelihoods.

Crop Patterns

Table 1 shows changes in cropping patterns since Owako's survey in 1965. The main points to note are an increase in the area planted to cash crops such as

Table 1

Cropping patterns in Ndueni

Crop	% Area 1965	% Area 1996
Maize, beans, peas	63	63
Grazing and waste land[1]	18	9
Coffee	4	13
Vegetables	4	4
Root crops	2	3
Millet, sorghum	2	negligible
Tobacco	2	none recorded
Tree lots	none recorded	3
Housing	none recorded	3
Others	5	2

Note: [1]Over 4% of the village is covered by expanses of exposed granite which are of little use for either grazing or cultivation, although are used as water collectors for downhill crops.

coffee, and an increase in tree cover (including a big rise in the number of fruit trees, often interplanted with food crops). Other changes seen include a big decrease in the amount of grazing land as it is given over to food and cash crops, and a virtual disappearance of traditional grains such as millet and sorghum.

These aggregate changes appear similar to those described by Tiffen *et al.* However, the household-level data collected in the current survey show how only 57 per cent of farmers have been able to afford the capital necessary for investment in cash crops. In contrast, the abandoning of sorghum and millet has not necessitated capital expenditure, and has been carried out by almost everyone in the village. Whilst the adoption of crops such as coffee has increased the diversity and the sum of many farmers' income, the move away from traditional grains has led to a reduction in the income diversity of poorer farmers, and increased dependence on only maize and bean cultivation on the smallest farms.

Livestock

Traditionally, it was thought that cattle were kept by the Akamba for the status they conferred, and only more lately for their meat. Stock were grazed in the open and looked after by young men and boys. The period since 1965, however, has seen an almost complete transition towards the zero grazing of cattle. This has allowed the stocking density of cattle per hectare to rise at the same time as land devoted to grazing has fallen. Despite this, cattle numbers have not grown as fast as human populations and there are thus fewer cattle per household. Zero grazing has been accompanied by a widespread transition from traditional breeds to exotic dairy cattle and half breeds, as farmers seek to maximize revenue from the sales of milk, which made up 34 per cent of agricultural income in the village.

These changes are much as described by Tiffen *et al.*, but the current household-level survey shows such changes to have been accompanied by a switching of gender roles, as those aspects of livestock care traditionally associated with women such as foddering and milking have become ever more important as male roles (herding and guarding) have declined. Women have guarded their roles carefully, with the result that livestock are now almost entirely under the care of women, who consequently often control the money from dairy sales.

Terracing

Unlike much of more lowland Machakos, terracing in Mbooni was almost complete by the time of Owako's survey in 1965. As early as 1955 the Divisional Officer for Mbooni was able to report that

> Bench terracing has proceeded at an ever greater pace during the year, and I believe it should be possible within only two years to ensure that all cultivation takes place on benches only.

The current survey is thus in many ways a study of changes occurring after the environmental transformation described by Tiffen *et al.* (1994). Despite this, terracing has continued to affect the relationship between society and its environment beyond the initial purpose of erosion control. For instance, terracing (together with the decreases in livestock holdings per family) now acts as a brake upon mechanization, and has led to a wide-scale abandonment of previously common technology such as ploughing, and a reversion to more labour intensive technologies such as hoeing. In 1965 over 25 per cent of farmers in the study area owned ploughs; in 1996 the figure was less than two per cent.

The trends outlined above show how, whilst *More People, Less Erosion* was able to overturn many received wisdoms regarding the relations between environmental degradation and population growth, household-level data have been able to reveal a more nuanced picture of change. Erosion control and agricultural intensification have not occurred simply as a result of societal processes, subsequently to exist in isolation from that society; but rather have reacted back in a dialectical way upon consequent economic and societal decisions. Such a process has been charted in sociology by Giddens and industrial geography by Massey, being only more recently applied to development studies by Leach *et al*. Landscape change is a fundamentally political process, and no assessment of environmental transformation is complete without a consideration of the accompanying economic and social change. The rest of the paper uses household-level data to describe more of the social and economic impacts of agricultural intensification.

Findings from Household-level Data Analysis

A significant polarization of wealth has occurred in the village in recent years, largely as a result of farmers' differential ability to tap into sources of urban capital.

In 1965 the poorest fifth of households owned eight per cent of the land whereas by 1996 the figure was three per cent and so the amount of land owned by poor families had decreased. In 1965 the richest quintile of farmers owned 40 per cent of the land, but the figure is now over 55 per cent, and so the amount of land owned by the rich has increased. In absolute terms this means that whilst average landholdings amongst the most landed 20 per cent are currently over 3.4 hectares, those with the smallest 20 per cent of holdings now own an average of only 0.2 hectares of land.

More uneven than the current distribution of land is the very unequal distribution of non-farm income in the village. Whereas the 20 per cent of families earning the most non-farm income controlled 67 per cent of the non-farm income of the village, earning an average of over $102 a month (US$1 = 58 Kenyan Shillings), over 30 per cent of the households in the village earn no non-farm income at all, except for the small amounts they can gain from occasional agricultural wage labour.

It is this differential access to non-farm income that has driven unfavourable changes in the distribution of land holdings and agricultural incomes. This is because land is very expensive in Ndueni and only people with non-farm income can afford to buy it. Irrigated land and land planted to coffee in the village is sold for prices in excess of $3400 per hectare, whilst other plots can expect to fetch around $1700 per hectare. By comparison, casual agricultural labour rates hover at around $0.90 a day when work is available. There is no tradition of rural money lending in Machakos in the same way as exists in parts of West Africa, and thus it is only people with access to income from non-farm employment, and more particularly, the soft loans and credit circles associated with formal sector work, that can afford to buy significant portions of land in the village.

For example, J- M- is one of the biggest landowners in the village, but has so far inherited nothing but an education from his father. He has purchased all his land from the proceeds of loans from his job as a Store Manager at the Ministry of Water in Machakos. Similarly, N- M- is a local entrepreneur who buys vegetables at the local market and then transports them down to Mombasa for sale. The profits from this trade have funded the acquisition of over 90 per cent of Nzasi's 16 hectares of land—the largest holding in the village. Thus whilst Owako thought that those taking up urban professions would be the rural poor, it appears that it is those households who have secured significant non-farm incomes who have been able to accumulate the most land in the countryside.

High earners are, therefore, climbing up the village land-holding scale from whatever position they started at because they have access to loans and credit to help them buy land. Indeed a limited reversal in the village wealth hierarchy has occurred over the last 50 years, as in the early colonial period it tended to be the sons of the poorest farmers (who had no cattle to provide a livelihood) who were the first to be sent to schools, and who then proceeded to buy land with the income from their urban jobs. For instance M- M-'s father was poor and had very little land or livestock. The clan (a form of extended family) sent M- to school, after which he secured a job driving road graders for the colonial administration. His wages enabled him to buy a lot of land before he retired.

In the past, farmers who made distress sales of land to people like M- were able, if things got too difficult in Mbooni, to migrate down to the lowlands where they could settle new land of lower agricultural potential. In this sense, many of the very poorest people from Ndueni are no longer there any more, but have been spun out of the upland system. However, such migration to escape poverty is no longer feasible for the poorest families in Mbooni, owing to the rising costs of migration and the closure of the land settlement frontier. Such people are now forced to remain in the uplands farming micro-scale holdings often less than 0.2 of a hectare.

The Productivity of Agriculture

According to the Boserup hypothesis, the decline in farm sizes occurring as a result of population growth and also, amongst the poorest families in Ndueni, owing to distress sales of land to richer farmers, should be offset by an increase in the productivity of agriculture per hectare, thereby preventing food availability decline. However, just as non-farm income has allowed farmers to invest in the quantity of land that they own, it is also becoming an increasingly necessary part of investment in the productivity of the land, with those farmers unable to purchase manure or artificial fertilizers struggling to maintain (let alone improve) the productivity of land per hectare with additional labour inputs alone. Average maize yields in Ndueni have fallen from 14.8 bags (90 kilograms per bag) per hectare in 1948 to only 12.3 bags per hectare in 1996. This decline has occurred despite the adoption of improved seed varieties and farming techniques such as planting before the rain, suggesting an inherent fall in the fertility of the soil.

Land in many parts of Mbooni has been cultivated for over 200 years. In the last 30 or so of these years, many fields have been continuously cultivated without breaks for fallow. Using the spatial analogue soil survey technique adopted by Tiffen *et al.*, a soil survey in the village showed that concentrations of soil nutrients in fields which had never received applications of manure or fertilizer were less than in non-cultivated areas or sites where such productivity investments regularly took place. Thus for poorer farmers in the village without access to livestock or non-farm income 'More People' may have meant 'Less Erosion,' but it did not guarantee 'More Productivity.' The families which were found to be most able to invest in the maintenance and improvement of soil productivity were those earning significant quantities of non-farm income, with the result that those families who were most prosperous off the farm, were also likely to be the most prosperous on the farm. This can be demonstrated by looking at the maize yields achieved by farmers in Ndueni, whereby a significant positive correlation ($r = 0.155$, significant $r = 0.123$ at 0.05 uncertainty) exists between non-farm income and maize yields.

More significant is the inverse relationship between maize yields and area planted ($r = -0.284$, significant $r = 0.231$ at 0.001 uncertainty), a relationship observed elsewhere by Sen and Hill. This inverse pattern reflects the more careful and labour intensive cultivation practices of farmers with microholdings, and has presumably existed in Mbooni ever since land shortages began. However, the trend is bucked by the higher yields obtained by the farmers planting the very largest amounts of maize and using more capital inputs. The five largest planters in the village achieved average yields of over 20 bags a hectare—raising the figure for the richest quintile, and creating a reverse 'J' distribution. What this suggests is that whilst Sen's theory is correct for the majority of smallholders, there now exists a small group of very commercial farmers who through their capitalization of production have managed to raise yields above those obtained by most of their more labour intensive peers. This breakdown of the inverse relationship between yields and area as a result of new technology and capital inputs has already been noted by Rahman in Bangladesh, and Lipton and Longhurst in India.

Social and Economic Trends

As a result of shrinking holdings of land due to the population growth and economic polarization described earlier, and, as a result of the static or declining productivity of agriculture on many poorer farms with no non-farm income, over 90 per cent of families are unable to feed themselves throughout the year from the food they harvest from the fields. Fortunately around 60 per cent of households in the village can afford to plant some cash crops, and so can make up a varying proportion of this subsistence gap by selling commodities such as coffee or tomatoes. However, over 40 per cent of farmers are, for between two and ten months of the year, dependent on selling their labour to buy food. For the majority who are unable to find permanent jobs in the towns, this results in participation in rural labour markets. These labour markets exist to supply the labour required for cash crops such as coffee and French beans grown by the richer farmers. However, when agricultural wage labour is unavailable locally, poor people in Mbooni go hungry.

A recent welfare survey in Makueni by a Danish International Development Agency (DANIDA) group found that malnutrition in the district was highest in Mbooni, despite the abundant greenery and horticultural exports. Food availability in Mbooni obviously does not equate easily with food entitlements. In contrast, the dryer but more sparsely populated lowlands enjoyed greater food security. This is counter-intuitive, even to people in other parts of Machakos and Makueni. Thus when a local councillor, T- M-, made a request for food aid for Mbooni at a District council meeting he was told 'You councillor, you just keep quiet!.' T- complained about people from other areas:

> Once they see all these trees (in Ndueni) they think that we are quite alright here—but trees are never eaten. At times the experts come and they go to (the local) market and they find a lot of sukumawiki (a local vegetable), cabbages and somebody goes away thinking that everybody has (food), while others living in the same area, unless they reach deeper into their pockets cannot even afford sukumawiki for a day.

Whilst these processes do show differentiation, only more recently has such differentiation been accompanied by class formation and diminishing social mobility. Before the late 1970s it was still relatively easy for members of even the poorest families to obtain an education and subsequently secure employment. Indeed, some of the richest people in the village today are old men from what were very poor families, who were sent to school by their clans, and who then accumulated land from the proceeds of their jobs, just as in the case of M- M- described earlier. However, as a result of recent economic reforms raising the costs of education and the concomitant decline of the clan as an institution of welfare support, poorer families are increasingly unable to afford an education for their children at the same time as educational requirements for jobs have been increasing. Consequently children from poor families are finding it more difficult to escape from poverty by working in the towns in the way that their forebears did in the past, and an undeniable element of structure is building up in society.

The Growing Importance of Capital in Agricultural Innovation and Environmental Conservation

The non-farm economy of Mbooni is largely focused around remittances from migrant labour in Machakos town and Nairobi. In a similar way, many of the other examples of high population density and sustainable intensive cultivation in Africa are to be found near major urban centres such as Kano. Without the widespread availability of such urban capital agricultural intensification may take an involutionary path of diminishing returns to effort. It is in this context that Haugerud acknowledges how the economic sustainability of rural Embu has only been maintained by becoming an 'outpost' of the wider Nairobi and Kenyan economy.

Agricultural intensification and conservation technology in Mbooni appear to have come in two waves, with 'second phase' technologies being increasingly dependent upon inputs of (urban derived) capital. Whilst almost everyone has adopted those first phase intensification practices that required no monetary inputs and only labour (such as terracing and mulching), second phase agricultural practices which require capital availability (such as manuring or the use of inorganic fertilizer) have been adopted only by those that can afford them. Families without access to such capital are, therefore, experiencing a detrimental and involutionary cycle of declining yields, declining soil fertility and diminishing returns to labour, as first phase conservation and productivity gains are overtaken by population growth. As one poorer respondent noted 'in the past there was a desire for more people, but now there is a fear.'

Thus Boserupian intensification on richer farms, and a form of Geertzian involution on poorer farms are seen to be proceeding side by side within the same village. At a village level, if the demands of population growth increase faster than the supply of capital to meet those demands through investments in productivity and new cash crops, then the village will tend further down the involutionary path of declining yields and returns to labour.

In such a situation, the importance of non-farm income becomes two-fold. Richer farms use non-farm income to make productive investments in agriculture, and so 'straddle' the urban and rural economies. Poorer households, on the other hand, are forced to use non-farm income to make up shortfalls in food needs, after which they often have very little investible surplus to put into agricultural production. Only greater access to land or non-farm income will enable escape from this trap. By contrast, continued population growth is increasing land hunger still further, and consequently inflating the numbers of people looking for non-farm employment in a slowing Kenyan economy. More people chasing a static number of jobs will only serve to push down the value of wages in real terms, and further reduce the ability of local farmers to 'straddle' the rural and urban economies.

Conclusions—Wealth Comes from the City

Having studied in the area of the ODI project 'More People Less Erosion,' this research can agree with many of the findings of the ODI project regarding the environmental effects of agricultural intensification and can also agree that it has come about as a result of people's profitable responses to market opportunities. However as a result of its focus on household-level processes, the new study is able to show that there has been a polarization of landholdings in the study village, driven largely by differential access to urban wages and capital. The application of such capital to production on the land, has resulted in those who earn the most off the land earning the most on the land as well. Recent changes have accelerated this process, and made it more difficult for the poorest farmers to enjoy upward social mobility.

Whilst these findings cannot be unproblematically extrapolated to other areas undergoing agricultural intensification, this study is not alone in observing increasing differentiation in such an area. A recent study in the Kabale region of Uganda by Lindblade *et al.* also noted a polarization of wealth, as did Haugerud in Embu, Kenya. In Rwanda, André and Platteau graphically chart the role of extreme land hunger and differentiation in fuelling the recent genocide. However, what is clear in Machakos at least, is that increasing inequality has not arisen, by and large, as a result of exploitation of the poorest farmers by a local bourgeoisie within rural labour markets; as has been shown to be the case in areas of India. Rather, such agricultural labour (generated by richer farmers growing new labour-intensive crops such as coffee and French beans) creates a lifeline to poorer households in the area, enabling them to cling onto their stake in the rural economy.

Polarization appears to have been driven by the application of, and differential access to, the 'investible surplus' from non-farm income. Whilst this is largely a function of non-farm income, it cannot be divorced entirely from the existing distribution of landholdings within the village. Farmers with the most land are likely to spend less of their income making up subsistence shortfalls, and thus will be able to utilize a greater proportion of their non-farm income to buy land and invest upon it. Obtaining non-farm income in this way helps smaller farmers to hold their economic ground in the face of increased pressures stemming from population growth. In contrast, those who fail to tap into new income streams are deriving diminishing returns from ever smaller pieces of land, lacking as they do the capital to intensify productively.

For many poorer families then, the environmental sustainability charted by Tiffen *et al.* has not been accompanied by a similar economic sustainability, with the implication that ecological and economic sustainability ought to be decoupled in the manner suggested by Mortimore and Munasinghe. Furthermore, the wealth and investment capital of many of the richer families in Ndueni has come from the city, with the implication that proponents of rural development ought to be looking to create not more rural 'Machakoses' but more urban 'Nairobies.' In the absence of greatly increased capital availability for investment in agricultural productivity, population growth will continue to drive many households in Ndueni along an involutionary spiral of poverty amidst green and terraced fields.

POSTSCRIPT

Is Food Production in Africa Capable of Keeping Up with Population Growth?

An interesting aspect of Mortimore and Tiffen's Machakos story was this rural community's ability to positively engage with the market. This is also a key point that Murton makes, as it is those households with nonfarm incomes who are most successful at agricultural intensification. This stands in contrast to a large amount of Africanist scholarship, which sees inter-action with the global market as extremely problematic. Boserup herself, whose theory was described in the introduction, and whose ideas drove both of these studies, saw the market as a potential stumbling block. Her concern was that employment opportunities elsewhere would draw off enough labor that the intensification process would be derailed. This con-cern is not entirely unfounded in the African context where labor (often more than land) is frequently the key constraint to agricultural production. Both Mortimore and Tiffen, as well as Murton, suggest that nonfarm income seems to more than compensate for any loss of labor in Machakos. What is crucial, however, is that households are willing to continually invest in this area. In other African contexts, labor constraints and lack of investment have been much more pronounced. See, for example, Jeffrey Alwang and P. B. Siegal's article in the August 1999 issue of *World Development* entitled "Labor Shortages on Small Landholdings in Malawi: Implications for Policy Reforms."

Murton is not alone in his criticism of Mortimore and Tiffen's analysis of the Machakos situation (but his study is an ideal match for this issue as it occurred in the same locale). Diane Rocheleau was actually the first scholar to articulate (in published form) the concern that Mortimore and Tiffen did not address properly the problem of economic differentiation over time (i.e., a growing gap between the rich and poor). For an excellent, and concise, articu-lation of this concern, see Rocheleau's September 1995 piece in the journal *Environment* entitled "More on Machakos." The same issue of the journal also carries a response to Rocheleau by Tiffen and Mortimore. Many other scholars have also sought to critically examine the relationship between population growth and agricultural change in African situations. For example, Thomas Conelly and Miriam Chaiken, in a 2000 article in *Human Ecology* entitled "Intensive Farming, Agro-Diversity, and Food Security Under Conditions of Extreme Population Pressure in Western Kenya," study an area in West-ern Kenya with similarly high population densities. Despite the wide variety of sophisticated practices that maintain a high level of agro-diversity, they conclude that intense population pressure has led to smaller land holdings, poorer diet quality, and declining food security.

Two final points about the Machakos case are worth considering. First, freehold tenure (or private property that may be bought and sold) now prevails in Machakos. This is not the case in many areas of Africa where only use, or usufruct, rights to land are allocated to community members (and, therefore, land may not be bought and sold). Some would view the lack of private property as a problem (because they believe this inhibits investment in land) while others would see the tradition of usufruct rights as beneficial because it may slow down the transfer of land from rich to poor households. Because there are local land and labor markets in Machakos, scholars should also consider how power differences between households might influence these economic transactions.

The second point to consider is that both authors in this issue agree that food production, on average, was able to keep up with population growth in this area. This is analogous to what is happening on the global scale over the same time frame, that is, global food production is keeping up with global population growth. The problem is that not all households are able to keep up and some may not even have the resources to buy available food on the market. This insight has important policy implications as it means that simply developing technologies to produce more food will not solve the problem for everyone (as these technologies, or the food produced, may be inaccessible to poor households). As such, finding ways for poor households to access food (through their own production or purchase) may be as or more important than simply producing more food.

For further reading on this topic, see a volume edited by Turner, Hyden, and Kates (*Population Growth and Agricultural Change in Africa,* University Press of Florida, 1993) that includes case studies from around the continent; or an article by Lamdin *et al.* in the December 2001 issue of *Global Environmental Change* entitled "The Causes of Land-Use and Land-Cover Change: Moving Beyond the Myths." For a more Malthusian perspective, see a book by Cleaver and Schreiber entitled *Reversing the Spiral: The Population, Agriculture, and Environment Nexus in Sub-Saharan Africa* (World Bank, 1994).

ISSUE 11

Does African Agriculture Need a Green Revolution?

YES: Kofi A. Annan, from Remarks on the Launch of the Alliance for a Green Revolution in Africa at the World Economic Forum, in Cape Town, South Africa (June 14, 2007)

NO: Carol B. Thompson, from "Africa: Green Revolution or Rainbow Revolution?" *Foreign Policy in Focus* (July 17, 2007)

ISSUE SUMMARY

YES: Kofi Annan, former UN Secretary General, deplores the fact that sub-Saharan Africa is the only region where per capita food production has declined. Annan is now leading a new organization that answers the call of many African leaders to build on the achievements and lessons learned from the Green Revolution in Asia and Latin America that began several decades earlier. He is spearheading an African Green Revolution that aims to increase African food production.

NO: Carol Thompson, professor of political economy at Northern Arizona University, suggests that increasing yields of a few targeted crops will not solve Africa's food problems. Rather, she argues that sustaining Africa's food crop diversity and indigenous ecological knowledge is the key to reducing hunger. She further eschews food security built on global market dependence in favor of food sovereignty.

Humans have long sought to increase agricultural output through the use of improved seed. Initially this was done by repeatedly (over generations) saving seeds from plants with the most desirable characteristics. African farmers have been found to maintain and utilize an amazing number of crop varieties. Even when fields are not intercropped (i.e., several different crops planted in the same field), African farmers will often plant different varieties of the same crop in accordance with soil and moisture conditions that may vary throughout the field. Preserving the rich genetic diversity among crops at the local level in Africa has been a concern of some environmentalists.

Formal plant breeding emerged at the end of the eighteenth century. Here, crops were systematically cross-fertilized in hopes of obtaining plants with a desirable mix of characteristics. A significant advance was the development of a technique known as hybridization in the early twentieth century. Hybridization involves cross-breeding two inbreds from desirable parentage, a technique that produces highly productive plants. In the 1960s, a concerted effort, known as the Green Revolution, was undertaken to boost food crop production in the developing world. The Green Revolution, involving highly productive hybridized crops in conjunction with pesticides and inorganic fertilizers, largely benefited Asia and South America because it devoted most of its attention to food crops prevalent in these regions, mainly rice, wheat, and maize to a lesser extent. Africa was bypassed by the Green Revolution for the most part, with a few significant exceptions such as maize in Zimbabwe. Although the Green Revolution of the 1960s and 1970s did boost food production, it has been criticized for not really resolving the hunger issues it was designed to address, tending to favor wealthier farmers and spawning a host of new environmental problems related to chemically intensive agriculture.

Now, some 40 years after the first Green Revolution, a group of policy leaders is calling for a new Green Revolution in Africa. The views of Kofi Annan, the former UN Secretary General, are presented in this issue, and he is also supported by the likes of Jeffrey Sachs, a prominent adviser to the UN and director of the Earth Institute at Columbia University. According to Annan, there is nothing more important than addressing poverty at its core, by increasing the ability of African farmers to improve their productivity, food security, and incomes. For Annan and others, Africa's basic problem is that it does not produce enough food to feed itself. As such, increasing agricultural yields is the best way to address hunger on the continent. Annan's organization, Alliance for a Green Revolution in Africa, is in the process of launching several programs, including those to develop improved seeds, to improve agricultural expertise, to improve the health of soils, to better manage water resources, and to improve crop storage, transport, and marketing.

Carol Thompson, a professor of political economy at Northern Arizona University, is concerned about the environmental sustainability and human health implications of a new Green Revolution in Africa. This concern extends to previous Green Revolution approaches that emphasized a few crops (and thereby limited crop biodiversity), the growing reliance on external sources for improved seeds, and the limited recognition of local technologies and approaches. She argues for an approach known as food sovereignty, which seeks to develop food security through local technologies and solutions rather than by reliance on the global market.

One question in this debate is whether food insecurity in Africa is a question of underproduction or maldistribution. Until quite recently, there was a consensus among specialists that hunger was more often the result of conflict, mismanagement, or poor distribution than underproduction. This former consensus was catalyzed, in large part, by the pioneering work of the Nobel laureate economist, Amartya Sen, who showed how national markets could be replete with grain, yet poor households might still not have the means to access this food.

YES

Kofi A. Annan

Remarks on the Launch of the Alliance for a Green Revolution in Africa at the World Economic Forum

Three years ago, as Secretary-General of the United Nations, I addressed the Africa's Green Revolution Seminar in Addis Ababa—a gathering of African leaders committed to achieving a goal that has eluded us for too long—lifting tens of millions of our children, parents, brothers and sisters out of poverty and hunger into a world of opportunity and hope.

Today, I have the high honour of accepting the position of Chairman of the Alliance for a Green Revolution in Africa. I am humbled, and yet excited.

I do this alongside all of you—our farmers, scientists, entrepreneurs, and elected leaders. And I accept this challenge with gratitude to the Rockefeller Foundation, the Bill & Melinda Gates Foundation, and all others who support our African campaign.

I do this because, for me, there is nothing more important. We must address poverty at its core. In Africa, this means enabling small-scale farmers to grow and sell Africa's food. Our goal is to dramatically increase the productivity, food security, incomes and livelihoods of small-scale farmers, many of whom are women.

All of us yearn for practical solutions to address the major cause of our continental poverty—an agricultural sector that has languished, but is now poised to be so much more productive and dynamic. We know that the path to prosperity in Africa begins at the fields of African farmers who, unlike farmers almost anywhere else, do not produce enough food to nourish our families, communities, or the populations of our growing African cities.

The facts are well known. Sub-Saharan Africa is the only region in the world where per capita food production has steadily declined. One-third of the continent's population is chronically undernourished. Most of our farmers lack access to productive crop varieties, adequate water resources and soil nutrients. In fact, our soils are the most depleted in the world. Good roads are scarce. Universities, and the research and talent they produce, are poorly staffed and underfunded.

We face many challenges. But, through this Alliance, we have reason for hope. But hope must be more than a dream—it requires the mobilisation of knowledge, capacity and resources to end the human misery that ravages our continent.

Many of our leaders have spoken about what is involved in launching a uniquely African Green Revolution—a revolution that looks to improved agricultural production as the basis of a larger effort to take Africa confidently into a new era of sustainable development; a revolution that improves the lives of farmers and delivers greater opportunity, enterprise and prosperity.

Such a vitalisation of African agriculture involves an ambitious agenda, and AGRA has such an agenda. Eager for results, AGRA has begun implementing its first programmes:

> In 2006, we began working with our partners to develop new seeds for small-scale farmers that are more productive and resilient varieties of Africa's major food crops. Also in 2006, we launched educational programmes to accelerate the development of African agricultural expertise, and to monitor and evaluate our work.

Over the next four years, we will systematically build on these programmes, adding initiatives to address other key aspects of the agricultural value chain.

> In 2007, we will launch a programme to improve the health of Africa's soils, now the most depleted in the world.
>
> In 2008, we will launch a water management initiative to help Africa's small-scale farmers get the most "crop for each drop."
>
> By 2009, we will address the key challenges facing off-farm systems and markets, such as improvements in market information systems, crop storage, processing, and transport.

Along with all of this work, AGRA will strongly advocate for policies that support small-scale farmers; those that promote rural development, environmental sustainability, and trade favourable to poor farmers in Africa.

Only with advocacy and policy change at national, regional, and global levels will small-scale farmers succeed in dramatically increasing yields, ending poverty and hunger, and lifting the economies of Africa.

Beyond this five-year time frame, the work of AGRA will continue to be informed by African farmers in the field, women's associations, and partners and leaders from all sectors of agriculture. We expect to see dramatic improvements in the livelihoods of small-holder farmers within 10 to 20 years.

We know that success depends on partnerships, and our approach is inclusive: all who share our goals are invited to this table. We are listening to others and learning about the approaches that are already improving food production in Africa. We launched our programmes only after extensive discussions with farmers in the field, as well as with our partner African institutions and experts who are carrying out this work.

In the coming months, AGRA will build upon and expand these partnerships. We will meet with farmer's unions, women's associations, networks of agro-dealers and civil society organisations. We will listen carefully to their needs and priorities, and learn from their perspectives and experience.

Many African heads of state have already committed to reaching the African Union's goal of reaching a 6 percent annual increase in farm production by 2015 and cutting food insecurity in half. AGRA fully endorses these goals, and will do all we can to help countries reach them.

While ambitious, ours is a practical campaign. We are focused on developing locally-driven and adapted solutions that address the full range of reforms required to ensure dramatically increased productivity on Africa's small-scale farms. Our farmers want better seeds, soils, and prices for what they sell. They want access to water, markets, and credit. They need to see national policies put in place that accelerate rural economic growth, investment, and job creation.

And, though we know this is a journey, that doesn't stop us from being in a hurry. We aim to make a concrete difference in our lifetimes. With respect to seeds, AGRA is already in the fields, working with African farmers and African agricultural scientists to breed new varieties of maize, cassava, rice, beans, sorghum and other major crops that will offer better resistance to disease and pests.

Our goal is to produce 100 new crop varieties in five years. And to ensure farmers have access to these seeds, we will also move to create a wider network of local seed distributors and agro-dealers to better serve remote rural areas.

Building on the Africa Fertiliser Summit held in Nigeria last year, we will launch a soil health initiative to improve soil management practices. We absolutely must improve the quality and health of our farm lands. We owe it to ourselves and future generations to enhance Africa's natural resource base and ensure sustainable production.

Water management is a cornerstone of our plans. A crucial challenge ahead will be to improve water use, especially where water is scarce. We will do this by working with farmers who are eager to develop low-cost, small-scale water management and irrigation techniques.

None of this will be possible without market improvements to increase access to credit for small-scale farmers. The ability of these farmers to sell their surplus crops is critical to any gains in African food production. So, we will work with existing and new organisations on this continent to make this uniquely African Green Revolution a lasting force for economic and social progress.

AGRA is answering the call of many African leaders to build on the achievements and lessons learned from the Green Revolution in Asia and Latin America that began more than a generation ago. That campaign—initiated by the Rockefeller Foundation—saved hundreds of millions of lives and more than doubled cereal production. There is much to be learned from these tremendous successes, as well as from their shortcomings.

That said, ours is a Revolution of the 21st century, one that we Africans will own, whose destiny we will shape, and which responds to the specific environmental challenges facing our continent. We will offer a wide range of innovative solutions. African nations and farmers will choose those that are best suited for our African cultures, climates, and economies.

We will move forward by empowering farmers and engaging rural communities to improve agriculture production in ways that reduce social and economic inequity. Our success in this regard will necessarily be linked with policies that support Africa's women farmers. Women do the lion's share of Africa's farming. It is they who grow, process, and prepare the continent's food. It is they who gather water and wood. Yet, women lack adequate access to credit, technology, training, and agricultural services.

We will be vigilant in protecting the environment that gives us our water, air, and land. We will conserve and make use of the natural biodiversity of Africa's crops and learn from the knowledge of our farmers.

The world remains captive to the old idea that we face a choice between economic growth and conservation. This is a false choice. Our fight against poverty is directly linked to the health of the earth itself. Let me be especially clear on this point: we will revitalise agriculture for Africa's small-scale farmers while protecting and enhancing the quality of our environment.

No doubt, ours is an ambitious agenda. And some would say that, in Africa, we are awash in ambitious agendas yet nothing changes. But I am convinced that the effort I am embracing today is moving us beyond a general commitment to help Africa's poor, towards specific solutions backed by the resources, talent and partnerships needed to produce tangible results.

You will be able to measure our progress by weighing the harvests as they come in from the field, by testing the soils on the farms, and by looking at the improvement in the livelihoods of farmers across Africa.

I have spent decades listening to people talk about Africa's problems, making promises to help. It's an experience that has left me thirsty for concrete action. AGRA is about taking action now, today, in a clear and meaningful way.

The final point I wish to make is this: no country or region of significant size has been able to lift itself out of poverty without raising productivity in its agricultural sector. This is our challenge. It is a long-term effort, but one that—with our partners—is within reach.

We in Africa know that, in the words of the Nigerian Chinua Achebe, it is always "morning yet on creation day." We now have an opportunity to create a new day, and with it, a new morning, for Africa.

I close today with the plea that I offered in Addis Ababa three years ago— let us all do our part to help Africa's small-scale farmers end chronic poverty. Let us generate a uniquely African Green Revolution—a revolution that will help the continent in its quest for prosperity and peace.

Africa: Green Revolution or Rainbow Evolution?

Kofi Annan has just agreed to head the Alliance for a Green Revolution in Africa, funded by the Bill and Melinda Gates Foundation and the Rockefeller Foundation.

The goals of these foundations are ambitious. "Our initial estimate is that over ten years, the program for Africa's seed systems (PASS) should produce 400 improved crop varieties resulting in a 50 percent increase in the land area planted with improved varieties across 20 African countries," reads the initiative's press release. "We have also initially estimated that this level of performance will contribute to eliminating hunger for 30–40 million people and sustainably move 15–20 million people out of poverty."

But can Africa afford this proposed "green revolution" in terms of human health and environmental sustainability? The foundation goals require resources that the continent does not have while derogating the incredible wealth it does possess. Although scientists, agriculturalists and African governments all agree that the continent has not remotely reached its agricultural potential, their advocated policies for food sovereignty drastically diverge from the high-tech, high-cost approach promoted by Gates and Rockefeller.

In 2002, while UN secretary general, Kofi Annan asked, "How can a green revolution be achieved in Africa?" After more than a year of study, the appointed expert panel of scientists (from Brazil, China, Mexico, South Africa and elsewhere) replied that a green revolution would not provide food security because of the diverse types of farming systems across the continent. There is "no single magic technological bullet . . . for radically improving African agriculture," the expert panel reported in its strategic recommendations. "African agriculture is more likely to experience numerous 'rainbow evolutions' that differ in nature and extent among the many systems, rather than one Green Revolution as in Asia." Now Annan has agreed to head the kind of project his advisors told him would not work.

Behind the Green Revolution

The green revolution of the 1970s promoted increased yields, based on a model of industrial agriculture defined as a monoculture of one or two crops, which requires massive amounts of both fertilizer and pesticide as well as the

From *Foreign Policy In Focus*, July 17, 2007. Copyright © 2007 by Carol B. Thompson. Reprinted by permission of the author.

purchase of seed. Although this approach to food production might feed more people in the short term, it also quickly destroys the earth through extensive soil degradation and water pollution from pesticides and fertilizers. It ruined small-scale farmers in Asia and Latin America, who could not afford to purchase the fertilizers, pesticides, and water necessary for the hybrid seed or apply these inputs in the exact proportions and at the exact times. To pay their debts, the farmers had to sell their land.

Increasing yields to provide food for the hungry remains the central justification for a green revolution. But as the expert panel above analyzed in great detail, increased yields of one or two strains of one or two crops ("monoculture within monoculture," as stated by a Tanzanian botanist) will not solve Africa's food problems. Africa's diverse ecological systems, and even more diverse farming systems, require multiple initiatives, from intercropping on to permaculture, from respecting and using traditional ecological knowledge to training and equipping more African geneticists. The UN Food and Agriculture Organization, for example, now promotes farmers' breeding seeds (*in situ*) as a better conservation measure than collecting seed for refrigeration in a few large seed banks (*ex situ*). The very best food seed breeders in Africa, the "keepers of seed," are women who often farm less than one hectare of land.

The key to ending hunger is sustaining Africa's food biodiversity, not reducing it to industrial monoculture. Currently, food for African consumption comes from about 2,000 different plants, while the U.S. food base derives mainly from 12 plants. Any further narrowing of the food base makes us all vulnerable because it increases crop susceptibility to pathogens, reduces the variety of nutrients needed for human health, and minimizes the parent genetic material available for future breeding.

Seeds are a key element in the equation. One figure not often quoted among the depressing statistics from the continent is that African farmers still retain control over this major farming input: of the seed used for food crops, 80% is saved seed. Farmers do not have to buy seed every season, with cash they do not have. They possess a greater wealth—their indigenous seeds, freely shared and developed over centuries. The proposed green revolution would shift the food base away from this treasure of seed. Instead, African farmers would have to purchase seed each season, thus putting cash into the hands of the corporations providing the seed. Is there a way of developing new varieties without further enriching Monsanto or DuPont by removing genetic wealth from African farmers?

Corporate development of new seed varieties, as promoted by the foundations, raises other questions. Will the new varieties be patented or protected by farmers' rights? Who will own and control the seed? One major reason for the decline of the World Trade Organization (WTO) is the global South's resistance to patenting life forms. In 1999, the African Union, representing all African governments, asked that its unanimous resolution rejecting any patenting on life be put on the agenda at the Seattle WTO meeting. The United States refused the request.

Another source of African wealth derives from indigenous ecological knowledge, reflecting centuries of adaptation to the different ecological zones, which values interspersing different plants to enrich the soil and deter pests from food crops. Shade trees, often cut down to open the land for monoculture farming, are not necessarily in the way of a plowing tractor. African farmers have the knowledge to use these trees as wind breaks, medicine, habitats for biodiverse insect communities, and food for all.

This wealth of knowledge raises another question whether the African continent needs newly manufactured varieties of food crops, or is the problem the lack of scientific recognition and market valuing of what African farmers have cultivated for centuries? Does the color green in this Green Revolution favor crops known and owned by the global North?

Sorghum is one example of a crop lost to markets in the global North but not to Africa. On the continent, it is planted in more hectares than all other food crops combined. As nutritious as maize for carbohydrates, vitamin B6, and food energy, sorghum is more nutritious in protein, ash, pantothenic acid, calcium, copper, iron, phosphorus, isoleucine, and leucine. One of the most versatile foods in the world, sorghum can be boiled like rice, cracked like oats for porridge, baked like wheat into flatbreads, popped like popcorn for snacks, or brewed for nutritious beer.

Although indigenous knowledge designed these diverse and rich uses of sorghum, most contemporary scientists have ignored its genetic wealth. "Sorghum is a relatively undeveloped crop with a truly remarkable array of grain types, plant types, and adaptability," concludes the National Research Council in the United States. "Most of its genetic wealth is so far untapped and even unsorted. Indeed, sorghum probably has more undeveloped genetic potential than any other major food crop in the world."

Engaging African scientists to discover the potential genetic wealth of sorghum would assist African food security. In a first glimpse of foundation expenditures, however, we see funds directed to the Wambugu Consortium (founded by Pioneer Hi-Breed, part of DuPont) for experiments in genetically modified sorghum. By adding a gene, rather than mining the genetic wealth already there, the consortium can patent and sell the "new" variety at a premium price for DuPont.

Toward Sustainability

Given the well-documented destruction of the previous green revolution, what if we decided that Africa's lack of use of fertilizer is a sign of *sustainable* development not of backwardness? Africa's use of chemical fertilizers is extremely low: nine kilograms per hectare in Sub-Sahara Africa, compared to 135 kilograms per hectare in East and Southeast Asia, 100 kilograms in South Asia, and an average of 206 kilograms in industrialized countries. Originating from excess nitrogen production left over after World War II, the massive use of chemical fertilizers defined industrial agriculture in the 20th century. Surely for the 21st century, yields can be increased without such a high cost of African environmental degradation.

The African continent also uses different terminology from that of the green revolution. Instead of food security, African voices articulate the goal of *food sovereignty.* Food sovereignty expresses resistance to the notion that food security can be provided by reliance on global markets, where price and supply vagaries can be as capricious as African weather. Experiencing political manipulation of global markets by the more powerful, African governments seek to control decisions about food sources, considering such choices as vital to national sovereignty.

African governments work to defend local, small-scale farmers from highly subsidized farmers in the United States or Europe. In most of Africa—with South Africa a notable exception—the majority of the population still lives in rural areas and still derives their incomes from farming. Dislocation of farmers to consolidate land for high-tech, green revolution farming is as serious a threat as chemical pollution of the environment.

Should the green wealth of ecological and farming knowledge among local small-scale farmers be destroyed for the cash wealth of much fewer large-scale farmers buying all their inputs from foreign corporations?

Each African government will answer the above questions about a green revolution differently. The diversity of policies matches the diversity of the continent. Yet they all reject patenting of life forms and strive to attain food sovereignty. High-tech answers to Africa's food crises are no answers at all if they pollute the environment with fertilizers and pesticides, destroy small-scale farming, and transform the genetic wealth of the continent into cash profits for a few corporations.

POSTSCRIPT

Does African Agriculture Need a Green Revolution?

The new Green Revolution for Africa, pushed by Kofi Annan and others, is both similar to and different from the Green Revolution of the 1960s and 1970s. It is similar to the extent that it views hunger as largely caused by an underproduction of food, that it emphasizes improved seed varieties and seed packages as a major thrust of the initiative, and that it has the support of some very large donors (the Rockefeller Foundation was/is a key player in both efforts). It is, however, also different in some respects. The more recent effort is not limiting itself to hybrid seeds as its leading technology, but also appears to be open to experimenting with genetically modified (GM) seeds. The key advance of GM seeds over hybridization is that plant breeders are now able to insert genes from other (unrelated) species to affect desirable characteristics in a food crop. Even though GM seeds are purported to have several advantages (e.g., higher yields, lower need for traditional pesticides), they also evoke a host of other concerns. These include unknown human health effects, the escape of genes from GM plants into other crops and plants, the need to continually invent and introduce new GM seeds in response to insect populations that become resistant to old GM seeds, and concern about undue reliance on the companies who develop such seeds. More information on Annan's group, Alliance for a Green Revolution in Africa, may be found at http://www.agra-alliance.org/.

Carol Thompson's emphasis on "food sovereignty" also deserves further explanation. At the Forum for Food Sovereignty in Sélingué, Mali, 27 February 2007, about 500 delegates from more than 80 countries adopted the Declaration of Nyéléni, which says "Food sovereignty is the right of peoples to healthy and culturally appropriate food produced through ecologically sound and sustainable methods, and their right to define their own food and agriculture systems." Food sovereignty is distinct from two closely related terms, *food security* and *food self-sufficiency*. Food security refers to access to enough food at all times for a healthy life (irrespective of the source of the food), whereas food self-sufficiency emphasizes the production of sufficient quantities of food within the borders of a country to feed its population. Food self-sufficiency was an explicit goal of many African nations in the 1970s and 1980s, whereas food security was the dominant parlance of the 1990s. Although food sovereignty is similar to food self-sufficiency in that it tends to emphasize local (or national) production of food, it differs with its stress on ecological sustainability and local participation. Food sovereignty is also quite different from food security in terms of how it views the market. Food security analysts often see the market as an acceptable means for acquiring additional food stuffs, whereas food sovereignty proponents typically view the market with great suspicion. They not only view the global market place as unfair to African producers, but as controlled by large corporate interests. For more information on food sovereignty, see, for example, a 2006 article by Hans Holmén in *The European Journal of Development Research* entitled "Myths about Agriculture, Obstacles to Solving the African Food Crisis," or the official Web site of the 2007 World Forum on Food Sovereignty in Mali (http://www.nyeleni2007.org/?lang=en).

ISSUE 12

Is Community-Based Wildlife Management a Failed Approach?

YES: **Peter J. Balint and Judith Mashinya**, from "The Decline of a Model Community-Based Conservation Project: Governance, Capacity and Devolution in Mahenye, Zimbabwe," *Geoforum* (2006)

NO: **Liz Rihoy, Chaka Chirozva, and Simon Anstey**, from "'People are Not Happy': Crisis, Adaptation, and Resilience in Zimbabwe's CAMPFIRE Programme," in Fred Nelson, ed., *Community Rights, Conservation and Contested Land: The Politics of Natural Resource Governance in Africa* (Earthscan, 2010)

ISSUE SUMMARY

YES: Peter Balint, of George Mason University, and Judith Mashinya, of the University of Maryland, found that the situation in Mahenye, Zimbabwe, has deteriorated significantly since an earlier time period when it was deemed to be a model community-based natural resources management (CBNRM) program. They do not blame this decline on political turmoil in Zimbabwe, but rather on a failure of leadership and the departure of outside agencies responsible for oversight and assistance. As such, they argue against full devolution of authority to the community level for wildlife management.

NO: Liz Rihoy, of the Zeitz Foundation, and Chaka Chirozva and Simon Anstey, of the University of Zimbabwe, arrive at a very different conclusion about the same community in Zimbabwe. Although they acknowledge that CBNRM could be viewed as a failure in Mahenye at certain moments in time, they see the situation as an ongoing process of development in which there are the seeds of opportunity. They claim that the CAMPFIRE program has had real impact in terms of empowering local residents, providing them with incentives, knowledge, and organizational abilities to identify and address their own problems.

A number of national parks were established in African countries during the colonial era. Although these parks were ostensibly established for the preservation of natural resources, they also served to sequester resources for the

European population. In the postcolonial era, national park systems have persisted and been expanded in many instances (particularly in East and southern Africa where relatively larger numbers of charismatic megafauna still reside). African park systems have been bolstered by a global environmental movement as well as a burgeoning ecotourism industry. In countries such as Kenya, Tanzania, Zimbabwe, Botswana, and South Africa, Western tourists flock to see the "Big Five," a term used to refer to the biggest, rarest, or most cherished animals traditionally sought after by trophy hunters (elephant, lion, rhinoceros, leopard, and African or Cape buffalo). Nature tourism has become big business.

African parks suffer from a poaching problem which is symptomatic of a much deeper set of social issues (numbering at least three). First, local people were often evicted without compensation from areas when national parks were established. This led to feelings of resentment and compromised livelihoods in many instances, that is, local people did not have the same resource base to rely on in their new location. Second, feelings of resentment and the inability of many national governments to effectively patrol park borders lead local people to encroach on parks in search of sustenance. Finally, big game animals rarely respect park boundaries and may wander onto the lands of communities abutting national parks. This is problematic because large animals may pose safety risks and destroy field crops. Elephants in particular are known for their ability to inflict a substantial amount of crop damage.

In lieu of the aforementioned problems, a number of conservation programs, often referred to as community-based natural resources management (CBNRM) programs, have been initiated that take into consideration the needs of local people. Part of the incentive for these programs is a genuine concern about the welfare of local people, but there also is recognition that many conservation initiatives, including parks, are doomed to failure unless they enlist the support of local people. A key way of securing such support has been to share ecotourism revenues with local people in exchange for their participation in protecting wildlife resources. Such programs typically work with the communities neighboring national parks and may even involve the establishment of buffer zones, that is, areas surrounding national parks where the activities of local people are restricted.

In this issue, the perceived success or failure of the CBNRM approach is explored. Peter Balint and Judith Mashinya found that the situation in Mahenye, Zimbabwe, has deteriorated significantly since an earlier time period when it was deemed to be a model CBNRM program. They do not blame this decline on political turmoil in Zimbabwe, but rather on a failure of leadership and the departure of outside agencies responsible for oversight and assistance. As such, they argue against full devolution of authority to the community level for wildlife management. In contrast, Liz Rihoy, Chaka Chirozva and Simon Anstey arrive at a very different conclusion about the same community in Zimbabwe. They claim that the CAMPFIRE program has had real impact in terms of empowering local residents, providing them with incentives, knowledge, and organizational abilities to identify and address their own problems.

YES

Peter J. Balint and
Judith Mashinya

The Decline of a Model Community-Based Conservation Project: Governance, Capacity, and Devolution in Mahenye, Zimbabwe

Introduction

In the late 1980s, the government of Zimbabwe instituted a program known as CAMPFIRE (Communal Areas Management Programme for Indigenous Resources) to promote community-based natural resource management (CBNRM) in its rural districts. Mahenye Ward in Chipinge District, in the southeast corner of the country, was an early site for implementation. Both CAMPFIRE and the Mahenye project have been followed closely. CAMPFIRE has been studied repeatedly because it was one of the first examples of national-level CBNRM and has served as a model for similar programs in other countries. Mahenye is of particular interest because it has frequently been cited as a strong CAMPFIRE project with a positive record.

This paper reports the results of a study of the Mahenye project that we conducted from late June to mid-August 2004. Our research focused on two sets of questions. First, we wished to examine whether the project has sustained previously reported gains despite Zimbabwe's severe post-2000 political and economic crisis. Second, we wished to explore the implications of outcomes in Mahenye for the commonly stated argument that a key flaw in CAMPFIRE design and implementation is the lack of full devolution of authority to the local level.

Arguments for devolution in natural resource management are commonly based both on political ecology theory and on empirical evidence from the field. Scholars bringing the perspective of political ecology to the study of CBNRM often build on foundational assumptions favoring the rights of communities to manage their own affairs. Bryant and Jarosz, for example, observe that diverse strands of political ecology share a point of view "that privileges the rights and concerns . . . of the poor over those of powerful political and economic elites." And researchers reporting the results of fieldwork note that domination of local resource management decision making by external authorities can lead both to social injustices and to conservation and development failures.

From *Geoforum*, vol. 37, no. 5, 2006, excerpts pp. 805–815. Copyright © 2006 by Elsevier Science Ltd. Reprinted by permission via Rightslink.

Yet the literature in political ecology also reflects considerable skepticism concerning the idealization of rural communities as harmonious entities that if left alone will naturally promote sustainable ecological and social development. For example, Gray and Moseley . . . note: "Much work is now focused on how local community structures are frequently unaccountable, inequitable and non-participatory. Programmes overlook the fact that village social relations are based on conflict and competition, which, in turn, can lead to negative environmental and equity outcomes." In related findings, empirical evidence from the field suggests that assumptions favoring local resource management regimes over state or multilevel management may be misplaced or oversimplified.

. . . In our fieldwork, we examined whether gains in Mahenye have been maintained despite Zimbabwe's ongoing national crisis and, consequently, whether outcomes in this project continue to support broader calls for devolution in CBNRM. The issues of the sustainability of gains and the appropriate extent of power sharing are linked in that arguments for full devolution of authority are at least partly contingent on the demonstrated capacity of local leaders to manage projects in the community interest despite difficult circumstances. . . .

Background
CBNRM

Human communities in rural areas of the developing world are often poor. Yet in some cases the territories where these communities are located have significant national or global conservation value. In eastern and southern Africa, where conservation efforts often focus on wildlife and habitat, this is particularly likely to be true in communal areas bordering national parks or game reserves. Protected areas generally do not encompass the entire ecosystems of concern, and wild animals rely on corridors through adjacent territories to reach other areas of their range. These protected areas and wildlife corridors are under increasing stress as external tourism and local populations expand and as current residents seek both to recapture or retain customary rights to nearby natural resources and to protect themselves and their property from wildlife predation.

In principle, this convergence of pressing development and conservation needs provides opportunities to integrate socioeconomic and environmental objectives. The idea is that if poor people who live near protected areas can earn significant income from the wildlife and habitat on their lands, local standards of living will improve and conservation threats will abate.

Projects designed to take advantage of these opportunities are often referred to generically as community-based natural resource management, although various related concepts and terms are found in the literature. CBNRM programs typically attempt to foster community development through revenue sharing programs established by park management agencies or more significantly through the promotion of independent ventures on communal lands, typically linked to sport hunting or wildlife viewing and cultural tourism. Conservation gains are expected as local residents then have incentives to limit poaching and maintain wildlife habitat on their territory.

Although straightforward in principle, CBNRM in practice faces a variety of obstacles and complications, and projects implemented in the field have an uneven record of success. Researchers examining individual projects or reviewing overall trends report significant structural challenges relating to historical patterns, current social and ecological conditions, stakeholder relationships, and project design and implementation practices. The problem of integrating conservation and development in rural communities appears to fall into the category of what are known as "wicked" problems, characterized by deep divisions among participants regarding social, economic, and environmental priorities; inherent complexity and uncertainty in predicting social and ecological outcomes; and the absence of optimal solutions.

CAMPFIRE

In 1975, the Rhodesian government devolved rights for management and commercial exploitation of game animals from the state to private landowners, particularly white farmers and ranchers. The new law did not apply to the rural black majority, who generally held land in common without secure land tenure. In 1982, the recently independent government of Zimbabwe amended the law to give the same rights to rural district councils (RDCs), sub-national government institutions with responsibility for development in the previously neglected communal areas. This amendment provided the statutory framework for CAMPFIRE. While communities continued to lack legal standing to institute CBNRM independently, the RDCs could apply for authority to implement projects on their behalf. The first CAMPFIRE projects, which focused on trophy hunting in the Zambezi River valley, were approved in 1988.

CAMPFIRE was one of the earliest national CBNRM programs. It garnered strong positive reviews and served as a model for similar efforts in Zambia, Botswana, Namibia, and elsewhere. Developing an overall judgment of the program is a complex task, however. On one hand, over the first decade of its existence CAMPFIRE absorbed more money in donor funding . . . than it produced from local projects. . . . Moreover, only about half the revenue directly benefited participating communities. The remainder of the income was withheld to fund activities of the RDCs and the CAMPFIRE Association, a national non-governmental organization (NGO) established to support local projects. Consequently, actual cash disbursements to households were modest. . . .

On the other hand, one estimate suggests that since its inception CAMPFIRE's contribution to Zimbabwe's gross domestic product may be on the order of five times greater than the direct revenue from constituent projects. While only a small portion of these gains accrued to the rural poor, the nation at large benefited as donor aid, project revenues, and visitors' additional non-CAMPFIRE spending flowed through the economy. Furthermore, households in communal areas with plentiful trophy animals and successful hunting concessions earned CAMPFIRE dividends significantly higher than the national average. Benefits other than direct income including roadwork, grinding mills, school buildings, and other rural infrastructure improvements related to CAMPFIRE also enhanced community life and provided employment and entrepreneurial opportunities that for some residents multiplied benefits associated with the small direct revenues.

As with CBNRM in general, however, once the initial enthusiasm of the early 1990s waned, scholars and practitioners began to review CAMPFIRE more critically. Some observers questioned the program's ability to meet either its conservation or development objectives. Others highlighted inherent conflicts among stakeholder groups—including local communities, donors, conservation and development NGOs, government agencies, and private sector firms—which often have sharply differing worldviews, interests, and incentives.

An additional concern regarding CAMPFIRE's structure and design is that, despite articulated ideals of devolution, communities do not have full authority for project management. Rural district councils retain the power to make and break contracts with hunting and tourism operators and to siphon off a significant portion of the proceeds through various taxes and levies. Lack of full devolution and continuing interference by the RDCs were the criticisms of CAMPFIRE that we heard most often as we talked to experts in Harare in preparation for our site visit.

History of the Mahenye CAMPFIRE Project

The central government of Zimbabwe authorized Chipinge District's CAMPFIRE program in 1991, making Mahenye one of the first officially recognized sites for implementation. But efforts at community-based natural resource management had been underway in Mahenye since the early 1980s and had served as an early model for development of the CAMPFIRE concept. . . .

From its inception through the worsening of Zimbabwe's political and economic crisis in 2000, the Mahenye project was consistently judged a model CAMPFIRE program both for its diversified sources of income and for its stable, participatory community leadership arrangements. For example, Matanhire evaluated local project management institutions in Mahenye for the first six months of 2001, the latest cycle for which data were available, as having an 89% performance rating, with 75% being the standard for model status.

National Conditions

For the past six years, Zimbabwe has suffered through a debilitating social, political, and economic crisis. The disturbances have a complex history, shaped by links between residual effects of colonial rule that only ended in 1980 and recent struggles over the country's political future. An essential component of the colonial legacy was a distorted pattern of land ownership in which the white minority continued to control most of the arable land. While the question of land reform had been on the agenda since independence, political and economic trends in the mid to late 1990s contributed to making it a primary focus of government policy by 2000.

Robert Mugabe, the leader of the movement in the 1970s to overthrow Ian Smith's white-minority government, became head of state at independence in 1980 and remains in power. . . . While broadly admired in Zimbabwe during the early years of his rule, his popularity fell during the 1990s as the economy weakened, adversely affected by recurring droughts, counterproductive structural adjustment programs, and ineffective governance. By 2000, his standing had declined to the point that in February of that year government-supported changes to the constitution were defeated in a referendum, and in June, despite widespread vote-rigging and intimidation, ZANU-PF lost substantial ground in parliamentary elections to the opposition Movement for Democratic Change (MDC). These unfavorable electoral outcomes contributed to Mugabe's implementation later in 2000 of an accelerated land redistribution program that in many areas degenerated into chaotic and often violent invasions of white-owned farms and ranches, and even of state-owned national parks and other protected areas. . . .

Negative international reactions to the contested election results, uncompensated land seizures, and associated domestic political violence caused extensive further damage to an already fragile economy. . . . In a self-reinforcing cycle, Mugabe's rule has become increasingly authoritarian and conditions in the country have continued to deteriorate.

The economic and political crisis in Zimbabwe has adversely affected CAMPFIRE in various ways. First, around the country project incomes have fallen. Negative international publicity dramatically reduced game viewing tourism, and, while sport hunters are less affected than conventional tourists by reports of political disturbances, hunting revenue has also declined. Bookings for trophy hunts are flat or down somewhat. More damaging, however, is the government's policy of significantly overpricing the Zimbabwe dollar in official exchange rates, thus devaluing CAMPFIRE receipts.

Moreover, in conjunction with its land reform program, the ruling party also worked to consolidate its position in the countryside by recruiting traditional leaders to enforce party discipline in preparation for the parliamentary elections of 2005. During the period of our research in 2004, for example, chiefs around the country, including in Mahenye, received valuable perquisites from the government, such as pickup trucks, boreholes, and electricity connections. This strengthening of the chiefs' authority through the backing of a government willing to use political violence could be expected to threaten the stability and durability of local participatory institutions established for management of CAMPFIRE projects.

Mahenye Case Study

Methods

. . . During an extended stay in Mahenye, we observed CAMPFIRE project activities and interviewed people living or working in the area. Both before and after visiting Mahenye, we interviewed representatives of relevant NGOs, government agencies, and private sector firms based in Harare and Chipinge District. . . .

We emphasize that we explored political and socioeconomic rather than ecological outcomes and acknowledge several additional limitations with our methods. For the most part we were not able to verify independently what our respondents told us about the history of the Mahenye project since 2000. We did not witness the events described, researchers have not published reports of the project's development over the past several years, and the project's records are incomplete and unreliable. In addition, several important respondents had reasons to dissemble, including both current insiders who may have wished to hide problems and former insiders who may have wished to exaggerate them. The impressions of other informants were clearly affected by anger or fear.

To address these potential weaknesses and distortions, we worked to take likely biases into account and to get multiple characterizations of important issues from as wide a variety of respondents as possible. In the end, the unusual unanimity of public opinion in Mahenye gave us confidence that we were capturing accurately both the substance of significant events and the perceptions of community members regarding project performance.

Results

. . . Most striking, we encountered broad and deep agreement among respondents that the project is no longer managed to benefit the community. No one unaffiliated with the leadership expressed satisfaction with current management practices, and even several interviewees with close ties to project leaders were critical of their performance. Regardless of gender, age, or education, and whether responding individually or in groups, local residents complained of bad management, corruption, nepotism, and intimidation. One respondent said, for example, "Let them steal a little. If I had CAMPFIRE money in my pocket and I was thirsty, I'd buy myself a beer, too. But it's not right to take it all."

Furthermore, we found no sense of community solidarity in the face of Zimbabwe's more general problems and the associated pressures on the CAMPFIRE project. Instead, community members not part of project management universally expressed various combinations of resignation, anger, and fear directed at their own local leaders. Reflecting the sentiments of many residents we interviewed, one woman when asked about CAMPFIRE replied, referring to the community leaders and their families, "It's for them, not for us."

In describing particular problems, our respondents repeatedly referred to several significant deviations from desirable governance and management practices. Perhaps most important, beginning in 2000 the democratic process for selecting CAMPFIRE project leaders was abandoned. Up through 2000, committee members and the chair were elected every two years as stipulated in the bylaws. In 2000, however, the chief, who has no formal authority over CAMPFIRE activities, ruled unilaterally that the sitting chairman could no longer serve because he had acquired property outside the village, thus raising questions about his residency. The chief then elevated his own brother, who was deputy committee chairman at the time, to the leadership post. Since then there have been no elections.

Procedures intended to promote community participation and maintain transparency and accountability have been undermined. The required annual general meeting at which the CAMPFIRE committee reports to the community at large was not held as scheduled in 2004. Also, the project's financial records have not been audited since the change in leadership in 2000. When at our request the current chairman showed us the records, we found them in obvious disarray. The most straightforward annual totals for income and expenditures could not be found. The chairman, apparently embarrassed, chastised the bookkeeper in our presence for what was clearly long-standing normal practice. While in the office, we also observed casual disbursements of CAMPFIRE funds. The amounts were relatively small, but there was no accounting.

We asked the current chairman about his accession to committee chairmanship, the lack of elections for the past four years, and the canceling of the annual general meeting for the current year. He acknowledged that our characterization of events was accurate. Regarding the change in leadership in 2000, he repeated that the previous chairman could no longer serve because of a change in residency. Regarding the lack of elections, he told us that the community had decided that stability in the leadership was important. Regarding the canceling of the annual general meeting, he said that villagers did not understand the issues and just liked to complain.

In addition to commenting on the lack of transparency and accountability, our respondents from the general community also reported evidence of misallocation of funds and mismanagement relating to the CAMPFIRE revenue-generating projects. For example, money promised in the past year for school construction and entered in the records as paid to the local school authorities has not been delivered. We observed that one of the two secondary school blocks continues to sit unfinished and deteriorating without a roof in place. We were told that no work has been done on the building since the change of CAMPFIRE committee administration in 2000. Also, the village store, established with CAMPFIRE funds as a cooperative to provide a convenient local outlet for common household goods and to generate funds for community improvements, has been given over to private merchants.

Interviewees in the community at large also expressed profound disillusionment and skepticism regarding the annual disbursements to families from the CAMPFIRE project. In the most recent cycle, for example, each of the approximately 1000 households in Mahenye eligible for benefits was to have received 6100 Zimbabwe dollars (Z$) as its share of CAMPFIRE revenues for the previous year. Before payment and without prior notification, however, clan leaders subordinate to the chief deducted Z$6000 from each family's payment to cover a community development tax. The RDC delegates this tax collecting authority to the local kraal heads, who are allowed to keep 10% of the funds they collect. Although this levy is unrelated to CAMPFIRE, villagers saw the manner and timing of its collection as one more means by which the traditional local leadership expropriates community CAMPFIRE benefits. In March 2004, after this tax was deducted, each household in Mahenye Ward received a payment of Z$100 in return for its participation in CAMPFIRE for 2003. As one respondent commented, this was not even enough for candy for the children, and it certainly was not sufficient to compensate families for the costs of living with wildlife.

Moreover, there is evidence that this amount did not reasonably account for project earnings. For example, while at the time of our research the project's records were poorly maintained, we did see documentation of a recent payment from the hotel firm. After withholdings to cover levies for the RDC and the CAMPFIRE Association, the committee in early 2004 received Z$28.9 million from the firm as the community's share of 2003 lodge receipts. This amount does not include project revenue from hunting safaris or from other sources, such as the operation of the grinding mill or the sale of elephant hides. Historically in Mahenye about half the income from CAMPFIRE is used for committee expenses and contributions to general infrastructure improvements in the village, and the other half is distributed directly to households. Yet following receipt of the payment from the lodges, the committee failed to distribute funds promised for school construction, and annual disbursements to households totaled only about Z$6 million, well below half of even this partial contribution to project receipts for the year. While this accounting of revenues and expenditures is far from comprehensive, it adds credibility to our respondents' skepticism regarding the CAMPFIRE committee's management practices.

Local leaders also monopolized equipment and employment opportunities meant to benefit the community. A pickup truck donated by the professional hunter to be used as an ambulance or for other local emergency services was co-opted for personal use by committee members. In the course of one private trip, the vehicle had been extensively damaged and as a consequence was not in working order at the time of our research. In addition, members of the chief's family filled jobs allocated for locals at the tourist lodges. Indeed, the chief's brother, the CAMPFIRE committee chairman, was himself on salary there as community tourism officer.

Moreover, at the time of our research, the CAMPFIRE committee, without authorization from the community or the RDC and without support from the CAMPFIRE Association or other outside agencies, was negotiating a new contract with lodge managers. Under the proposed arrangement, the community would receive about 15% of the lodges' profits rather than 12% of gross revenues, as currently stipulated. Because of the decline in tourism in Zimbabwe, however, the lodges are unlikely to be profitable for some time, and as a consequence, the proposed new arrangement would clearly benefit the firm at the community's expense. Yet, in these negotiations the asymmetry of business expertise unfairly favored the firm over the community, and the employer/employee relationship between the lodge management and the CAMPFIRE committee chair created a conflict of interest.

Perhaps most disturbing, we heard allegations of intimidation during several credible, independent interviews. One respondent told us, referring to the current CAMPFIRE committee chairman, "He's my uncle, but I'll still tell you he's a bad man." The respondent went on to say, speaking of the community leaders, "If you speak against them, the sun will not set on you." We were also told that because the chief retains his traditional authority to determine land-use patterns in the communal area, those who might criticize the current CAMPFIRE leadership are silent for fear of losing rights to the plots on which they depend for subsistence livelihoods.

It might be reasonable to suppose that Zimbabwe's national political unrest is largely responsible for the collapse of participatory project management processes in Mahenye. Yet this is an isolated community with little history of violence linked to the national struggle for power. Even during the 2000 parliamentary election cycle, Mahenye was spared the attacks by war veterans and youth militia loyal to ZANU-PF that occurred in many parts of the country. . . . Indeed, none of our respondents reported any interference or intimidation from factions linked to ZANU-PF, and we observed no activity by either the ruling party or the opposition MDC.

We have focused on issues of governance in Mahenye to this point, but some of our respondents also reported significant problems with management and oversight at other levels as well. Community leaders, for example, expressed frustration with the Chipinge RDC. They told us that the current professional hunter secured the Mahenye hunting concession through RDC favoritism and that his performance has been unsatisfactory. They noted that in the previous year the hunter led safaris taking four elephants. They reported that he did not take the full quota of seven animals, and thus generate maximum revenue for the community, because he was overextended with other concessions elsewhere in the country. They also complained that he has often been unavailable when the community needs him to deal with problem animals that threaten village residents and their property. Community leaders further asserted that under current arrangements between the hunter and the RDC they have no representative present when trophies are weighed and their values assessed.

In response to criticism of RDC management of the hunting concession, a Chipinge district councilor explained that the council selected the present hunter because the previous hunter is also a partner in the Mahenye Lodge management firm. The councilor told us that RDC officials felt this gave the previous hunter too much involvement in the affairs of the local CAMPFIRE project and led to conflicts of interest.

Thus, relations between the RDC and the Mahenye CAMPFIRE committee remain adversarial rather than cooperative, continuing a pattern dating back to the early 1980s when CBNRM was first implemented in Mahenye. This is not unexpected since the two institutions have differing incentives. Nevertheless, several residents of Mahenye that we spoke to, particularly the primary school teachers, spoke against the idea of full devolution of authority from the RDC to the community. While acknowledging that the RDC is not an ideal custodian of the community's interests, these respondents maintained that under current local conditions, characterized by mismanagement, misallocation of funds, and intimidation, the district council serves as an essential check on the power of local leaders.

The NGOs responsible for guidance and oversight are part of a third layer of project management subject to criticism. At the beginning of the CAMPFIRE program, the CAMPFIRE Association was formed to provide administrative support to the communities and the RDCs, which lacked experience in CBNRM. Other national and international NGOs, which had acted as implementing agencies when donor money flowed in to underwrite CAMPFIRE development, also had capacity building responsibilities. Yet, during the time of our research, our respondents reported a complete absence of external support for the Mahenye CAMPFIRE project. Moreover, NGO staff members whom we interviewed in Harare were unaware of the project's collapse. As mentioned earlier, their primary concern and recommendation was a change in the law to allow full devolution of authority to the community.

Discussion

Our study reveals the decline of a promising CBNRM program. We found that the Mahenye community no longer receives the flow of significant social and economic benefits reported in earlier studies. In this section we summarize the problems, consider reasons for the deterioration in outcomes, and discuss the implications of our findings for devolution in community-based natural resource management.

The central failure of the Mahenye CAMPFIRE project is that participatory decision-making processes have broken down. Following the undemocratic takeover of the committee in 2000 by the chief's immediate family, there have been no elections and no outside audits of receipts and expenditures. Moreover, progress on school construction and other community infrastructure improvements has stalled, and households have received only insignificant annual disbursements that fall well below the 50% share of project revenue that they received before the change in administration. Given the sharp decline of benefits and the lack of transparency and accountability, community members no longer trust the CAMPFIRE leadership or feel any sense of ownership in the project. Residents have seen the ruling clan fill jobs at the lodges, use project vehicles for personal purposes, and privatize the community's cooperative store. These overt violations of CAMPFIRE principles fuel what our findings suggest are residents' legitimate suspicions regarding the management of project revenues and the motives of the leadership.

Zimbabwe's broader crisis has contributed to these adverse outcomes. We found evidence of significant negative impacts from the national turmoil, including lower revenues from the lodges, challenging problems of financial management in a time of hyperinflation, and both reduced services and increased pressure for higher shares of project income from the RDC. Yet we also found that these national political and economic disruptions are not sufficient to explain the full extent of the collapse of Mahenye's CAMPFIRE project. Our research suggests that there were opportunities for resilience and survival.

First, on the political front, none of our respondents, whether members of the CAMPFIRE leadership, other community residents, or outsiders, suggested that Mahenye had been directly affected by violence and intimidation linked to national politics. . . .

Second, on the economic front, the project's two primary sources of income, hunting revenue and lodge receipts, are both to some degree buffered against external shocks. As mentioned, trophy hunters are less likely than other types of tourists to avoid countries with political troubles. . . . Also, while game-viewing tourism in Zimbabwe is down sharply, the project still received substantial revenue from the lodges because language in their current contract guaranteed payment of a percentage of gross receipts rather than a percentage of now non-existent profits.

Thus, while we observed damaging effects from the national crisis, our findings suggest that local failures in governance and capacity contributed significantly to the decline in community benefits and the near universal distrust and disillusionment voiced by our respondents. Indeed, it appears that problems noted more generally in community-based programs may ultimately have undermined outcomes in Mahenye as well. For example, as mentioned in the introduction, the tendency of local elites to expropriate benefits and the instability of local participatory processes linked to community projects that our respondents in Mahenye described have both been reported elsewhere. In addition, our results indicate that in Mahenye, as noted in other projects in times of economic stress, the incentives of private firms—the professional hunter and the lodge operator in the Mahenye case—came to conflict with community development goals. We also found that the Mahenye project experienced the pattern reported in other cases that when donor funding comes to an end, as it did nationally for CAMPFIRE in 2000, essential outside support for local projects drops off, thereby undermining success.

In Mahenye, following the withdrawal of NGOs and government agencies responsible for oversight and capacity building, the traditional community leaders usurped power from the elected CAMPFIRE committee and then co-opted benefits and otherwise mismanaged project activities. These outcomes highlight both the importance and the fragility of good governance and adequate capacity in CBNRM. Our findings thus add weight to arguments for caution in promoting full devolution of authority, particularly in the absence of safeguards to protect the broader community interest.

Liz Rihoy, Chaka Chirozva, and Simon Anstey

"People Are Not Happy": Crisis, Adaptation, and Resilience in Zimbabwe's CAMPFIRE Programme

Introduction

In the early 1990s Mahenye Ward, located in southeast Zimbabwe, was a leading local reference point for the widely heralded CAMPFIRE Programme (Communal Areas Management Programme for Indigenous Resources), which was in turn a leading influence on wider experimentation with community-based natural resource management (CBNRM) across southern Africa. Regional and international analyses of CAMPFIRE held Mahenye up as a leading functional example of the programme's aspirations to forge new links between local democracy, rural development and wildlife conservation.

Key factors in the relative success of local people in Mahenye to sustainably manage and derive benefits from their natural resources included 'the insights, ingenuity and commitment of socially dedicated individuals in positions of influence or leadership . . . which has been balanced in its sources of traditional and popular legitimation;' an 'enlightened private sector;' a capacity for flexibility and acceptance of innovation; and particularly local intra-communal cohesiveness. . . .

More recently, and again both reflecting and informing changed national and regional discourse around CBNRM, the narrative emerging from Mahenye has shifted to one of crisis and collapse and a questioning of the merits of devolving rights over natural resources to the local level. Scholars report how local élites have undermined the formerly flourishing CAMPFIRE system and formerly democratic local institutions.

The narratives and counter-narratives about Mahenye and its CBNRM initiative matter not only because of the centrality of natural resources to the people of the Ward, to their survival and their future livelihoods. They also matter because, as in the 1990s, Mahenye has an impact and reach far beyond a peripheral zone of Zimbabwe. The recent narrative of crisis in CAMPFIRE in Mahenye Ward, in questioning the merits of devolution of natural resource governance, in local élite capture of benefits and decision-making, in the strivings for participatory democracy, and in resilience and adaptability all have their reflections and relevance at other scales. These range from academic or policy debates on the 'crisis' in CBNRM in southern Africa, on the contested evolutions of democracy in Zimbabwe or the region and on natural resource management and human livelihoods more widely. . . .

The National Context

Zimbabwe has undergone significant and far-reaching political, economic and social upheavals since the mid-1980s when CAMPFIRE was first introduced, and since 2000 has descended into a state of protracted crisis. Its relatively strong economy has been reduced to the weakest in the region. Once reasonably stable political conditions are now characterized by civil unrest and political repression and a previously well-functioning bureaucracy is in tatters. Respect for basic democratic principles, the rule of law and human rights are limited in their observation. Zimbabwe, once a darling of the international donor community, has become a pariah and exhibits many of the attributes of 'disorder as a political instrument' in which political actors and élites seek to maximize their returns from conditions of confusion and uncertainty. This decline has had significant impacts on many different elements of the CAMPFIRE programme, including the process of policy-making, the economic benefits available from wildlife and tourism, donor or private investment, governance arrangements and implementation capacities of both NGOs and government agencies.

Economic Conditions

The negative macro-economic and political environment in the post-2000 period presents major challenges for local communities to generate revenue from wildlife. . . .

The political and economic turmoil has led to the collapse of the tourism sector. Nemarundwe highlights the negative impacts of this economic climate on CAMPFIRE, compromising not only its income-generating potential through tourism but also undermining community investment projects. Inflationary changes in prices make a mockery of budgeting, erode financial benefits and value, and given the cycle in which payments of household cash dividends from CAMPFIRE revenue activities takes place six months to a year after activities have occurred, the loses to inflation of cash benefits are massive. Finally, in the absence of many other income or taxable options, the current situation is further increasing the RDCs' dependence on CAMPFIRE wildlife revenue for survival, presenting a disincentive for fiscal or other devolution.

Political Climate

The extreme economic and political problems that now face Zimbabwe can best be analysed and understood in the context of its history. Zimbabwe emerged from almost a century of white rule, following a long and violent liberation war that ended in 1980, fought largely over land. Since 1980, the political priorities of the government have been dominated by reversing decades of racially-biased inequalities in land, resource and asset distribution. As the ruling party slogan 'the land is the economy, the economy is the land' implies, struggles over land have been at centre stage throughout the colonial and post-colonial period. This struggle over land and natural resources is central to understanding the political dimensions of natural resource management in Zimbabwe, and explaining why it receives such a high degree of political prominence. . . .

In broad political terms Zimbabwe can no longer be described as an ordered political polity in which political opportunities and resources are formally defined and codified by legislation or precedent. Whereas in the 1980s Zimbabwe had a relatively well-functioning bureaucracy, at present informal political relationships have come to play a much greater role in policy formulation and implementation. Powerful ruling party politicians have assumed leading roles within the wildlife management industry in Zimbabwe and overt political influence on government decision-making is now prevalent.

Civil Society

Throughout the 1980s and 1990s, Zimbabwe witnessed the growth of a strong and vibrant civil society. NGOs received generous support from donors and effectively collaborated with many government programmes. CAMPFIRE exemplified this, with the CAMPFIRE Collaborative Group (CCG), a joint facilitating structure of both government agencies, NGOs and academic institutions, playing a key role in implementation until 2000. However, the shift in the political landscape of Zimbabwe immediately prior to 2000 resulted in major opposition by civil society organizations to a government-led constitutional amendment referendum. From 1999, some segments of civil society began to challenge the government on land, electoral and human rights issues. This challenge was treated as a sign of political defiance warranting the repression of NGOs, and the government introduced the 2005 NGO Bill which considerably curtailed NGO functions and independence. This volatile political climate translated into a difficult operational environment for civil society, particularly in any area of governance or involvement in rural development.

The impact of this marginalization of civil society on CAMPFIRE has been profound. Members of the CCG formerly played a key role in capacity building at grass-roots level. Members of the CCG also fulfilled a critical role as third-party brokers providing neutral arbitration in instances where community-level polarization stalled progress in programme implementation. As of 2003, because of the political backlash against civil society, NGOs have been prevented from playing any significant role in implementing CAMPFIRE. Compounding this operational marginalization has been the loss of access to funding that has been experienced by NGOs throughout Zimbabwe as a result of donor withdrawal arising from the political situation. . . .

CAMPFIRE Evolutions in Mahenye, 2000–2005
Institutions, Management and Local Governance

Since 2000 there have been significant shifts of power within and between different actors and institutions in Mahenye, as well as the major shifts in macro-economic and national political context that have occurred in Zimbabwe as a whole. One outcome of these shifts has been the dramatic demise of CAMPFIRE in the view of the overwhelming majority of local inhabitants interviewed, and summed up as follows by one woman: 'CAMPFIRE used to be for all the people, now it's a family business.'

The demise of CAMPFIRE in Mahenye, its core local institution (the MCC) and dramatic falls in the value of household dividends coincide with, and have been strongly influenced by, four related local events:

1. the death of the highly respected old Chief Mahenye in 2001 and replacement by his son, who is the current Chief;
2. on the explicit instructions of the new Chief, the complete change in MCC office bearers following the flawed MCC elections of 2001, including the direct appointment (not election) of the Chief's younger brother as Chairman;
3. the election of a new Councillor for the Ward;
4. the re-tendering of the sport hunting concession which has led to ongoing conflict and the widespread belief among most local stakeholders that the operators are currently un-transparently bidding for the concession and are competing amongst each other in their attempts to illicitly 'buy off' the Chief and MCC to ensure preferential treatment.

These changes have effectively removed the strong local leadership whose commitment and accountability were formerly such a distinctive feature of Mahenye. These included the Chief, headmen and respected elders, the school headmaster and other teachers and an elected leadership including the Ward Councillor and members of the MCC. Collectively these provided a leadership structure that was balanced in its sources of traditional and popular legitimacy.

Local power and authority have shifted away from the delicate balance established between traditional and elected democratic institutions and the leadership of these structures, and concentrated into the hands of a core local élite concentrated within the traditional leadership. 'Honest brokers' in local dynamics, whether of the private sector, NGO, state, RDC or other have become rare, ineffectual or sidelined. As many people in Mahenye said, the result is that they now have their own 'dictator.' An important point in the following discussion is the premise that it is not the institution . . . of either the MCC or customary authority that is the root source of these governance problems, but the *distortion* of the rules governing both by particular forces since 2000 that have permitted élite capture and perpetuated stalemate, contrary to the past existing delegation and accountability mechanisms. . . .

Shrinking Incomes and Incentives

The earnings in Mahenye from CAMPFIRE declined dramatically from 2000 to 2005, as a result of both local misappropriation and leakages arising from national economic distortions. These leakages primarily result from:

- the loss in value occurring when converting foreign exchange to the massively over-valued Zimbabwe dollar;
- the loss in value resulting from annual inflation rates as high as 650 per cent to over 800 per cent (as of 2004–2005) when revenues remain stored in bank accounts for periods of six months to up to a year before household dividend payments are made.

Throughout the 1990s annual allocations to household dividends were consistently around 50 per cent of total budgets in Mahenye. Since 2001 there has only been one allocation for household dividends. . . . As a proportion of the overall stated revenues, this sum of 'actual cash in hand' dividend represented less than 1 per cent (0.2 per cent) compared to the 50 per cent averages in the 1990s. . . .

The simple facts are that the households in Mahenye are getting no meaningful economic dividends from CAMPFIRE, in stark contrast to the 1990s. The outcome of this situation is that there is no longer any independent local body that represents the interests of the people or to which the grievances of the people can be aired. All discussions and decisions now take place at the Chief's *Dare* (assembly meeting). This is the context of changed local governance and economic incentives against which the following section of local narratives are set.

Local Narratives and Perceptions

Vanhu varwadziwa, havana kwavanochemera

(People are not happy, but they don't know where to complain.)

Given the competing interests at stake it is perhaps not surprising that the narratives surrounding CAMPFIRE in Mahenye differ amongst the various stakeholders and that different scenarios for change are identified by these groups. In very broad terms the stakeholder groups can be identified as follows:

- the traditional leadership and current MCC members;
- general community members;
- external stakeholders such as the RDCs and NGOs such as the CAMPFIRE Association.

However, as the following discussion indicates, this simplistic breakdown of disparate actors hides an overlapping and constantly shifting array of perceptions, alliances and networks. . . .

Traditional Leadership

The traditional leadership in Mahenye consists of the Chief, two headmen and 29 kraal heads. Given the thorough co-optation of the MCC by the Chief and his immediate family—in 2005 every member of the 12-person MCC was a relative of the Chief—we combine the traditional leadership and the MCC here as falling within the same stakeholder group, even though there are very clear fault lines developing amongst various individuals and sub-groups. Despite this close association of the Chief with the programme, he claims to have no direct relationship with it, although he is outspoken in his support, noting that:

> CAMPFIRE has been here a long time and brought many good things but it needs changes. The main problem is that money from hunting goes to the RDC first, it should come directly to Mahenye; also the RDC want to interfere in who we select as our hunter.

The narrative constructed by both the Chief and the MCC Chairman is one of a successful CAMPFIRE programme that has brought development to Mahenye, whilst protecting the natural resource base and upholding local culture and traditions. They identify some problems with the programme but consider that these are brought about by external agents and technical deficiencies with the implementation process, what they portray as the greed and inefficiency of the current safari operator, coupled with the unwillingness of the RDC to commit to fiscal devolution and local-level decision-making regarding the selection of safari operators.

However, with the exception of these two individuals, the other members of the traditional leadership and MCC interviewed presented a different story by identifying failures in leadership, financial management and governance—including detailing several instances of abuse and misuse of funds and MCC assets by the Chairman—coupled with the technical and administrative problems identified by the Chief and Chair as being the most significant impediment to the programme. . . .

General Population

The story told by people in the general community . . . had at its centre disappointment and disillusionment with the current situation, but also a sense that events were still unfolding and that they collectively had at their disposal means to address the current problems. This group unanimously identified poor leadership, governance issues and the misappropriation of power by the MCC as the root cause of their problems but there was also considerable concern and confusion articulated about the private sector tourism operations, the role of NGOs and the role of the RDC.

CAMPFIRE was described as a source of local pride and confidence as well as development for over 10 years. It was considered to have been a genuinely representative process about which the majority of ward residents had considerable information concerning the nature and extent of their rights and technical details relating to wildlife management, and in which they enthusiastically participated and benefited. People articulated trust in and respect for their leaders during that time, who they credited with having brought about this success. Specifically mentioned on many occasions were the (former) Chief, (former) Councillor, (former) MCC members, the private sector partner, as well as NGOs formerly active in the area.

There is universal agreement over the cause of the problems that subsequently emerged:

> Our troubles started when the old Chief passed and . . . [the former MCC Chairman] and the others were pushed out of the committee and . . . was made Chairman for life.

There was also widespread acknowledgement that there are constraints to what they can do about this because 'people fear to challenge the Chairman, this is challenging the Chief and would result in losing land or even being chased from the area.' . . . There is also a common view that 'the RDC has more power, they should do something.' . . .

Thus there is a remarkable level of agreement on the basic situation and the way to resolve it amongst the majority of those in Mahenye. . . .

This strategy involves appeals to the RDC, as the only institution with the authority, legitimacy and mandate, to intervene and assist in the restoration of local structures that are accountable and representative of the community. Thus the collective local demand is for the RDC to accept its responsibilities as the agency granted Appropriate Authority (AA) for wildlife in the district and act accordingly to ensure that the CAMPFIRE 'Constitution' (the bylaws of the MCC) and democratic local institutions (the MCC under the rules of the bylaws) are in place. . . .

The most striking clement of the local community's narrative is the level of agreement on the nature of the problem and how it can potentially be solved through RDC intervention to restore earlier local democratic institutions. Despite considerable problems (and dangers), the people of Mahenye continue to demonstrate the remarkable level of 'intra-communal cohesiveness' and capacity for expressing 'constituency demands' identified in the past.

The Rural District Council (RDC)

The role of the RDC includes formal awarding of the hunting concession following an established process of advertising and competitive tendering. As well as having a legal obligation in this regard, they also have a financial incentive to ensure that the process is efficiently managed as they are recipients of 20–35 per cent of income as an administrative fee or tax. . . .

Following a written request from the Mahenye Ward Councillor, backed up by anonymous letters from Mahenye residents, the RDC undertook an independent audit of the MCC in 2004. This audit clearly revealed the validity of accusations of mismanagement and misappropriation of CAMPFIRE funds by the élite within Mahenye.

According to the RDC Chief Executive Officer (CEO) the situation in Mahenye is thus 'a big mess' which has largely occurred because 'one individual is no longer accountable' which is bringing the RDC into disrepute. . . .

The RDC's chosen strategy has been to analyse what they see as the two elements of the problem: lack of accountability, and conflicts between the broader community and the safari operator. . . .

The story according to the RDC is that they are aware of problems and are in the process of making a measured and responsible determination of how to proceed, which will respond to the demands and needs of their constituency. Given such a reasonable response it is fair to speculate why action has been so slow in forthcoming. The audit—which clearly illustrates fraud and corruption—was carried out in August 2004, whilst the Commission of Inquiry took place in May 2005. And yet by October 2005, despite the CEO acknowledging that it was a priority for the RDC, no action had been taken. This may simply be a result of bureaucratic ineptitude, but once again it is possible to identify alternative reasons.

Chief Mahenye's position provides him with networks linked to politically powerful national factions that may have an influence on the strategies adopted by the RDC. . . . These personal national networks and political affiliations provide an additional level of complexity in local power struggles which impact on the balance of power between the RDC and traditional authorities, and this may at least partially account for the reluctance of the RDC to take any decisive action.

Non-Governmental Organizations (NGOs)

The marginalization of civil society from policy-making and implementation in Zimbabwe's politically contested rural areas has had significant impacts on CAMPFIRE. The consequence of this marginalization is that those former CAMPFIRE Collaborative Group (CCG) members (particularly NGOs such as WWF and Zimbabwe Trust) who formerly played key roles in institutional development within Mahenye are no longer able to do so.

Some scholars have criticized NGOs for this, but this glosses over the reality that Zimbabwe's national political context since 2000 has served to marginalize and exclude those NGOs from the local governance arena. This has occurred by denying NGOs access to funds but also by removing their mandate. NGOs formerly active in Mahenye have been aware of the problems there but have no means or resources with which to address the problem, and also felt intimidated to try to do so. . . .

The one NGO that is still highly active in CAMPFIRE implementation is the CAMPFIRE Association (CA). They are familiar with the current situation in Mahenye and are involved with the RDC in seeking a solution to the problems based on their understanding that:

> There are a lot of undeclared interests at play in Mahenye. There's a need to identify the root cause of the problem and sort the institutional problems. We strongly felt as a commission there was used for changes in tenure of office, to elect a new committee.

As in the case of the RDC narrative, there is also a sense of some deadlock in taking actions or decisions in the discourse of the CA; particularly given this is precisely the institution taxed (literally, given that the CA membership fees are deducted from Mahenye revenue) with the task of linking the producer communities of CAMPFIRE with district and national agencies and with the overall coordination of the programme.

Discussion

In discussing contemporary CAMPFIRE and natural resource governance evolutions in Mahenye, a good place to begin is to recognize the complexity of the current situation both in Zimbabwe and in Mahenye, but also the extent to which there is remarkable congruence and depth in the narratives of local, district and national scales about existing challenges and the most urgent next steps to take. At the crux of these stories is a multi-tiered and interrelated set of politically and socially constructed stalemates inhibiting those steps from being taken and governance problems being addressed. As noted in the previous section by one interviewee: 'people are not happy, but they don't know where to complain.'

Local Governance, CBNRM Institutions and Historical Precedent

One of the paradoxes and strengths of the case of Mahenye is the degree of adaptation and cross-scale linkages that characterize local governance dynamics over the course of the past two decades. Mahenye had, by the mid- to late 1990s, developed a complex set of multi-tiered natural resource governance linkages involving upward delegation and downward accountability depending on political agency and ecological and social scale requirements. It had in that decade moved beyond the 'chicken and egg' structural dilemma of full devolution as prerequisite for CBNRM versus fragile local common property regimes as a cause of failure of CBNRM. The egg had produced the chicken and chicken produced the egg in a context, as described earlier, of happy congruence where the strengths of the local society . . . was linked to higher scale organizations of the state, private sector and NGOs, with powerful economic incentives and political capital supporting these evolutions. The challenge was to come from 2000 with the series of connected local and national events which generated the dramatic distortions to economic incentives, political dynamics and local leadership. The informal and precedent basis of the Mahenye 'constitution' was inadequate to counterbalance these profound changes. In simple terms, the devolution-jurisdictional egg was hatching out in a much rougher neighbourhood.

But it is important to stress, as do the majority of the local narratives from Mahenye, that this does not preclude local ability to react or adapt. The fact that the precedent of tackling significant challenges from 1982 to 1991 existed, and the widely agreed strengths of the institution then until 2000 were established. provides hope that the scenarios and strategies for change envisaged by most Mahenye people can engage with contemporary crises.

National to Local Links, Mirrors and Influences

Whilst the past decade's problems in CAMPFIRE in Mahenye do indeed reflect 'local failures in governance and capacity,' those changes in local governance are fundamentally shaped by developments at the national scale. The situation in Zimbabwe, where political trends since 2000 have resulted in the promotion of those institutions and individuals associated with the ruling party, whilst those affiliated in any way with opposition parties and polities have been marginalized, has been comprehensively documented by many analysts, both Zimbabwean and foreign. The Mahenye situation in this regard mirrors that of the nation; the impact has been profound in determining the balance of power between various local actors in Mahenye, as well as determining which individuals continue to play active roles within institutions, based upon their political affiliations.

One of the most significant legislative changes promoting shifts in the institutional dynamics and balance of power within Mahenye has been the Traditional Leaders Act (TLA) of 2001, which has strengthened the power of traditional authorities nationally whilst also bringing them under the influence of the ruling party, ZANU-PF. Until the passing of this Act, policy since independence had strengthened the role of elected RDCs at the expense of traditional authorities. The TLA is a significant shift in direction, empowering traditional leaders not least in terms of natural resource management. A widespread interpretation of the TLA is that it aims to co-opt traditional leadership to ensure political penetration of the ruling party into rural areas. This Act has not only enhanced the authority of chiefs locally but has also changed the nature of the relationship between chiefs, the RDC and the private sector. . . .

But changes in the national context have not been limited to legislative or administrative changes. The year 2000 saw a dramatic and public shift in the political dynamics of Zimbabwe, culminating in an increase in politically motivated violence and in the collapse of the rule of law. This situation was underlain by a racial and populist moral discourse about the return of 'African soil' to Africans adopted by the ruling party, which served to marginalize and vilify whites and, by inference, political opponents of the ruling party. At local levels this often translated into the violent persecution and marginalization of MDC supporters and introduced greater suspicion of wildlife management as it was considered to be 'a ploy of whites to forestall land acquisition and justifying multiple and extensive land holdings.' Many of those interviewed noted that the impact in Mahenye has been to marginalize key figures who were known to be opposition supporters and a further reinforcement of the powers of the Chief.

Thus the relationship between the traditional and ruling party institutions has fundamentally changed, with the result that the power and influence of traditional authorities has been enhanced but at the expense of increased dependency on the ruling party. In Mahenye this has allowed for the creation of one institution within which all power is vested: that of the Chief. . . . This has occurred because of the mutually beneficial relationship and endorsement from ZANU-PF and the other newly created or co-opted institutions such as the MCC under the current Chairman. The new roles acquired by the Chief and his family translate into real power over and above that traditionally extended to them.

The national context has enabled the Chief to translate his newly enhanced legal position as regards natural resources and his new position as powerful ZANU-PF representative to divert the claims of others and validate his own claims over these resources, thus expanding his control over development in Mahenye. One of the first actions undertaken by the Chief on his ascendancy in 2001 was to ensure that CAMPFIRE and its benefits were brought under his control.

Networks, Patronage and Power

By effectively capturing CAMPFIRE operations in Mahenye, the traditional authorities have essentially created a powerful patronage tool for themselves through which they can construct and reproduce power relationships and perpetuate their authority. CAMPFIRE provides the means by which to develop a strong network of loyal supporters. . . . The Chairman and Chief have ensured that these people are beholden to them. By extending participation in certain key meetings to include all members of the traditional authorities and other party-endorsed positions, this network has been extended further to all those in positions of authority in the village. . . . By consolidating their positions of power in other institutions outside the MCC, the Chairman and Chief can threaten retribution to any who question their decisions, not just in the form of losing the benefits that have been forthcoming from being part of their network but also through the potential loss of access to food aid, land or being labelled an opposition supporter. This last threat can also be extended to private sector operators and the RDC through the manipulation of national political networks.

Thus the Chief and Chairman would appear to have built themselves an unassailable position of power and authority. Yet this is clearly not the case. There is unanimous condemnation of the Chairman—although many, particularly the traditional authorities, were careful to draw a distinction between the Chief and the Chairman—and on the need to find a solution to the current problems, even though such a solution would probably lead to some people losing privileged positions as network beneficiaries. However, whilst there is unanimous discontent, the situation within Mahenye is effectively a socially and politically constructed stalemate with no local means of sufficient agency or power to break the deadlock. Therefore, people have identified alternative mechanisms to assist them to solve their problems. The long and successful history of CAMPFIRE in the area has ensured that there is considerable local knowledge about the process, including a widespread understanding of the roles and responsibilities of various governance bodies. Thus whilst the RDC is widely distrusted on the grounds that it has its own agenda in relation to the safari operations and securing its own revenue, there is nevertheless clear recognition within Mahenye that it has a legal responsibility to step in to break the local stalemate and the—albeit so far latent—political agency and state-party linkages to do so.

It is generally recognized locally that improving the existing situation involves two different but interconnected activities. First, addressing issues of local governance and second, addressing fiscal accountability and use of revenue. Only once these issues have been resolved do the majority of people in Mahenye want to see greater fiscal devolution occurring. That is to say, their scenario for change is a sequence of events in rebuilding a process of devolution based on local responsibilities and authority but also with strategic linkages and practical politics to get there.

Building Accountability from Local to District Scale

The situation in Mahenye suggests that one of the most significant impacts of CAMPFIRE over the last 20 years has been to empower local people by making them aware of the value of the natural resources in their areas and their (albeit restricted) rights to these, whilst raising awareness of mechanisms through which they can exercise those rights.

Mahenye illustrates that community members can have the knowledge, confidence and organizational awareness to counter local élites who are usurping power and undermining democratic decision-making, and to articulate demands to their political representatives at the district level to assist in resolving the problem. Thus despite the fact that local political mobilization has had to be largely covert in recent years due to fear of reprisal, it has nevertheless created space for political negotiation between the local and district level and catalysed two external and damning investigations. This could ultimately lead to greater accountability of the RDCs to their local constituents. Allied with a strategy of practical politics in a win–win approach to the revenue and economic incentives for the residents of the Ward, the RDC and the private sector, the potential for breaking the current stalemate certainly exists.

Conclusions

The foremost lesson from the experiences of Mahenye is that CBNRM is a process of applied and incremental experiments in local democracy and most valuable in this because it involves not a single idealized state of full devolution but the interaction of tiers of governance over time in adaptive processes. What could be construed as a 'failure' or 'crisis' at any one moment is in reality part of an ongoing process of development which, in this case, contains the seeds of opportunity through which rural people can develop organizational mechanisms and abilities to voice their demands. The analysis, as drawn out in the narratives presented here, demonstrates that CAMPFIRE has had a real impact in terms of empowering local residents, providing them with incentives, knowledge and organizational abilities to identify and address their own problems, recognize the constraints that they are operating within and identify where external interventions are required.

It is apparent that alliances and boundaries are formed throughout these processes, and when situations change these alliances and boundaries shift and reconfigure the landscape of governance and politics of natural resource management. The situation currently facing Mahenye is, in the stories of the residents themselves, just a snapshot of a moment in time. Their eye is on the future and how to effect an outcome that is favourable to all people in Mahenye, not just temporarily powerful local élite. Thus what an observer may view as a crisis is viewed by many local inhabitants as part of an ongoing contest for control over resources within which lie opportunities for positive change. This has been readily apparent in other CAMPFIRE locales during the past decade, notably the community of Masoka which used the crisis brought about by unprecedented RDC appropriation of revenues in 2004 to force a process of renegotiation which led to record benefits being realized by the community only two years later.

The situation in Mahenye illustrates the centrality of political dynamics at multiple scales to natural resource governance outcomes. A core concern of CBNRM is therefore working towards the recognition and translation of political capital into a political tool for mobilizing power and bringing local demands to bear on relevant authorities in order to support communities to capture and enlarge the political and policy spaces fundamental to local partic-ipation. While in Zimbabwe RDCs are notoriously associated with 'capturing' CAMPFIRE revenues, in the current context of Zimbabwe RDCs could provide a system of checks and balances at the local level which can prevent capture of the process by local élites.

But our argument goes further than simply acknowledging the vital role of local government and addresses the broader issue of democratization. Local government has a vital role to play in ensuring democratic outcomes. Mamdani argues that emphasizing local participation or empowerment in an isolated or autonomous fashion, at the expense of cross-scale alliances and representative forms of democracy, can serve to reinforce authoritarian local structures. He concludes that 'to create a democratic solidarity requires join-ing the emphasis on autonomy with the one on alliance, that on participatory self-rule with one on representational politics.'

Put simply, a properly democratic system requires the effective linking of the local and national. CBNRM provides a means and incentives by which this can be done. Mahenye, albeit based on informally legitimized institutional foundations (the bylaws) was in the process of doing this in the late 1990s; now in more complex times it retains the potential to do so again and to provide continually evolving political and structural applications of CBNRM.

POSTSCRIPT

Is Community-Based Wildlife Management a Failed Approach?

The debate presented in this issue tugs at a couple of deeper philosophical questions, including the chasm between social and physical sciences in thinking about biodiversity protection and the ongoing debate within the environmental community regarding preservationist versus conservationist approaches.

Although it may be an obvious point, the natural sciences have tended to prioritize wildlife in their design and evaluation of conservation programs, whereas the social sciences have generally emphasized the welfare of people. Although some advocates of the sustainable development paradigm argue that their approach integrates the two concerns (development leading to better conservation and sustainable use of resources leading to enhanced development), the reality is that sustainable development is conceptualized in very dissimilar ways by different constituencies, for example, economists versus deep ecologists. Some African intellectuals complain that environmental concerns have been used to unfairly constrain their development prospects, a problem sometimes referred to as "green imperialism." In his 1997 article (January/February issue) in *The Ecologist,* titled "The Authoritarian Biologist and the Arrogance of Anti-Humanism: Wildlife Conservation in the Third World," Ramachandra Guha, an Indian scholar, asserts that biologists have been overly influential in environmental policy making. According to Guha, "biologists have a direct interest in species other than humans. . . . This interest in other species, however, sometimes blinds them to the legitimate interest of the less fortunate members of their own." In contrast, many environmentalists complain that the World Bank, for example, operationalizes a view of sustainable development that is overly anthropocentric.

Within the environmental community itself, there is an ongoing debate about the effectiveness of preservationist versus conservationist approaches to biodiversity protection. Preservation calls for the total nonuse or nonconsumptive use of natural resources, an approach typically associated with parks. Conservation allows for the use of natural resources to meet human needs within certain biological limits. This debate is directly relevant to discussions in Africa regarding protected areas. Some countries, such as Kenya, have opted for a more park-based approach, whereas others rely on more of an integrated conservation and development model, such as Namibia or Zimbabwe.

Following a first generation of CBNRM programs widely heralded as successful (including CAMPFIRE in its early years), there was a wave of studies that more critically examined this approach. These critiques ranged from those who argued that CBNRM was bad for local people (e.g., Roderick P. Neumann, *Imposing Wilderness: Struggles over Livelihood and Nature Preservation in Africa,* University of California Press, Berkeley, 1998) to others who questioned the ability of these programs to protect biodiversity. One element of the latter is a resurgent fortress conservation (or park) movement led by scholars such as John Terborgh (*Requiem for Nature,* Island Press, 1999). A third group acknowledges the concerns of the previous two, suggesting that we must continue to experiment with the CBNRM approach. This group points to promising new CBNRM programs in countries such as Namibia and Botswana (e.g., Wolfram Dressler et al., "From Hope to Crisis and Back Again? A Critical History of the Global CBNRM Narrative," *Environmental Conservation,* 2010).

Internet References . . .

Minnesota State University's E-Museum, African Cultures Section

This site offers useful background information on over 60 ethnic groups in Africa. It includes information on each group's geographic location, language, history, religion, and typical livelihood strategies.

http://www.mnsu.edu/emuseum/cultural/oldworld/africa.html

Population Council: Africa

The Population Council's Africa page explains population and family planning issues, reproductive health, HIV/AIDS, and other issues by country and for the African continent as a whole.

http://www.popcouncil.org/projects/projectsbycountry.asp

Washington Post: AIDS in Africa

The Washington Post has an ongoing "Special Report" on the HIV/AIDS issue that contains news on the HIV/AIDS epidemic from African countries, debate and information on U.S. actions to fight the epidemic, and world HIV/AIDS information.

http://www.washingtonpost.com/wp-dyn/world/issues/aidsinafrica/index.html

World Health Organization

The World Health Organization offers information on diseases and epidemiological facts for African countries (as well as other areas of the world).

http://www.who.int/en/

The University of Pennsylvania's African Studies Center, Women's Issues

The women's issues page of this Web site offers links to several other reputable sites dealing with women and gender.

http://www.africa.upenn.edu/About_African/ww_wmen.html

Social Issues

*P*erhaps more than any other set of contested African issues, those pertaining to the social sphere tend to provoke deep-seated emotional responses. This is also an area where differences in perspective between Africanists and non-Africanists tend to be more apparent. Such a degree of contestation is not surprising as these are, after all, deeply personal and culturally specific issues dealing with sexuality, gender roles, intra-household dynamics, and customs. Despite the highly private nature of some of the topics, this is also a realm that has come under incredible public scrutiny given concern about the global AIDS pandemic and the increasingly global nature of the human rights and feminist movements.

- Should Female Genital Cutting Be Accepted as a Cultural Practice?

- Are Women in a Position to Challenge Male Power Structures in Africa?

- Is the International Community Focusing on HIV/AIDS Treatment at the Expense of Prevention in Africa?

ISSUE 13

Should Female Genital Cutting Be Accepted as a Cultural Practice?

YES: Fuambai Ahmadu, from "Rites and Wrongs: Excision and Power among Kono Women of Sierra Leone," in B. Shell-Duncan and Y. Hernlund, eds., *Female "Circumcision" Africa: Culture, Controversy, and Change* (Lynne Reiner, 2001)

NO: Liz Creel et al., from "Abandoning Female Genital Cutting: Prevalence, Attitudes, and Efforts to End the Practice," A Report of the Population Reference Bureau (August 2001)

ISSUE SUMMARY

YES: Fuambai Ahmadu, an anthropologist at the London School of Economics, finds it increasingly challenging to reconcile her own experiences with female initiation and circumcision and prevailing (largely negative) global discourses on these practices. Her main concern with most studies on female initiation is the insistence that the practice is necessarily harmful or that there is an urgent need to stop female genital mutilation in communities where it is done. She suggests that "the aversion of some writers to the practice of female circumcision has more to do with deeply imbedded western cultural assumptions regarding women's bodies and their sexuality than with disputable health effects of genital operations on African women."

NO: Liz Creel, senior policy analyst at the Population Reference Bureau, and her colleagues argue that female genital cutting (FGC), while it must be dealt with in a culturally sensitive manner, is a practice that is detrimental to the health of girls and women, as well as a violation of human rights in most instances. Creel et al. recommend that African governments pass anti-FGC laws, and that programs be expanded to educate communities about FGC and human rights.

When examining the issue of female genital cutting in Africa (also known as female circumcision, female genital mutilation, or female genital alteration), it is difficult for many Westerners not to have an emotional reaction. In order to carefully evaluate the topic, the reader should try to keep as open a mind as possible.

This issue tugs at a deeper debate between those who believe there are certain universal rights and wrongs, and that female genital cutting is just wrong, irrespective of the cultural context, and cultural pluralists who believe we need to evaluate a practice within its own cultural context. Advocates of the universality of certain norms often depict female genital cutting as a violation of basic human rights. They may further disparage defenders of female genital cutting as cultural relativists. Cultural relativism is often cast as problematic because it may be used as an excuse to say that anything goes. For example, some individuals have and continue to argue that slavery is appropriate in some cultural contexts.

Others would argue that, despite one's personal objections to the practice, female genital cutting must be viewed within the context of cultural pluralism. Cultural pluralists assert that there are separate and valid cultural and moral systems that may involve social mores that are not easily reconcilable with one another. In contrast to cultural relativists, however, cultural pluralists would argue that everything does not go, and that there are certain universal norms (e.g., murder is wrong). The challenge for cultural pluralists is to determine if a practice violates a universal norm when it is viewed in its proper cultural context (rather than in the cultural context of another). The result of this deep philosophical divide is that we often see Western feminists pitted against multiculturalists (two groups that frequently function as intellectual allies in the North American context) over this controversial African issue.

In this issue, Fuambai Ahmadu, an anthropologist at the London School of Economics, reflects on her own position as a Western trained academic, as a member of an ethnic group that practices female initiation, and as someone who underwent the procedure herself. She is troubled by the ethnocentric insistence of outsiders that the practice is necessarily harmful or that there is an urgent need to stop female genital cutting in communities where it is done. Research and personal experience lead her to conclude that the positive aspects of initiation outweigh the negative ones. She argues that "initiation was the 'acting out' and celebration of women's preeminent roles in history and society." She believes the practice should not be banned, but medicalized. Liz Creel et al. argue that female genital cutting (FGC), while it must be dealt with in a culturally sensitive manner, is a practice that is detrimental to the health of girls and women, as well as a violation of human rights in most instances. They recommend that African governments pass anti-FGC laws and that programs be expanded to educate communities about FGC and human rights. They also believe that the use of medical professionals to perform the procedure should be discouraged.

Fuambai Ahmadu

Rites and Wrongs: Excision and Power among Kono Women of Sierra Leone

The issue of female initiation and circumcision is of significant intellectual and personal interest to me. Like previous anthropologists, I am fascinated by social, ideological and religious/symbolic dimensions of these rituals, particularly from indigenous perspectives. I also share with feminist scholars and activists campaigning against the practice a concern for women's physical, psychological and sexual well-being, as well as with the implications of these traditional rituals for women's status and power in society. Coming from an ethnic group in which female (and male) initiation and circumcision are institutionalized and a central feature of culture and society and having myself undergone this traditional process of becoming a "woman," I find it increasingly challenging to reconcile my own experiences with prevailing global discourses on female circumcision.

Most studies on FGC [female genital cutting] in Africa have been conducted by "outsiders" or individuals who are not from the societies they analyze and who have no personal experience of any form of the operation. The limited number of African women who have written about FGC either come from ethnic groups where female genital operations are not practiced (i.e. Thiam, Dorkenoo, Koso-Thomas) or have never undergone the procedures themselves (Toubia, El-Nadeer). There is an unfortunate and perturbing silence among African women intellectuals who have experienced initiation and circumcision. This reticence however is understandable given the venomous tone of the "debate" and unswerving demand that a definitive stance be taken—evidently, if one is educated—*against* the practice. However, "insider" voices from initiated/circumcised African women scholars can go a long way in providing fresh approaches to our understanding of these practices and their continued significance to the bulk of African women.

This essay is an attempt at reconciling "insider" representations with "outsider" perspectives. I seek to contextualize my own experience within the broader framework of initiation in Sierra Leone's Kono society and then contrast dominant Kono paradigms with conflicting international debates which focus on female circumcision as a peculiar manifestation of womens global subordination.

My main quarrel with most studies on female initiation and the significance of genital cutting relates to the continued insistence that the latter is necessarily "harmful" or that "there is an urgent need to stop female genital mutilation in communities where it is done." . . . Both of these assertions are based on the alleged physical, psychological and sexual effects of female genital cutting. I offer, however, that the aversion of some writers to the practice of female circumcision has more to do with deeply imbedded western cultural assumptions regarding women's bodies and their sexuality than with disputable health effects of genital operations on African women. For example, one universalized assumption is that human bodies are "complete" and that sex is "given" at birth. A second assumption is that the clitoris represents an integral aspect of femininity and has a central erotic function in women's sexuality. And, finally, through theoretical extension, patriarchy is assumed to be the culprit—that is, women are seen as blindly and wholeheartedly accepting "mutilation" because they are victims of male political, economic and social domination. According to this line of analysis, excision is necessary to patriarchy because of its presumed negative impact on women's sexuality. Removal of the clitoris is alleged to make women sexually passive, thus enabling them to remain chaste prior to marriage and faithful to their husbands in polygynous households. This supposedly ensures a husband sole sexual access to a woman as well as certainty of his paternity over any children she produces. As victims, then, women actively engage in "dangerous" practices such as "female genital mutilation" to increase their marriageability . . . , which would ultimately enable them to fulfill their honored, if socially inferior, destiny of motherhood.

When attempting to reconcile Kono practice with dominant anti-"FGM" discourses, a number of problems arise, starting with the alleged physical harm resulting from the practice. Part of the problem . . . is the unjustified conflation of varied practices of female genital cutting and the resulting over-emphasis on infibulation, a rare practice which is associated with a specific region and interpretation of Muslim *purdah* ideology. Kono women practice *excision*, removal of the clitoris and labia minorae. . . . The purported long-term physical side-effects of this procedure may have been exaggerated. It can be argued, as well, that although there are short-term risks, these can be virtually eliminated through improved medical technology. . . .

Furthermore, among the Kono there is no cultural obsession with feminine chastity, virginity nor with women's sexual fidelity, perhaps because the role of the biological father is considered marginal and peripheral to the central "matricentric unit." Finally, Kono culture promulgates a dual-sex ideology, which is manifested in political and social organization, sexual division of labor and, notably, the presence of powerful female and male secret societies. The existence and power of *Bundu*, the women's secret sodality, suggest positive links between excision, women's religious ideology, their power in domestic relations, and their high profile in the "public" arena.

The Kono example makes evident underlying biases of such culturally loaded notions as the "natural" vagina or "natural female body." The word "natural" is uncritically tossed around in the Female Genital Mutilation (FGM) literature to describe an uncircumcised woman when actually it needs definition and clarification. Kono concepts of "nature" and "culture" differ significantly from western assumptions and it is these local understandings which compel female (and male) genital cutting. In essence, what this paper amounts to is a critique of a profound tendency in western writing on female circumcision in Africa to deliver male-centered explanations and assumptions. Scholars must be wary of imposing western religious, philosophical and intellectual assumptions which tend to place enormous emphasis on masculinity and its symbols in the creation of culture itself. In traditional African societies, as is the case with the Kono, womb symbolism and imagery of feminine reproductive contributions form the basis of meanings of the universe, of human bodies, of society and its institutions—social organization, the economy, and even political organization can be viewed as extensions of the "matricentric core" or base of society. Female excision, I propose, is a negation of the masculine in feminine creative potential, and in the remainder of this essay I will show how the Kono case study demonstrates this hypothesis.

This paper is a culmination of several years of informal inquiry as well as formal research into the meaning of female circumcision and initiation, particularly among my parental ethnic group, the Kono, in northeastern Sierra Leone. This study constitutes an analysis of five stages: (1) my subjective experience of initiation in December/January 1991–92 which lasted just over month; (2) indigenous interpretations from other participants, mainly ritual leaders and their assistants, recorded at the time; (3) later academic study, when I returned to Kono for an additional 2 months in December 1994 and December 1996; (4) a total of nine months conducting formal and informal interviews among Kono immigrants in and around the Washington, D.C. area; and finally, (5) approximately three months spent between January and July of 1998 traveling back and forth between Conakry and Freetown talking to Kono refugees and women activists, mainly about their more immediate survival concerns but also about circumcision, initiation, and the future of women's secret societies. These discussions included informal interviews, as well as formal semi-structured interviews with three ritual officials: two traditional circumcisers, or *Soko* priestesses, and one *digba*, or ranking assistant to the *Soko*.

The cumulative data is drawn from interviews with a broad range of Kono men and women: young, old, university educated professionals in Freetown and in the U.S., as well as illiterate villagers and traditional rulers in Kono. If I have sacrificed quantification, it has been for the benefit of collecting detailed qualitative data which would enable my search for meaning and significance, both of which I felt could be best obtained through carefully selected, knowledgeable informants. What this study attempts to explain are the views, beliefs, and rationales of *supporters* of initiation and circumcision. The extent to which these attitudes reflect those of all or the majority of Kono women is left open for future research. . . .

The "Debate": Physical, Sexual and Psychological Effects

Anthropologists have not been the only ones interested in initiation and female genital cutting. In the last decade or so many others—feminists, politicians, international aid organizations, international medical community etc. within Africa and without—have produced a plethora of literature and convened conferences and the like on the subject of the effects of various forms of genital cutting on women's bodies, their sexuality and psychological well-being. My intention in this section is to interrogate some of the major assumptions in prevailing international discourses on female circumcision in light of my own experiences and the data collected from other Kono women, primarily but not limited to immigrants residing in the U.S.

First, as regards the health implications of excision, several short and long-term risks have been associated with the practice. . . . I have personally interviewed several male and female Sierra Leonean gynecologists who profess that although they regard excision as "medically unnecessary" the practice does not pose any significant adverse long-term effects to women and that, moreover, traditional circumcisers are on the whole "very well trained" and are "experts" at what they do. None had personally treated women with long-term problems related to excision but all stated that they had come across "reports" of horror cases.

Each of the doctors I have spoken to, irrespective of their position on the legitimacy of the practice, agree that short-term risks can be significantly reduced if not altogether eliminated through the use of antiseptic instruments, anesthetics to reduce pain, and skilled traditional officials. Also, it must be noted that most Kono women I have spoken to maintain that excision has existed in their society for hundreds of years and the practice has not adversely affected their fertility nor has it been giving their womenfolk the types of gynecological or obstetrical problems which have been over the past decade become associated with the operation. Thus, if some medical practitioners are saying that safe excisions are possible under the right conditions and if many Kono women do not attribute gynecological/obstetrical problems to their operations and choose to continue to uphold their tradition, a genuine case for limited medicalization can be made on this basis. . . . Such steps may reduce the immediate risks of the operation for young girls, until such a time that women are collectively convinced to give up the practice.

Second, my research and my experience contradict received knowledge regarding the supposed negative impact of removing the clitoris on women's sexuality. Much of this taken for granted information comes perhaps from popular misconceptions about the biological significance of the clitoris as the source of female orgasms. It is probable that such myths evolved as a result of the heightened focus on female clitoris during the 1960s sexual revolution and subsequent discourses regarding women's sexual autonomy. The clitoris has come to be seen in western societies as not only the paramount organ responsible for women's sexual pleasure but it has also been elevated as *the* symbol of women's sexual independence—the latter suited women's objectives in asserting their sexual agency and rejecting previous constraining notions of their roles as wives and mothers.

However, the presumptions which inform Kono women's values regarding female sexuality, as in other aspects of socio-cultural life, emphasize sexual interdependence and complimentarity, principles which are profoundly heterosexual. Western women's notions of the importance of the clitoris to female sexual autonomy can be contrasted with *Bundu* officials' stress on vaginal stimulation which implies male penetration, and this glaringly suggests heterosexual intercourse which, because the latter leads to reproduction, is considered the socially ideal form of sexual relations. My informants consider vaginal orgasm as independent of the clitoris but still fundamental to a woman's sexuality. Perhaps because women believe that the "internal" vagina is the appropriate locus of women's sexual pleasure, they profess that the clitoris is redundant and leads to excessive "sexiness." Also, as the clitoris is associated with androgyny and "nature," its presence signifies lack of self-control or self-discipline which are attributes of "culture." *Bundu* officials insist that the clitoris is "no good" and that it leads to uncontrolled masturbation in girls and sexual insatiability in adult women. It is believed to be a purely superfluous erotic organ, unlike a "proper" adult penis, its sex-corollary, which at least has reproductive functions. It is thus understandable, even if one does not agree, how some Kono women can claim that while excision curbs a woman's desire for sex, the operation itself does not reduce her enjoyment of sexual pleasure.

. . . There can be no way to "objectively" test the evidence regarding the impact of excision on women's sexuality as the latter is subjective and individually variable. Notwithstanding, an interesting finding in the Hite report is that the external clitoris constitutes a small fraction of total nerve endings which account for sensations for the entire appendage. . . . This suggests that excision leaves uncut most of the nerve endings which are beneath the vaginal surface. Thus, paradoxically, even according to "objective" biological science, it is possible for a woman's sensitivity to remain for the most part undiminished after excision. This would probably explain how it is that many women who had sexual experiences *prior* to excision, the author included, perceive either no difference or increased sexual satisfaction following their operation. In any case, most contemporary, urban educated as well as rural Kono women are just as interested in their sexuality as are their counterparts in western countries, and they do not perceive of excision as inhibiting them in any way. Also it is worth noting, especially since it is usually omitted, that significant numbers of western women, despite having their clitorises intact, experience their own difficulties in achieving any kind of orgasm, clitoral or vaginal.

Finally, as regards to the psychological well-being of young girls and women who have undergone initiation and excision, more research is needed before any credible generalizations can be made. A small but growing number of African female activists against various forms of circumcision have detailed the pain and trauma they underwent and the lasting impact such negative experiences have had on their lives and thus, they campaign against what they rightfully believe to be an affront to their human rights and womanhood. I have spoken to a few young Kono women who are adamantly opposed to initiation because of their experiences of pain, abuse and maltreatment by female elders in the "bush." However, the bulk of women I have interviewed fervently support the practice and my ground observations confirm that not only do most girls continue to look forward to their initiation but, further, they demonstrate their on-going support for the practice by actively participating in later ceremonies involving younger female friends and relations.

Conclusion

The question is often put to me: "How can a Western-bred and educated African woman *support* a practice which degrades women and deprives them of their humanity?" Notwithstanding the ethnocentrism in this remark, and the fact that I prefer to consider myself "neutral" in terms of the continuation of the practice, I am aware of many educated, professional circumcised African women gracefully negotiating their way through culturally distinct settings. There are those, the author included, who refuse to privilege one presumably objective, scientific model of personhood over supposedly "misguided" local interpretations but, rather, seek to juggle "modern" and "traditional" identities according to appropriate cultural context. Educated circumcised African women, like most people of multicultural heritage, maneuver multiple identities depending on the specific circumstances in which they find themselves. Personally, I do not see any conflict or contradiction in being educated and being circumcised, as the contexts which require each of these cultural idiosyncracies are separate and distinct.

For me, the negative aspect of excision was that it was a physically excruciating experience, for which, given my relatively cushioned Western upbringing, I was neither emotionally nor psychologically prepared. This is in contrast to most of the prepubescent Kono girls with whom I was "joined." As with the young Mandinka girls in the Gambia among whom I am currently conducting fieldwork, they "took" excision "bonically" (Krio term used to describe sheer human strength, strength of the flesh) and in a few hours were up, laughing and playing. After one or two days, they were jumping up and down dancing the "bird dance" to the rhythm of makeshift drums in preparation for their big "coming out" dance. To impose on my research what was my own experience of "pain" would be a gross distortion of the experiences of most of the other novices and thus, a certain disservice to anthropological knowledge in general.

The positive aspects have been much more profound. Initiation was the "acting out" and celebration of women's preeminent roles in history and society. Although I could not at the time put together all the pieces, I felt I was participating in a fear-inspiring world, controlled and dominated by women, which nonetheless fascinated me because I was becoming a part of it. In the years since, I have managed to make sense of much of the ritual symbolism and acting-out, enough to understand that women claim sole credit for everything from procreation to the creation of culture, society and its institutions, and, most important, they maintain a "myth of male dominance" so that their fundamental prerogatives are not threatened by increasing masculinization of religion, culture and society in Africa.

One such prerogative is the virtual deification of mothers among the Kono (and most African societies for that matter). This is not only symbolically important for women, but it gives them "real" power in inter- and intra-domestic lineage and immediate family relations, by virtue of the moral privilege women have over sons and daughters. I agree with Diop's assertion that the greatest abomination of any African, male or female, high or low, is the curse of his or her mother. This could also explain the findings of Skramstad and Hernlund indicating that Gambian Mandinka women continue "circumcision" first and foremost out of respect for their mothers and grandmothers. Even for the few Kono women who have second thoughts about "joining" their own daughters, the idea of eradication never comes up. It is not so much that an unexcised woman is unfathomable to them, but the public defiance and condemnation that abolition campaigns require would constitute a most unfathomable "insult" against their mothers and grandmothers.

Another feature of excision is the way in which the scar itself symbolizes women's sameness or common female identity. In effect, the operation rite is what defines and, thus, essentializes womanhood. Unlike in Western society, there is no confusion or fruitless intellectualizing about the definition of "woman." Among the Kono, a woman is a woman by virtue of the fact that she has been initiated and nothing else. But initiation also creates a hierarchical ordering of women in society. At the apex is, of course, the *Soko,* the mother of the community, then an individual's mother, after which is her mother-in-law, and then all other older women in the community. A woman's equals are her age-mates, those with whom she was initiated and/or those falling within the same age group. Thus, sameness is not always tantamount to equality, and neither does it imply strict conformity to dominant values of womanhood, such as motherhood. For example, my grandmother often nags me about not yet having children and about the importance of motherhood to a woman but, for her, as for the entire community, it was my initiation which "made" me into a "woman." This is perhaps how it is possible for some circumcised African women to be educated, westernized and yet not view the practice as an affront to their womanhood. In short, initiation/excision has the positive value of creating sameness among all women and maintaining equality within age-groups as well as a general hierarchy of female authority in society.

Other advantages of initiation include beliefs about women's esoteric knowledge and their monopoly over powerful "medicines." Although excision cannot be said to be a marker of ethnicity today (most ethnic groups in Sierra Leone practice excision), what does distinguish Kono *Bundu* from those of other groups are the "medicines" which are used. According to a high-ranking *Bundu* official, even more important than excision itself is women's "medicine" which is used in ritual. The more powerful a sodality's "medicine" is reputed to be, the more feared, and thus, influential are the women leaders of such a group. Kono women often assert the power of their own "medicine" and claim that this is how they dominate their men and "keep them home." Also, it is believed that the medicine which is used for the novices during initiation will protect them against all sorts of witchcraft and other malevolent supernatural practices which may be aimed at them throughout their lives.

Finally, and more subjective, is my shared view of the aesthetics of excision and (male) circumcision. I propose that the basis of Kono appreciation of male/female genital modifications is the latters compatibility and harmony with basic principles of complimentarity and interdependence. These ideals underpin cosmological beliefs regarding sex and gender difference and are manifested in the dual-sex organization of culture and society. As long as there are deeply implanted mental associations of the clitoris with masculinity, then the former will continue to be regarded as dirty, abnormal, unclean, and harmful by a culture which sees "male" and "female" as fundamentally separate and distinct moral categories.

To Cut or Not to Cut?: The Future of Excision

While location and identity may establish who is an "outsider" as opposed to an "insider" with respect to studies on FGC, these factors do not automatically determine the position of any writer regarding abolition. For example, not a few anthropologists—who are by discipline western scholars and often by nationality "outsiders"—have been bitterly criticized for their attempts to represent the cultural viewpoints and values of their informants. Conversely, indigenous African female activists or "insiders," fighting against "FGM" often promulgate the same messages contained in global discourses that link the practice to women's social, sexual and psychological oppression. What is certain is that the future of FGC will depend on the extent to which "insiders" themselves are convinced of purported negative effects of the practice.

The medical evidence as well as the speculations regarding adverse effects on women's sexuality do not tally with the experiences of most Kono women. It is the immediate physical pain and risk of infection which concern most mothers and both of these hazards can be reduced, if not eliminated, through medicalization, education and general modernization of the operation. A compelling point has been made, however, that all the eradication mechanisms, such as policies of international organizations and local NGOs devoted to change peoples attitudes and behaviors, have already been set in place and that most likely there can be no going back. . . . But the virtually universal resistance to change after several decades of international and internationally sponsored local campaigning, conferencing and legislating suggests that what is seriously needed is a re-thinking of previous eradication strategies and a deeper appreciation of the historical and cultural relevance of this ancient practice and its symbolically dynamic and fluid links to women's changing sources and notions of power.

In my opinion, if eradication has become an irreversible "international" political compulsion, then the ideal of "ritual without cutting" . . . seems to be a reasonable middle-ground. The ritual without cutting model positively values many cultural aspects and beliefs underlying female genital operations and initiation while attempting to eliminate the actual physical cutting. Perhaps what *is* needed to replace the physical act of cutting is an equally dynamic symbolic performance which will retain the same fluidity in associated meanings, that is eschewal of masculinity, womanhood, fertility, equality, hierarchy, motherhood, and sexual restraint. However, for rural Kono women in particular, the "cutting" and "medicine" are all-important. Also, as I discussed earlier, "ritual without cutting" can be very dangerous when taken out of context such as in the recruitment and "training" of child rebels in Kono.

I continue to support, however, the goal of medicalizing and modernizing initiation and "circumcision"—not necessarily in a full sense of institutionalizing female circumcision or transferring the practice from the "bush" to hospitals, as in male circumcisions today, because this would reduce the power and authority of female ritual leaders and female elders—but rather by making available basic, modern hygienic equipment and medications to traditional officials to use during rituals. I support change which will promote safe, sanitary environments, so that initiates are given adequate, modern medical assistance to reduce pain and risks of infection. The position that this only legitimizes the practice is dangerously arrogant: the practice is already seen as legitimate by its proponents, who have themselves undergone excision, and denying them the benefits of medicalization only continues to endanger the health and lives of innocent young girls. Modernization should also include impartial, neutral education within primary and secondary schools. Such education should entail both positive historical and cultural significance of initiation/circumcision as well as possible negative health effects. Emphasis should be on preparing young girls to make informed choices about their futures and the futures of their own girl-children.

What direction individual women take should be left to them and their immediate family members. Just as much as die-hard "traditionalists" must relinquish their insistence that uncircumcised women are not socially and culturally "women" who therefore must be denied legal rights and dignity within society, hard-line efforts by abolitionists to coerce women against the practice and stigmatize those who uphold their ancestral traditions as "illiterate," backward, and against "women's rights" and "progress" are unacceptable. It is the bulk of circumcised African women who are unfortunately caught between a rock and a hard place, as the adage goes: Either break traditional customary laws and face the consequences of "not belonging," or ignore increasing efforts to ban the practice and face possible legal penalties instigated by eradicators at the national and international level. Today, it seems that the pressure on circumcised African women, educated or not, is to choose between these two extremist positions—to be either "anti-culture" or "anti-progress."

Change may indeed be occurring gradually, but I do not believe this is necessarily a direct result of "anti-FGM" campaigning. In my grandmother's days, excision was a universal *rite-de-passage*. For my educated, Christian mother who has spent over thirty years in the United States, initiating her daughters was a matter of judgment, an expedient choice to enable us to easily navigate between worlds. My generation is faced with a dramatically different and greater complexity of issues and other priorities (i.e. the complete destruction of Kono through civil war) and, as a result, initiation and excision can hardly be said to be the most pressing preoccupation of young, contemporary Kono women. In the event that I ever have a daughter I would like her to be well-informed about the socio-cultural and historical significance of the operation as well as its purported medical risks so that she can make up her own mind, like I had the opportunity to do. Mbiti has noted regarding female initiation and circumcision in Africa: "If they are to die out, they will die a long and painful death." However, through more culturally sensitive and appropriate "education" as well as limited medicalization strategies, the "death" of female circumcision could be more gradual, more natural, and a lot less painful for millions of future African women and girls. . . .

Liz Creel et al.  **NO**

Abandoning Female Genital Cutting: Prevalence, Attitudes, and Efforts to End the Practice

Introduction

More than 130 million girls and women worldwide have undergone female genital cutting [FGC]—also known as female circumcision and female genital mutilation—and nearly 2 million more girls are at risk each year. The practice often serves as a rite of passage to womanhood or defines a girl or woman within the social norms of her ethnic group or tribe. The tradition may have originated 2,000 years ago in southern Egypt or northern Sudan, but in many parts of West Africa, the practice began in the 19th or 20th century. No definitive evidence exists to document exactly when or why FGC began. FGC is an ancient practice but has also been recently adopted, for example, among adolescents in Chad.

FGC is generally performed on girls between ages 4 and 12, although it is practiced in some cultures as early as a few days after birth or as late as just prior to marriage, during pregnancy, or after the first birth. Girls may be circumcised alone or with a group of peers from their community or village. Typically, traditional elders (male barbers and female circumcisers) carry out the procedure, sometimes for pay. In some cases, it is not remuneration but the prestige and power of the position that compels practitioners to continue. The practitioner may or may not have health training, use anesthesia, or sterilize the circumcision instruments. Instruments used for the procedure include razor blades, glass, kitchen knives, sharp rocks, scissors, and scalpels. A discouraging trend is the use of medical professionals (physicians, nurses, and midwives) in some countries (e.g., Egypt, Kenya, Mali, and Sudan) to perform the procedure due to growing recognition of the health risks associated with FGC and heightened concern regarding the possible role of FGC in HIV transmission. WHO [World Health Organization] has strongly advised that FGC, in any of its forms, should not be practiced by any health professional in any setting—including hospitals and other health centers.

FGC has health risks, most notably for women who have undergone more extreme forms of the procedure (see Box 1). Immediate potential side effects include severe pain, hemorrhage, injury to the adjacent tissue and organs, shock, infection, urinary retention and tetanus—some of these side effects can lead to death. Long-term effects may include cysts and abscesses, urinary incontinence, psychological and sexual problems, and difficulty with childbirth. Obstructed labor may occur if a woman has been infibulated. This involves cutting off the external genitalia and sewing together the two sides of the vulva, leaving a small hole for urination and menstruation. If the woman's genitalia is not cut open (defibulated) during delivery, labor may be obstructed and cause life-threatening complications for both the mother and the child, including perineal lacerations, bleeding and infection, possible brain damage to infants, and fistula formation.

TYPES OF FEMALE GENITAL CUTTING

Female genital cutting (FGC) refers to a variety of operations involving partial or total removal of female external genitalia. The female external genital organ consists of the vulva, which is comprised of the labia majora, labia minora, and the clitoris covered by its hood in front of the urinary and vaginal openings. In 1995, the World Health Organization classified FGC operations into four broad categories described below:

> **Type 1 or Clitoridectomy:** Excision (removal) of the clitoral hood with or without removal of the clitoris.
>
> **Type 2 or Excision:** Removal of the clitoris together with part or all of the labia minora.
>
> **Type 3 or Infibulation:** Removal of part or all of the external genitalia (clitoris, labia minora, and labia majora) and stitching and/or narrowing of the vaginal opening, leaving a small hole for urine and menstrual flow.
>
> **Type 4 or Unclassified:** All other operations on the female genitalia including
>
> - pricking, piercing, stretching, or incising of the clitoris and/or labia;
> - cauterization by burning the clitoris and surrounding tissues;
> - incisions to the vaginal wall; scraping or cutting of the vagina and surrounding tissues; and introduction of corrosive substances or herbs into the vagina.

Note:
1. World Health Organization, *Female Genital Mutilation: Report of a Technical Working Group* (Geneva: WHO, 1996): 9.

All of these possible side effects may damage a girl's lifetime health, although the type and severity of consequences depend on the type of procedure performed (see Box 1). Infibulation or Type 3 is the most invasive and damaging type of FGC. Operations research studies conducted in Burkina Faso and Mali have shown that women who were infibulated were nearly two and a half times more likely to have a gynecological complication than those with a Type 2 or Type 1 cut. Risks during childbirth also increased according to the severity of the procedure. For instance, in Burkina Faso, women with Types 2 or 3 cutting had a higher likelihood of experiencing hemorrhaging or perineal tearing during delivery.

While it is difficult to determine both the number of women who have undergone FGC and how many have undergone each type of circumcision, WHO has estimated that clitoridectomy, which accounts for up to 80 percent of all cases, is the most common procedure. Fifteen percent of all circumcised women have been infibulated—the most severe form of circumcision.

FGC is practiced in at least 28 countries in sub-Saharan and north-eastern Africa but not in southern Africa or in the Arabic-speaking nations of North Africa, with the exception of Egypt. It is practiced at all educational levels and in all social classes and occurs among many religious groups (Muslims, Christians, animists, and one Jewish sect), although no religion mandates it. For countries presented here with DHS [Demographic and Health Surveys] data, prevalence varies from 18 percent in Tanzania to nearly 90 percent or more in Egypt, Eritrea, Mali, and Sudan. According to WHO estimates, 18 African countries have prevalence rates of 50 percent or more. Through migration, the practice has also spread to Europe, North and South America, Australia, and New Zealand. Although doctors, colonial administrators, and social scientists have documented the adverse effects of FGC for many years, governments and funding donors have become increasingly interested in the practice because of the public health and human rights implications.

Global efforts to end FGC have used legislation to provide legitimacy for project activities, to protect women, and to discourage circumcisers and families who fear prosecution. In the 1960s, WHO was the first United Nations (UN) specialized agency to take a position against female genital cutting. It began efforts to promote the abandonment of harmful traditional practices like FGC in the 1970s, focusing largely on gathering information about FGC's epidemiology and health consequences and speaking out about FGC at international, regional, and national levels. In 1982, WHO issued a formal statement to the UN Commission on Human Rights and recommended several actions:

- Governments should adopt clear national policies to end FGC, and educate and inform the public about its harmful aspects.
- Anti-FGC programs must consider the practice's association with difficult social and economic conditions and respond to women's needs and problems.
- Women's organizations at the local level should be encouraged to take action.

In 1988, WHO began to integrate FGC into the development context of primary health care. Over the intervening years, WHO shifted its position on FGC from addressing the practice only in terms of health to acknowledging it as both a health and human rights issue. In the 1990s, FGC gained recognition as a health and human rights issue among African governments, the international community, women's organizations, and professional associations. The 1993

Vienna Human Rights Convention, the 1994 International Conference on Population and Development, and the 1995 Fourth World Conference on Women called for an end to the practice. When performed on girls and non-consenting women, FGC violates a number of recognized human rights protected in international conventions and conferences, such as the Convention on Children's Rights, the Convention on the Elimination of All Forms of Discrimination against Women, and recommendations of the Committee on the Elimination of Discrimination against Women (CEDAW). These conventions explicitly recognize harmful traditional practices such as FGC as violations of human rights, including the right to nondiscrimination, the right to life and physical integrity, the right to health, and the right of the child to special protections.

Respect for international human rights law does not require that every culture use an identical approach to abandoning FGC. One Muslim scholar suggested that respecting different cultures means accepting "the right of all people to choose among alternatives equally respectful of human rights," and that human rights must include life, liberty, and dignity for every person or group of people.

In Africa, 10 countries–Burkina Faso, the Central African Republic (CAR), Côte d'Ivoire, Djibouti, Ghana, Guinea, Niger, Senegal, Tanzania, and Togo—have enacted laws that criminalize the practice of FGC. The penalties range from a minimum of six months to a maximum of life in prison. In Nigeria, three of 36 states (as of 2000) had also enacted legislation regarding FGC. In Burkina Faso, Ghana, and Senegal, these laws are enforced and circumcisers are imprisoned. In these countries, various groups educate the public about the law, use a variety of strategies (e.g., public service announcements and watchdog committees) to denounce FGC, and stop circumcisers by going to the police. Several countries also impose fines. In Egypt, the Ministry of Health issued a decree declaring FGC unlawful and punishable under the Penal Code. There have been several prosecutions under this law, which include jail time and fines. In addition, seven more developed countries that receive immigrants from countries where FGC is practiced—Australia, Canada, New Zealand, Norway, Sweden, the United Kingdom, and the United States—have passed laws outlawing the practice. Enforcement of these laws, however, is extremely uneven. France, on the other hand, consistently enforces general penal code provisions against providers of FGC but has not adopted specific legislation regarding FGC.

Why Is FGC Performed?

The traditions surrounding FGC vary from one society to another. In some communities, FGC is a rite of passage to womanhood and is performed at puberty or at the time of marriage. In other communities, it may be performed on girls at a younger age for other reasons such as a celebration of womanhood, preservation of custom or tradition, or as a symbol of ethnic identity. The ritual cutting is often an integral part of ceremonies, which may occur over several weeks, in which girls are feted and showered with presents and their families are honored. It is described as a joyous time with many visitors, feasting, dancing, good food, and an atmosphere of freedom for the girls. The ritual serves as an act of socialization into cultural values and an important connection to family, community, and earlier generations. The ceremonies often involve three interrelated aspects:

- **Educational** A girl learns her place in society and her role as woman, wife, and mother.
- **Physical** A girl must undergo physical pain to prove she is capable of assuming her new role courageously without showing suffering or pain; the pain is experienced both through the actual cutting and through punishment received by girls in complete submission throughout weeks of initiation.
- **Vow of silence** Each girl must make a solemn pledge not to speak about her experience during the ceremony.

The reasons for performing FGC differ, but many practicing communities believe that it preserves the girl's virginity and protects marital fidelity because it diminishes her sexual desire. Practicing communities cite reasons such as giving pleasure to the husband, religious mandate, cleanliness, identity, maintaining good health, and achieving good social standing. At the heart of all this is rendering a woman marriageable, which is important in societies where women get their support from male family members, especially husbands. A circumcised woman will also attract a favorable bride price, thus benefiting her family. The practice is perceived as an act of love for daughters. Parents want to provide a stable life for their daughters and ensure their full participation in the community. Many girls and women receive little formal education and are valued primarily for their role as sources of labor and future producers of children. For many girls and women, being uncircumcised means that they have no access to status or a voice in their community. Because of strong adherence to these traditions, many women who say they disapprove of FGC still submit themselves and their daughters to the practice.

Understanding Why the Practice Continues

FGC is a cultural practice. Efforts to end it require understanding and changing the beliefs and perceptions that have sustained the practice over the centuries. Irrespective of how, where, and when the practice began, those who practice it share similar beliefs—a "mental map"—that present compelling reasons why the clitoris and other external genitalia should be removed. The details of these mental maps vary across countries, and there are distinctive features to each culture that providers, community workers, and others involved with anti-FGC campaigns need to take into consideration.

Figure 1 provides a conceptual framework for understanding the role of FGC in society. This mental map shows the psychological and social reasons, and the religious, societal, and personal (hygienic and aesthetic) beliefs that contribute to the practice. These beliefs involve continuing long-standing custom and tradition; maintaining cleanliness, chastity, and virginity; upholding family

Figure 1

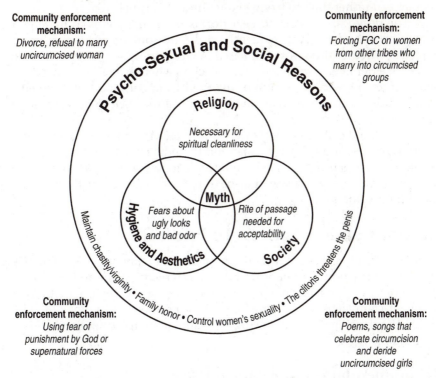

Why the Practice of FGC Continues: A Mental Map

Community enforcement mechanism:
Divorce, refusal to marry uncircumcised woman

Community enforcement mechanism:
Forcing FGC on women from other tribes who marry into circumcised groups

Psycho-Sexual and Social Reasons

Religion
Necessary for spiritual cleanliness

Myth

Hygiene and Aesthetics
Fears about ugly looks and bad odor

Rite of passage needed for acceptability
Society

Maintain chastity/virginity • Family honor • Control women's sexuality • The clitoris threatens the penis

Community enforcement mechanism:
Using fear of punishment by God or supernatural forces

Community enforcement mechanism:
Poems, songs that celebrate circumcision and deride uncircumcised girls

Source: Asha Mohamud, Nancy Ali, Nancy Yinger, World Health Organization and Program for Appropriate Technology in Health (WHO/PATH), *FGM Programs to Date: What Works and What Doesn't* (Geneva: WHO, 1999):7.

honor (and sometimes perceived religious dictates); and controlling women's sexuality in order to protect the entire community. In the countries surveyed by Demographic and Health Surveys, good custom/tradition is the most frequently cited reason for approving of FGC. Bad custom/tradition is also mentioned as one of the primary reasons for discontinuing the practice.

To encourage abandonment of FGC, health care providers, community workers, and others involved with anti-FGC programming need to understand the mental map in the communities where they are working. Communities have a range of enforcement mechanisms to ensure that the majority of women comply with FGC. These include fear of punishment from God, men's unwillingness to marry uncircumcised women, insistence that women from other tribes get circumcised when they marry into the group, as well as local poems and songs that reinforce the importance of the ritual. In some cases, women who are not circumcised may face immediate divorce or forced excision. Girls who do undergo FGC sometime receive rewards, including public recognition and celebrations, gifts, potential for marriage, respect and the opportunity to engage in adult social functions. In other instances, girls and women are cut without an accompanying ceremony; thus, importance is attached to being circumcised rather than to having gone through a ritual.

The desire to conform to peer norms leads many girls to undergo circumcision voluntarily, yet frequently girls (and sometimes infants) have no choice in whether they are circumcised. A girl's family—typically her mother, father, or elder female relatives—often decides whether she will undergo FGC. Due to the influence of tradition, many girls accept, and even perpetuate, the practice. In Eritrea, men are more likely than women to favor ending the practice.

FGC could continue indefinitely unless effective interventions convince millions of men and women to abandon the practice. Many African activists, development and health workers, and people following traditional ways of life recognize the need for change but have not yet achieved such an extensive social transformation. . . .

Recommended Actions to End FGC

Data on attitudes, practices, and prevalence can provide important background information on opportunities for intervention. In addition, lessons from program experiences provide an important context for formulating abandonment campaigns. PATH [Program for Appropriate Technology in Health] and WHO developed the following recommendations for policymakers and program managers.

Recommendations for Policymakers

Policymakers are those who are in a position to influence policies and provide funding related to FGC.

> 1. *Governments and donors need to support the groundswell of agencies involved in FGC abandonment with financial and technical assistance.*

An increasing number of agencies, especially NGOs [nongovernmental organizations], are involved in efforts to end FGC. However, programs tend to be small, rely heavily on volunteers and funds from foreign donors, and reach a small proportion of the people in need. Additional support is needed to make the growing network of agencies more effective and expand their reach.

In Egypt, 15 NGOs, including the Egyptian Fertility Care Society, the Task Force Against FGM, the Cairo Institute for Human Rights Studies, and CEOSS [Coptic Evangelical Organization for Social Services] have been instrumental in advocating the abandonment of FGC. In order to become more effective, these groups need to collaborate with one another, enhance training in advocacy and communications skill building, and evaluate the impact of their programs.

2. Governments must enact and use anti-FGC laws to protect girls and edu-cate communities about FGC and human rights.

Passing anti-FGC legislation is one of the most controversial aspects of the FGC abandonment movement. It is extremely difficult to enforce anti-FGC laws. There is fear that heavy-handed enforcement may drive the practice underground. In fact, this has occurred in some countries. Still, most program planners and activists agree that anti-FGC legislation can demarcate right from wrong, provide official legal support for project activities, offer legal protec-tion for women, and ultimately, discourage circumcisers and families for fear of prosecution. The key is to use the law in a positive fashion—as a vehicle for public education about and community action against FGC.

3. National governments need to be active both in setting policy and in expanding existing programs.

A key role for governments is to "scale up" successful community-based FGC abandonment activities. To date, most governments have provided sup-port in the form of in-kind contributions to NGOs working in the communi-ties. The excellent program models that NGOs have carried out on a pilot basis need to be expanded, either by direct government interventions or by increased support for the NGO networks.

4. To sustain programs, governments need to institutionalize FGC abandon-ment efforts in all relevant ministries.

Currently, none of the anti-FGC efforts underway are sustainable over the long run, in part because they have failed to change the social norms underlying FGC. The integration of FGC issues into government programs, however, has met with some success. In Burkina Faso, the National Committee to Fight Against the Practice of FGC has effectively promoted FGC abandon-ment through participation in national events such as the international day of population, by integrating FGC abandonment into all of the relevant min-istries, and through training and awareness raising activities.

Efforts in other countries, such as Mali, have encountered more difficul-ties. While various ministries have expressed their support for anti-FGC activities, they have not been integrated into the relevant ministries, particularly in pro-grams carried out by Mali's Ministry of Health. The primary nursing and medical schools in Mali do not include FGC as an adverse health practice in their cur-ricula. Presently, PRIME II, a partnership of U.S.-based organizations, is working with the Ministry of Health to develop a national curriculum integrating FGC.

Governments have a responsibility to make political decisions and place FGC abandonment in the mainstream of reproductive health and development programs. Limited success has been achieved through increased fundraising, greater integration of anti-FGC activities into government and civil society programs, and through solicitation of community support.

5. Health providers at all levels need to receive training and financial support to treat FGC complications and to prevent FGC.

A key foundation for FGC abandonment is to make health providers aware of the extent and severity of FGC-related complications and to give them the skills and resources to treat these problems. Health providers often encounter women and girls suffering from FGC-related complications, yet they are often not prepared to treat and counsel women, or to prevent recurrence of the circumcision practice. Because there is limited training on clinical treatments for circumcised women or counseling women suffering from psychological or sexual problems, women lack access to high quality, relevant services in most countries.

A 1998 operations research study in Mali sought to assess the use of health personnel to address FGC. The study, which was conducted by an NGO, Association de Soutien au Developpement des Activités de Population (ASDAPO), and the Ministry of Health, evaluated the effectiveness of a three-day training course on identifying and treating medical complications related to FGC and counseling patients about the problem. The study focused on 14 urban and rural health centers in Bamako and the Ségou region and included 107 health providers from experimental and control sites. Results indicated that the course was highly effective in changing provider attitudes toward FGC. After receiving training, three in four trained providers knew at least three immediate and long-term complications of FGC. The study also indicated that providers felt they had limited competence in caring for FGC complications (even after receiving training) and needed further training in how to discuss FGC with their clients.

6. Governments, donors, and NGOs working on FGC abandonment should continue to coordinate their efforts.

Findings from field assessments reveal an impressive array of cooperative efforts and exchanges of information and resources among NGOs, government institutions, and donors. Agencies typically invite each other to meetings and training activities and coordinate at program sites to avoid duplication of efforts. Although occasional conflicts arise over funding and strategies, they should not discourage agencies from continuing to coordinate and build on each other's strengths.

7. International agencies should assist staff of NGOs and government to develop their advocacy skills.

Advocacy is essential to ensuring that FGC abandonment programs are established and maintained until the practice of FGC ceases. Agencies involved in abandonment efforts increasingly use advocacy for public education and to influence legislation, but they need to improve their skills.

POSTSCRIPT

Should Female Genital Cutting Be Accepted as a Cultural Practice?

After reading the different arguments in this issue, there are at least five different points that one may want to ponder before taking an informed stance on this topic. First, female genital cutting covers a very broad range of practices in the African context. In some instances, apparently contradictory statements from either author may have as much to do with the specific practices they have in mind as with the fact that they have differing views on the general topic.

Second, the whole notion of choice, the prospect of giving girls and women the option of undergoing or not undergoing the procedure, implies that they are aware that there is a choice. Many girls and women may just assume that this is what is done, or what is normal.

Third, if a girl or young women opts not to undergo female genital cutting, there could be serious social consequences in some settings. Within the context of the educational programs advocated by Creel et al., is there an obligation to make sure individuals are made aware of the medical dangers of the procedure as well as the social consequences of not being initiated?

Fourth, some African countries (e.g., Kenya) have a mix of ethnic groups that may or may not practice female genital cutting. In other nations, such as Mali or Eritrea, the vast majority ethnic groups, and the population in general, practice female genital cutting. This level of homogeneity or heterogeneity could have implications for people's exposure to different practices (particularly in urban areas where ethnic groups tend to mix) as well as the chances of success or failure of education programs in this domain.

Finally, the different authors have very different views on medicalizing FGC in Africa. Ahmadu contends that this will minimize health problems whereas Creel et al. fear that this may help perpetuate the practice. How likely is making the procedure safer going to contribute to its spread?

For more information on this debate, see Richard Scweder's article "What about 'Female Genital Mutilation?'" and "Why Understanding Culture Matters in the First Place" in *Daedalus* (Fall 2000) or Monica Antonazzo's paper "Problems with Criminalizing Female Genital Cutting" in *Peace Review* (2003). For a publication questioning the medical evidence against female genital surgeries, see an article by Carla M. Obermeyer in *Medical Anthropology Quarterly* (no. 13, 1999) with the title "Female Genital Surgeries: The Known, The Unknown, and the Unknowable." Finally, for more of an anti-FGM perspective, see *Eradicating Female Genital Mutilation: Lessons for Donors* (by Susan Rich and Stephanie Joyce, Wallace Global Fund for a Sustainable Future, 1990).

ISSUE 14

Are Women in a Position to Challenge Male Power Structures in Africa?

YES: Richard A. Schroeder, from *Shady Practices: Agroforestry and Gender Politics in The Gambia* (University of California Press, 1999)

NO: Human Rights Watch, from "Double Standards: Women's Property Rights Violations in Kenya," A Report of Human Rights Watch (March 2003)

ISSUE SUMMARY

YES: Richard A. Schroeder, an associate professor of geography at Rutgers University, presents a case study of a group of female gardeners in The Gambia who, because of their growing economic clout, began to challenge male power structures. Women, who were the traditional gardeners in the community studied, came to have greater income-earning capacity than men as the urban market for garden produce grew. Furthermore, women could meet their needs and wants without recourse to their husbands because of this newly found economic power.

NO: Human Rights Watch, a nonprofit organization, describes how women in Kenya have property rights unequal to those of men, and how even these limited rights are frequently violated. It is further explained how women have little awareness of their rights, that those "who try to fight back are often beaten, raped, or ostracized," and how the Kenyan government has done little to address the situation.

$\mathbf{A}$s is the case in other parts of the world, African women suffer from discrimination and inequality. According to the World Bank, the female adult illiteracy rate in Africa in 2001 was 46 percent as compared to 38 percent for the general population. Girls also continue to attend primary school in lower numbers than boys in many African countries (although this ranges from near equality in nations such as South Africa, Zimbabwe, and Namibia to great disparities in countries like Benin, Chad, and Guinea). Despite these disadvantages, women are the backbone of the rural economy in many African settings where it is estimated that they produce, on average, 70 percent of the food supply.

The inequities faced by many women in the African context led to the rise of the women in development (WID) movement in aid circles in the late 1970s and 1980s. These WID programs were also instigated because of a general recognition that many aid programs had not addressed the needs of women or had excluded them entirely. Many agricultural development programs catered almost exclusively to men. In many instances, such programs exacerbated economic disparities between men and women. WID programs were specifically designed to counteract these problems, including a number of initiatives related to gardening, income generation, health care, and education.

In promoting these initiatives, development agencies occasionally exploited the image of African women as a downtrodden class of people who undertake a disproportionate share of the work, yet are severely disadvantaged in terms of access to education, health care, land, and legal protection. While there may be some truth to this generalization, it is problematic because it denies African women "agency." In other words, it could negate or understate the ability of African women to change their situation. This is not to say that African women should not form alliances with outside groups to work for transformation, but those peddling the assistance need to be careful that they are not trafficking images and stereotypes that may be disempowering.

In this issue, Richard A. Schroeder presents a case study of a group of female gardeners in The Gambia. These women, who were the traditional gardeners in the community, benefited from outside funding for fencing and wells during the heyday of WID programming in the 1980s. They eventually came to have greater income-earning capacity than men as the urban market for garden produce grew, and they adeptly intensified production. As a result, men were often forced to turn to their wives for loans. This allowed women to "purchase" freedom of movement and social interaction. Furthermore, women with growing economic clout could challenge male power structures because they were capable of meeting their needs and wants without recourse to their husbands. They also were less susceptible to the threat of divorce (which historically implied the nearly impossible obligation of repaying one's bride-price) because women were now capable of repaying their bride-price with their gardening income.

In contrast, the selection from Human Rights Watch contains the assertion that in Kenya "discriminatory property laws and practices impoverish women and their dependents, put their lives at risk by increasing their vulnerability to HIV/AIDS and other diseases, drive them into abhorrent living conditions, subject them to violence, and relegate them to dependence on men and social inequality." The Kenyan government is castigated for having done little to address this situation.

YES

Richard A. Schroeder

Shady Practices: Agroforestry and Gender Politics in The Gambia

Introduction

Some sixty kilometers upriver along the North Bank of The River Gambia lies the Mandinka-speaking community of Kerewan (ke´-re-wan). The dusty headquarters of The Gambia's North Bank Division is located on a low rise overlooking rice and mangrove swamps and a ferry transport depot that facilitates motor vehicle transport across Jowara Creek (Jowara Bolong), one of The River Gambia's principal tributaries. Since the Kerewan area was dominated by opposition political parties throughout the nearly thirty-year reign of The Gambia's first president, Al-Haji Sir Dawda Jawara (1965–1994), it became something of a developmental backwater. Before 1990, Kerewan town had no electricity or running water beyond a few public standpipes. For a community of 2,500 residents, there were no restaurants and only a poorly stocked market that lacked fresh meat. Indeed, from the standpoint of the civil servants assigned to the North Bank Division, Kerewan was considered a hardship post. Mandinka speakers sarcastically referred to the divisional seat as "Kaira-wan," a place where "peace" (Mandinka: *kaira*) reigned to the point of overbearing stagnation. Neighboring Wolof speakers, meanwhile, disparaged the community by dubbing it "Kerr Waaru"—"the place of frustration."

Kerewan's reputation was only partially deserved, however, for the community was actually the center of a great deal of productive economic activity. Over two decades beginning in the mid-1970s, the town's women transformed the surrounding lowlands into one of the key sites of a lucrative, female-controlled, cash-crop market garden sector. A visitor to Kerewan as recently as 1980, when I made my first trip to The Gambia, would have found that vegetable production on the swamp fringes ringing Kerewan on three sides was decidedly small-scale. Most gardeners, virtually all of whom were women, worked single plots that were individually fenced with local thorn bushes or woven mats. Outside assistance in obtaining tools, fences, and wells was minimal. Seed suppliers were not yet operating on a significant scale, and petty commodity production was largely confined to tomatoes, chili peppers, and onions. The market season, accordingly, stretched only a few weeks, and sales outlets were all but nonexistent. Most Kerewan produce was sold directly to end users in the nearby Jokadu District by women who transported their fresh vegetables by horse or donkey cart and then toted them door to door on their heads (a marketing strategy known as *kankulaaroo*).

By 1991, when I completed the principal phase of research for [my] book, large gardens on the outskirts of Kerewan had come to dominate the landscape. Each morning and evening during the October–June dry season, caravans of women plied the footpaths connecting a dozen different fenced perimeters to the village proper. Over the course of nearly twenty years, the number of women engaged in commercial production rose precipitously from the 30 selected to take part in a pilot onion project in the early 1970s to over 400 registered during an expansion project in 1984, and some 540 recorded in my own 1991 census. The arrival of the first consignments of tools and construction materials donated by developers for fencing and wells in 1978 initiated an expansion period which saw the area under cultivation more than triple in size, growing from 5.0 ha to 16.2 ha in ten years. Between 1987 and 1995, a second wave of enclosures nearly doubled that area again. At least a dozen separate projects were funded by international NGOs, voluntary agencies, and private donors. These donations were used toward the construction of thousands of meters of fence line and roughly twenty concrete-lined irrigation wells. In addition, there were some 1,370 hand-dug wells and nearly 4,000 fruit trees incorporated within Kerewan's garden perimeters. Growers purchased seed, fertilizer, and other inputs directly from an FAO [Food and Agriculture Organization]-sponsored dealership in the community and sent truckloads of fresh produce to market outlets located up and down the Gambia-Senegal border, which thrived on the vegetable trade. In sum, the Kerewan area developed over two decades into one of the most intensive vegetable-producing enclaves in the country. . . .

Theories Connecting Gender, Development, and the Environment

The image most widely used to capture the "plight" of Third World women is that of an African peasant woman toting an improbably large and unwieldy bundle of firewood on her head. She may or may not have a young child tied to her back, but the image is always meant to convey that she has traveled a great distance to gather her load. As a metaphor, this feminine icon suggests the incredible burdens women shoulder, and the great lengths they go to, to satisfy the multiple and competing demands society and their families place on them. The implication is that women suffer these conditions universally, as a class, and that pure, selfless motives drive them to undertake routinely dull, repetitive, and ultimately thankless tasks. At the same time, the graphic portrayal of firewood collectors is meant to underscore the idea that close connections exist between women and the natural environment. It suggests that women forced to gather wood from the countryside lead a hand-to-mouth existence, where knowledge of the landscape is bred of necessity and deep personal experience, and where the vagaries of climate and ecology have profound and immediate implications for human well-being. Thus, by virtue of their collective lot in a singular division of labor, women mediate the relationship between nature and society, and they feel the brunt of natural forces as a consequence.

Such images convey a stark reality: life for peasant women is often filled with considerable toil and drudgery. Yet if these women suffer a common plight, it resides not in any particular niche in some all-encompassing division of labor but in the countless ways the range and variety of their lived experiences are distorted in the words and images conveyed by outsiders. The wood-gathering icon represents Third World women as Africans, African women as peasants, and peasant women as a single type. There is no geographical detail at either localized or macropolitical scales that might serve as an explanation for the plight thus portrayed. Moreover, to render such women as beasts of burden, dumb, stolid, unwavering in their support of their families, unstinting in their service of same, is to acquiesce in the notion that they are perpetual victims, steeped in need, and incapable or disinclined to contest their lot creatively.

This tension between images of women as victims and women as autonomous actors traces back to the earliest efforts of developers to promote Women in Development (WID) programs in the Third World. The United Nations–sponsored convocation in Mexico City in 1975 proclaimed an International Decade for Women and initiated efforts within the major development agencies to address a broad agenda of issues deemed especially pertinent to Third World women. . . .

Gone to Their Second Husbands: Domestic Politics and the Garden Boom

One of the offshoots of the surge in female incomes and the intense demands on female labor produced by the garden boom was an escalation of gender politics centered on the reworking of what [Anne] Whitehead once called the "conjugal contract." In Kerewan, the political engagement between gardeners and their husbands can be divided into two phases. The first phase, comprising the early years of the garden boom, was characterized by a sometimes bitter war of words. In the context of these discursive politics, men whose wives seemed preoccupied with gardening claimed that gardens dominated women's lives to such a degree that the plots themselves had become the women's "second husbands." Returning the charge, their wives replied, in effect, that they may as well be married to their gardens: the financial crisis of the early 1980s had so undermined male cash-crop production and, by extension, husbands' contributions to household finances, that gardens were often women's only means of financial support during this period.

As the boom intensified, so, too, did intra-household politics. The focus of conflict in the second phase—which extended into the mid-1990s—was the role of garden income in meeting household budgetary obligations. Several studies have examined "non-pooling" households in Africa, that is, households in which men and women tend to engage in distinctly different economic activities and control their own incomes from these enterprises. The garden boom offers a case study in which women, by virtue of their new incomes, entered into intra-household negotiations over labor allocation and income disposition with certain economic advantages. The upshot of these negotiations was not, however, quite so simple. In terms of budgetary obligations, women in the garden districts assumed a broad range of new responsibilities from their husbands. Moreover, they frequently gave their husbands part of their earnings in the form of cash gifts. This outcome appears in some respects as a capitulation on the part of gardeners. I argue, . . . however, that it can also be read as symbolic deference designed to purchase the freedom of movement and social interaction that garden production and marketing entailed. In effect, gardeners used the strategic deployment of garden incomes to win for themselves significant autonomy and new measures of power and prestige, albeit not always at a price of their own choosing.

. . . Before the garden boom, men in Mandinka society had powerful economic levers at their disposal which they could, and did, use to "discipline" their wives. They controlled what little cash flowed through the rural economy due to their dominant position in groundnut production and were able to fulfill or deny a range of their wives' expressed needs at will. These included such basic requirements as clothing, ceremonial expenses (naming ceremonies, circumcisions, and marriages for each individual woman's children), housing amenities, and furnishings. The power vested in control over cash income was only enhanced by polygamous marital practices and the opportunities they afforded to play wives off against one another. A second advantage was derived from the husband's rights in divorce proceedings. In the event of a divorce, Mandinka customary law requires that the bride's family refund bridewealth payments. Consequently, when marital relations reach an impasse, divorce is not automatic; the financial arrangement between the two families must first be undone. Typically, the woman flees or is sent back to her family so that they can ascertain to their own satisfaction whether she has made a good faith effort to make her marriage work. The onus is on the woman to prove her case, however, and she is not infrequently admonished by her own family to improve her behavior before being returned to her husband.

The advent of a female cash-crop system reduced the significance of both these sources of leverage, not least because women's incomes had outstripped their husbands' in many cases. A rough comparison of the garden incomes of women in Kerewan and Niumi Lameng and the earnings their husbands reported from groundnut sales showed that 81 percent and 47 percent of women in the Niumi Lameng and Kerewan samples, respectively, earned more cash than their husbands from sales of these crops. This reversal of fortunes changed fundamentally the way male residents of the garden districts dealt with their wives:

> Before gardening started here, if you saw that your wife had ten dalasis you would ask her where she got it. At that time, there was no other source of income for women except their husbands. . . . But nowadays a woman can save more than two thousand dalasis while

the husband does not even have ten dalasis to his name. So now men cannot ask their wives where they get their money, because of their garden produce.

—Gardener's husband

Indeed, the garden boom reduced male authority ("If she realizes she is getting more money than her husband, she may not respect him"), and the extent of gardeners' economic influence expanded proportionately. The simple fact that women could largely provide for themselves ("If we join [our husbands] at home and forget [our gardens in] the bush, we would all suffer. . . . Even if he doesn't give you [what you want], as long as you are doing your garden work, you can survive") constituted a serious challenge to the material and symbolic bases of male power. In the first phase of conflict brought on by the boom, men openly expressed their resentment in pointed references to female shirking and selfishness. Their feelings were also made plain in actions taken by a small minority who forbade their wives to garden, or agitated at the village level to have gardening banned altogether. In the second phase, men dropped their oppositional rhetoric, became more generally cooperative, and began exploring ways to benefit personally from the garden boom. Sensing the shift in tenor of conjugal relations, women, accordingly, began a prolonged attempt to secure the goodwill necessary to sustain production on a more secure basis.

. . . Survey data show that both senior members of garden work units and women working on their own took on many economic responsibilities that were traditionally ascribed to men. Fifty-six percent of the women in the Kerewan sample, for example, claimed to have purchased at least one bag of rice in 1991 for their families. The great majority bought all of their own (95%), and their children's (84%), clothing and most of the furnishings for their own houses. Large numbers took over responsibility for ceremonial costs from their husbands, such as the purchase of feast day clothing (80%), or the provision of animals for religious sacrifice. Many paid their children's school expenses. In a handful of cases, gardeners undertook major or unusual expenditures such as roofing their family's living quarters, providing loans to their husbands for purchasing draught animals and farming equipment, or paying the house tax to government officials. There are, once again, unfortunately no baseline data that could be used to gain historical perspective on this information. Nonetheless, several male informants stated unequivocally that, were it not for garden incomes, many of the marriages in the village would simply fail on the grounds of "non-support.". . .

Gone to Their Second Husbands

It is fair to say that domestic budgetary battles did not originate with the garden boom in Mandinka society; nor are they wholly unique to either The Gambia or Africa. Nonetheless, the Gambian garden boom clearly produced dramatic changes in the normative expectations and practices of marital partners in the country's garden districts. . . .

The price of autonomy notwithstanding, women in The Gambia's garden districts succeeded in producing a striking new social landscape—by embracing the challenges of the garden boom, they placed themselves in a position to carefully extricate themselves from some of the more onerous demands of marital obligations. Indeed, in a very real sense, they won for themselves "second husbands" by rewriting the rules governing the conjugal contract. Thus the product of lengthy intra-household negotiations brought on by the garden boom was not the simple reproduction of patriarchal privilege and prestige; it was instead a new, carefully crafted autonomy that carried with it obligations and considerable social freedoms.

Double Standards: Women's Property Rights Violations in Kenya

Summary

Shortly after Emily Owino's husband died, her in-laws took all her possessions—including farm equipment, livestock, household goods, and clothing. The in-laws insisted that she be "cleansed" by having sex with a social outcast, a custom in her region, as a condition of staying in her home. They paid a herdsman to have sex with Owino, against her will and without a condom. They later took over her farmland. She sought help from the local elder and chief, who did nothing. Her in-laws forced her out of her home, and she and her children were homeless until someone offered her a small, leaky shack. No longer able to afford school fees, her children dropped out of school.

—Interview with Emily Owino, Siaya, November 2, 2002

When Susan Wagitangu's parents died, her brothers inherited the family land. "My sister and I didn't inherit," said Wagitangu, a fifty-three-year-old Kikuyu woman. "Traditionally, in my culture, once a woman gets married, she does not inherit from her father. The assumption is that once a woman gets married she will be given land where she got married." This was not the case for Wagitangu: when her husband died, her brothers-in-law forced her off that homestead and took her cows. Wagitangu now lives in a Nairobi slum. "Nairobi has advantages," she said. "If I don't have food, I can scavenge in the garbage dump."

—Interview with Susan Wagitangu, Nairobi, October 29, 2002

Women's rights to property are unequal to those of men in Kenya. Their rights to own, inherit, manage, and dispose of property are under constant attack from customs, laws, and individuals—including government officials—who believe that women cannot be trusted with or do not deserve property. The devastating effects of property rights violations—including poverty, disease, violence, and homelessness—harm women, their children, and Kenya's overall development. For decades, the government has ignored this problem. Kenya's new government, which took office in January 2003, must immediately act to eliminate this insidious form of discrimination, or it will see its fight against HIV/AIDS (human immuno-deficiency virus/acquired immune deficiency syndrome), its economic and social reforms, and its development agenda stagger and fail.

This report recounts the experiences of women from various regions, ethnic groups, religions, and social classes in Kenya who have one thing in common: because they are women, their property rights have been trampled. Many women are excluded from inheriting, evicted from their lands and homes by in-laws, stripped of their possessions, and forced to engage in risky sexual practices in order to keep their property. When they divorce or separate from their husbands, they are often expelled from their homes with only their clothing. Married women can seldom stop their husbands from selling family property. A woman's access to property usually hinges on her relationship to a man. When the relationship ends, the woman stands a good chance of losing her home, land, livestock, household goods, money, vehicles, and other property. These violations have the intent and effect of perpetuating women's dependence on men and undercutting their social and economic status.

Women's property rights violations are not only discriminatory, they may prove fatal. The deadly HIV/AIDS epidemic magnifies the devastation of women's property violations in Kenya, where approximately 15 percent of the population between the ages of fifteen and forty-nine is infected with HIV. Widows who are coerced into the customary practices of "wife inheritance" or ritual "cleansing" (which usually involve unprotected sex) run a clear risk of contracting and spreading HIV. The region where these practices are most common has Kenya's highest AIDS prevalence; the HIV infection rate in girls and young women there is six times higher than that of their male counterparts. AIDS deaths expected in the coming years will result in millions more women becoming widows at younger ages than would otherwise be the case. These women and their children (who may end up AIDS orphans) are likely to face not only social stigma against people affected by HIV/AIDS but also deprivations caused by property rights violations.

A complex mix of cultural, legal, and social factors underlies women's property rights violations. Kenya's customary laws—largely unwritten but influential local norms that coexist with formal laws—are based on patriarchal traditions in which men inherited and largely controlled land and other property, and women were "protected" but had lesser property rights. Past practices permeate contemporary customs that deprive women of property rights and silence them when those rights are infringed. Kenya's constitution prohibits discrimination on the basis of sex, but undermines this protection by condoning discrimination under personal and customary laws. The few statutes that could advance women's property rights defer to religious and customary property laws that privilege men over women. Sexist attitudes are infused in Kenyan society: men that Human Rights Watch interviewed said that women are untrustworthy, incapable of handling property, and in need of male protection. The guise of male "protection" does not obscure the fact that stripping women of their property is a way of asserting control over women's autonomy, bodies, and labor—and enriches their "protectors."

Currently, women find it almost hopeless to pursue remedies for property rights violations. Traditional leaders and governmental authorities often ignore women's property claims and sometimes make the problems worse. Courts overlook and misinterpret family property and succession laws. Women often have little awareness of their rights and seldom have means to enforce them. Women who try to fight back are often beaten, raped, or ostracized. In response to all of this, the Kenyan government has done almost nothing: bills that could improve women's property rights have languished in parliament and government ministries have no programs to promote equal property rights. At every level, government officials shrug off this injustice, saying they do not want to interfere with culture.

As important as cultural diversity and respecting customs may be, if customs are a source of discrimination against women, they—like any other norm—must evolve. This is crucial not only for the sake of women's equality, but because there are real social consequences to depriving half the population of their property rights. International organizations have identified women's insecure property rights as contributing to low agricultural production, food shortages, underemployment, and rural poverty. In Kenya, more than half of the population lives in poverty, the economy is a disaster, and HIV/AIDS rates are high. The agricultural sector, which contributes a quarter of Kenya's gross domestic product and depends on women's labor, is stagnant. If Kenya is to meet its development aims, it must address the property inequalities that hold women back.

Unequal property rights and harmful customary practices violate international law. Kenya has ratified international treaties requiring it to eliminate all forms of discrimination against women (including discrimination in marriage and family relations), guarantee equality before the law and the equal protection of the law, and ensure that women have effective remedies if their rights are violated. International law also obliges states to modify discriminatory social and cultural patterns of conduct. Kenya is violating those obligations.

With a new government in office and a new draft constitution containing provisions that would enhance women's property rights set for debate, this is a pivotal time for Kenya to confront the deep property inequalities in its society. It must develop a program of legal and institutional reforms and educational outreach initiatives that systematically eliminates obstacles to the fulfillment of women's property rights.

Conclusion

Women's property issues touch deeply the ways people live, think, and organize their social and economic lives. It's not just a matter of getting a few women in parliament. People feel threatened.

—Professor Yash Pal Ghai, chairman, Constitution of Kenya Review
Commission, Nairobi, October 23, 2002

Property rights abuses inflicted on women in Kenya should be recognized for what they are: gross violations of women's human rights. Discriminatory property laws and practices impoverish women and their dependents, put their lives at risk by increasing their vulnerability to HIV/AIDS and other diseases, drive them into abhorrent living conditions, subject them to violence, and relegate them to dependence on men and social inequality.

Despite the slow recognition that property rights violations harm not just women and their dependents but Kenya's development as a whole, little has been done to prevent and redress these violations. Averting these abuses in a country where dispossessing women is considered normal will be difficult. A concerted effort is needed not just to improve legal protections, but to modify customary laws and practices and ultimately to change people's minds. With extreme poverty, a moribund economy, rampant violence, and catastrophic HIV/AIDS rates, Kenya can no longer afford to ignore women's property rights violations. Eliminating discrimination against women with respect to property rights is not only a human rights obligation; for many women, it is a matter of life and death.

POSTSCRIPT

Are Women in a Position to Challenge Male Power Structures in Africa?

In many ways, the viewpoints presented in this issue get at a deeper debate about social change and the best way to improve the situation of women in Africa. The selection by Human Rights Watch presents the local situation for women in Kenya as deplorable and intractable, suggesting that a top-down, legislative solution is the best course of action. Critics of this approach might argue that, while this is all well and good, it is largely ineffectual as the reach of government is fairly limited in many African contexts. The case study presented by Schroeder about women in The Gambia provides ammunition for those who suggest that a bottom-up approach that is focused on economic empowerment is the best avenue to greater gender equality in Africa. Imagine, for example, what type of social change might occur in the United States if women earned more on average than men (the situation in Kerawan). This can be compared with U.S. Bureau of Labor Statistics survey results showing that American women earned 77 percent of their male counterparts' salaries in 1999. However, it should be noted that a reading of Schroeder's entire volume (of which a small portion was excerpted for this issue) reveals that the situation was later constrained because women's access to land for gardening was somewhat tenuous. Many men who had temporarily loaned land to women for gardening began reasserting their rights to these plots in the 1990s. As such, it may be that both top-down (i.e., legislative) and bottom-up approaches are needed in order to improve the situation of women in Africa.

The case presented by Schroeder is not an isolated incident of economically empowered women in Africa. Another classic example concerns the "Nanas-Benz" of Togo who are wealthy cloth merchants. They are emblematic of how successful women can be in the West African marketplace. *Nanas* means "established woman" or "woman of means." Benz refers to the type of auto preferred by these market women. The most successful of these merchants can turn over about $600,000 in cloth per month. They act as agents between importers and wide-ranging clientele in West Africa. Successful Nanas-Benz make sure that their children attend a university. The girls study economics, management, and administration while the boys become architects, teachers, and bankers. The business is often passed down to a woman's female children.

While WID programs still exist today, there has been an effort to move beyond stand-alone programs focused on women to attention and awareness of the situation and needs of women in all types of programs and policy initiatives. This broader approach is often simply referred to as "gender" or "gender and development." For examples of the WID and gender approaches, see relevant sections of the Web sites of the United States Agency for International Development, http://www.usaid.gov/wid/links.htm, and the World Bank, http://www.worldbank.org/gender/.

ISSUE 15

Is the International Community Focusing on HIV/AIDS Treatment at the Expense of Prevention in Africa?

YES: Andrew Creese, Katherine Floyd, Anita Alban, and Lorna Guinness, from "Cost-Effectiveness of HIV/AIDS Interventions in Africa: A Systematic Review of the Evidence," *The Lancet* (2002)

NO: Philip J. Hilts, from "Changing Minds: Botswana Beats Back AIDS," in *Rx for Survival: Why We Must Rise to the Global Challenge* (Penguin Books, 2005)

ISSUE SUMMARY

YES: Andrew Creese and his colleagues, who work for the World Health Organization (WHO) and European universities, suggest that cost-effectiveness is an important criterion for deciding how to allocate scarce health care funding. A case of HIV/AIDS can be prevented for $11 by selective blood safety measures and targeted condom distribution with treatment of sexually transmitted diseases. In contrast, antiretroviral treatment for adults can cost several thousand dollars. They argue that a strong economic case exists for prioritizing preventive interventions and TB treatment.

NO: Philip J. Hilts, who teaches journalism at Boston University, describes a comprehensive HIV/AIDS program in Botswana. This program offered not only preventive care, but sophisticated triple drug AIDS treatments to all people of the nation, free of charge. By 2005, the program was treating 43,000 people and the cost of treatment is one-tenth of what it is in the United States.

Slowing the spread and impact of HIV/AIDS is one of the greatest challenges facing contemporary Africa. Unlike some other diseases, AIDS is particularly problematic because it strikes the working-age population and thus has serious economic and social consequences. In 1999 AIDS became the leading cause of death in Africa, overtaking malaria. As of 2007, the Joint United Nations

Programme on HIV/AIDS (UNAIDS) reported that 22.5 million adults and children were living with the HIV virus in Sub-Saharan Africa (SSA), which accounts for 68 percent of the infected population worldwide. SSA also accounted for more than three-quarters of AIDS related deaths in 2007, as well as 68 percent of new infections (1.7 of 2.5 million). Unlike other world regions, the majority of people living with HIV in SSA are women (61%). The overall infection rate in SSA is estimated at 5 percent as compared to .8 percent worldwide.

Southern Africa is the worst affected region with eight countries with a national adult prevalency rate that exceeded 15 percent in 2005 (including Botswana, Lesotho, Mozambique, Namibia, South Africa, Swaziland, Zambia, and Zimbabwe). South Africa is the country in the world with the most HIV infections, but like many other countries in Sub-Saharan Africa, its adult prevalency rate is stable or declining.

The spatial pattern of the disease has changed over time. Initially, it was distributed along major highways and in the urban centers of eastern and southern Africa. In the 1990s, the epicenter of the virus moved to southern Africa. In Africa, the disease is largely transmitted through unprotected heterosexual sex and unsafe medical practices. Truck drivers, sex workers, and military personnel all have above-average infection rates and are believed to play a significant role in the spread of the virus.

Until the 2000s, much of the HIV/AIDS work in Africa focused on prevention as opposed to treatment. Uganda was touted as an example of a country that aggressively pursued prevention and was able to dramatically reduce its rate of new infections. The exorbitant costs of antiretroviral drugs also seemed to make such expenditures unrealistic. But increasingly, access to such drugs at affordable prices began to be framed as a human rights issue. It was also argued that African countries could not afford to lose so much human capital to early death.

In this issue, Andrew Creese and coauthors, who work for the World Health Organization (WHO) and European universities, suggest that cost-effectiveness is an important criterion for deciding how to allocate scarce health care funding. A case of HIV/AIDS can be prevented for $11 by selective blood safety measures and targeted condom distribution with treatment of sexually transmitted diseases. In contrast, antiretroviral treatment for adults can cost several thousand dollars. They argue that a strong economic case exists for prioritizing preventive interventions and TB treatment. Cost-effectiveness criteria could also be used to decide between different types of treatment.

Philip Hilts, who teaches journalism at Boston University, describes a comprehensive HIV/AIDS program in Botswana. This program offered not only preventive care, but sophisticated triple drug AIDS treatments to all people of the nation, free of charge. By 2005, the program was treating 43,000 people and the cost of treatment is one-tenth of what it is in the United States. This program benefited from substantial levels of external funding from the Merck pharmaceutical company.

YES ⬏   Andrew Creese, Katherine Floyd, Anita Alban, and Lorna Guinness

Cost-Effectiveness of HIV/AIDS Interventions in Africa: A Systematic Review of the Evidence

Introduction

HIV/AIDS accounts for about 20% of all deaths and disability-adjusted life-years (DALYs) lost in Africa, which makes it the biggest single component of the continent's disease burden. The epidemic has reduced life expectancy in the worst affected countries by more than 10 years, and its social and economic consequences have been devastating.

Substantial new resources are becoming available for prevention, care, and support. The European Commission is committed to a major increase in spending on the diseases of poverty, including HIV/AIDS. A global fund to fight AIDS, tuberculosis, and malaria became operational in January, 2002; so far pledges are in the region of US$2 billion. . . .

To ensure that any new resources have the maximum possible effect on the epidemic, cost-effectiveness should be considered in the design of strategies for prevention, care, and support. As Kahn and Marseille have pointed out, the scale of the HIV/AIDS epidemic combined with scarcity of resources makes cost-effectiveness especially important in developing countries. Up to now, however, cost-effectiveness has been well documented only for industrialised countries. For low-income and middle-income countries, we could identify only one detailed review, which addressed interventions to reduce mother-to-child transmission. For Africa, investigators focused on individual HIV/AIDS-related interventions. We could not identify any published report that brought together the evidence base in a standardised way that allowed comparison among interventions.

We report a critical assessment of studies of the cost-effectiveness of HIV/AIDS interventions in Africa, and present their results in a standard form.

Methods

Review of Published Work

We searched Medline, Popline, and EconLit databases for 1984–2000 using the key words HIV, AIDS, and HIV/AIDS in combination with each of the terms: costs; cost-effectiveness; cost-benefit analysis; economics; and Africa. Citations and reference lists were then reviewed to identify any additional relevant studies. Abstracts from international conferences were searched but were not included because they provided insufficient detail. Unpublished data were obtained through contact with experts in HIV/AIDS. A total of 57 studies and nine reviews were identified, including several unpublished reports and presentations.

Criteria for Inclusion and Exclusion of Identified Studies

We assessed each study using a standard checklist. We then decided in three stages about inclusion in our review. First, we included any study that met all these five criteria: (i) the report contained data for Africa; (ii) it measured both cost and effectiveness; (iii) it seemed to use standard methods for estimating costs and outcomes, (iv) it seemed to include all major cost items, and (v) it allowed a generic measure of outcome (either HIV infections prevented or DALYs gained) to be calculated. We focused on studies in which investigators had analysed costs and effects together, rather than reviewing evidence on costs and effects separately, because the two items are not independent of each other.

Second, studies that met these inclusion criteria were excluded if (a) they were about regimens that are now out of date, such as long-course zidovudine for prevention of mother-to-child transmission; (b) they had estimated the effectiveness of an intervention before clinical trial results were available, and subsequent cost-effectiveness studies had used clinical trial results in their effectiveness estimates; or (c) drug prices had altered substantially since publication. We therefore excluded three studies of interventions to reduce mother-to-child transmission (table 1).

Third, we identified interventions not covered by studies meeting the five initial inclusion criteria, but for which some cost and effectiveness data existed. We identified two such interventions—highly active antiretroviral treatment (HAART) for HIV-positive adults, and promotion of female condoms. In view of the current importance of antiretroviral treatment, we decided to include a study that used only drug costs, even though drug prices have fallen since its publication, and we could only calculate a cost per life-year gained rather than a cost per DALY gained from the data presented. To provide a more recent estimate of cost-effectiveness, we used laboratory test costs for antiretroviral therapy in adults enrolled in the HIV drugs access initiatives in Uganda and Côte d'Ivoire, and the cost of drugs cited by Médecins Sans Frontières in 2001. An unpublished study of promotion of female condoms was included only after written communication with its authors.

Standardisation of Studies Meeting Inclusion Criteria

Thus, we included 24 of the initial 66 studies identified. Data from these studies spanned 13 years (1988–2000), and differed widely in their methods and assumptions. A few studies had primary data for both costs and outcomes, but most used epidemiological models to estimate effectiveness. In the modelling, some studies included analysis of the secondary infections prevented by an

Table 1

HIV/AIDS Intervention Groups, Individual Interventions, and Standardised Cost-Effectiveness Results, US$ for Year 2000

Intervention groups (numbered) and individual interventions	Place and year of publication	Cost per HIV infection prevented	Cost per DALY gained
Prevention			
1. Condom distribution			
Condom distribution plus STD treatment for prostitutes	Sub-Saharan Africa, 1991	11–17	1
Female condoms targeted to:			
Prostitutes	Kenya, 1999	275	12
High-risk women	Kenya, 1999	1066	48
Medium-risk women	Kenya, 1999	2188	99
2. Blood safety			
Hospital based screening	Tanzania, 1999	18	1
	Zambia, 1995	107	5
Strengthening blood transfusion services through:			
Defer high risk donors	Zimbabwe, 1995	18–107	1–5
Test and defer high risk donors	Zimbabwe, 1995	48–74	2–3
Rapid test	Zimbabwe, 1995	62	3
Improved transfusion safety with outreach	Zimbabwe, 2000	208–256	10–12
Improved blood collection and transfusion	Tanzania, 1999	950	43
3. Peer education for prostitutes	Cameroon, 1998	79–160	4–7
4. Prevention of mother-to-child transmission			
Single dose nevirapine-targeted	Sub-Saharan Africa, 2000	20–341	1–12
	Uganda, 1999	308	10
Single-dose nevirapine-universal	Uganda, 1999	143	5
	Sub-Saharan Africa	268	9
Petra regimen	South Africa, 1999	268	9
ZDV CDC Thai regimen	South Africa, 2000	949–2198	33–75
	South Africa, 1999	2356	81
Formula recommendation	South Africa, 1999	3834	131
Breastfeeding 3 months	South Africa, 1999	5006	171
Formula provision	South Africa, 1999	6355	218
Breastfeeding 6 months	South Africa, 1999	21 355	731
5. Diagnosis and treatment of STIs	Tanzania, 1997	271	12
6. Voluntary counselling and testing	Kenya and Tanzania, 2000	393–482	18–22

(Continued)

Table 1   (Continued)

Intervention groups (numbered) and individual interventions	Place and year of publication	Cost per HIV infection prevented	Cost per DALY gained
Treatment and care			
7. Short-course tuberculosis treatment for new sputum-smear positive pulmonary patients			
Ambulatory care	Malawi, Mozambique, Tanzania, 1991	n/a	2–3
	Uganda, 1995	n/a	2–4
	South Africa, 1997	n/a	8–16
IUATLD model (involves 2 months hospitalisation at treatment outset followed by monthly visits to a health clinic to collect drugs)	Uganda, 1995	n/a	3–4
	Malawi, Mozambique, Tanzania, 1991	n/a	4–8
	South Africa, 1997	n/a	34–68
Community-based DOT	South Africa, 1997	n/a	14–21
8. Co-trimoxazole prophylaxis for HIV+ tuberculosis patients	Hypothetical low income country, sub-Saharan Africa	n/a	6
9. Home-based care			
Community-based programme	Tanzania, 2000	n/a	77
	Zambia, 1994	n/a	99
Health facility based programme	Zambia, 1994	n/a	681
	Tanzania, 2000	n/a	786
	Zimbabwe, 1998	n/a	469–1230
10. Preventive therapy for tuberculosis			
Isoniazid for 6 months	Uganda, 1999	n/a	169
Rifampicin plus pyrazinamide, 2 months	Uganda, 1999	n/a	282
Isoniazid plus rifampicin, 3 months	Uganda, 1999	n/a	288
11. Antiretroviral therapy for adults	Senegal and Côte D'Ivoire, 2000	n/a	1100
	South Africa, 2000	n/a	1800

intervention, whereas others did not; and different values were used for some variables (e.g., the efficiency of HIV transmission) that determine effectiveness. Several studies included an analysis of treatment-cost savings but most did not; others also included savings from averting loss of productivity in their calculations. In most studies, investigators focused on costs from a provider perspective only, but a few also looked at costs incurred by patients. Different prices were used, particularly for antiretroviral drugs, whose prices and regimens have changed substantially in the past 5 years. Discount rates, effectiveness measures, the reporting of costs and effects, assumed life expectancy at birth, and the year in which costs were assessed also varied.

To ensure the widest possible comparability among interventions, we standardised both cost and effectiveness data; therefore, the figures we report differ from the results shown in the original publications. Standardisation of cost data included the year of prices, the price of 1 year of triple combination therapy, how costs were assessed, and savings related to averted treatment costs and productivity losses. For effectiveness, we undertook no new modelling. However, we standardised the discount rate used to estimate the present value of future health gains; life expectancy at birth; average age at HIV infection; assumptions for tuberculosis treatment, including years of life gained through cure and death rates in the absence of treatment; the disability weighting associated with years of life lived with AIDS; and the frequency of home-based care visits.

For all studies we calculated unit costs and effectiveness. Once both had been standardized, we calculated two measures of cost-effectiveness: (1) cost per HIV infection averted (for the preventive interventions) and (2) cost per DALY gained (for all interventions). Sensitivity analyses were excluded if they were based on variable measurements (e.g., life expectancy at birth, discount rate) for which we had already standardised results, or if there was too little detail to allow recalculation of figures. . . .

Results

Cost per HIV Infection Prevented

There was a wide range in the cost per HIV infection prevented. Costs for condom distribution ranged from as little as $11 to over $2000. Measures to improve blood safety cost between just under $20 and about $1000 to prevent one case of HIV. There was especially large variation in the different strategies to reduce mother-to-child transmission. Breastfeeding and formula-feeding interventions cost from around $4000 to over $20 000 per infection prevented, whereas single-dose nevirapine cost much less—about $20–341. Diagnosis and treatment of sexually transmitted infections cost just over $270 per infection prevented, and the figure for voluntary counselling and testing (VCT) was higher, at around $400–500.

Cost per DALY Gained

The cost per DALY gained by interventions ranged from around $1 for a combined treatment of sexually-transmitted disease (STD) and condom promotion programme and for blood screening, to well over $1000 for HAART in adults. Blood safety measures, and single-dose nevirapine for prevention of mother-to-child transmission, cost as little as $10 per DALY gained. Tuberculosis treatment could also be less than $10 per DALY gained, but as high as $68 when inpatient care was involved. VCT and co-trimoxazole prophylaxis for HIV-positive patients with tuberculosis cost around or below $20. Home-based care varied from around $100 to $1000, with community based care programmes having a lower cost per DALY than programmes organised from health facilities.

Discussion

Our results show that there are few studies of the cost-effectiveness of HIV/ AIDS prevention, treatment, and care interventions in Africa, and there is considerable variability in the cost-effectiveness of such interventions. The most cost-effective interventions are for prevention of HIV/AIDS and treatment of tuberculosis, whereas HAART for adults, and home based care organised from health facilities, are the least cost effective. For some interventions, such as prevention of mother-to-child transmission, tuberculosis treatment, and home based care, there are particular strategies that provide the best value for money (best buy).

This review has several limitations. For five interventions, only one study was identified, and the maximum number of studies—for mother-to-child transmission—was four. In no one country were all interventions assessed, which made unbiased comparison of interventions difficult. Cost data were not always comprehensive, and were sometimes too few for standardised sensitivity analysis. The cost of HAART was underestimated, because data for only a very restricted subset of costs were considered. There were no data for the costs of use and strengthening of general health services necessary for provision of HAART. The effect of some interventions on HIV prevention might have been underestimated because some potential effects that are difficult to measure—such as reduced stigma arising from increased knowledge of status—were not accounted for. None of the studies on interventions to reduce vertical transmission looked at the effect of VCT on horizontal transmission. The effectiveness of HAART might have been underestimated because we had insufficient data to measure its effect on transmission through lowering viral loads. It could also have been overestimated. First, its use might increase transmission since risky behaviour by HIV-positive people with improved life expectancy could be encouraged. Second, side-effects mean that the value of 1 year of life is likely to be less than the 1 DALY assumed here. Some studies are based on project implementation at only a few sites (for example the study of VCT), or on theoretical analyses of interventions (e.g., some studies of mother-to-child transmission). Thus, costs and effects in practice and on a large scale might be different from those shown. Finally, some interventions may complement each other in ways that are missed in analyses of individual interventions.

These limitations mean that both generalisability and interpretation should be viewed with caution. Ideally, we would have data for every intervention from several studies in similar settings—both income levels and prevalence rates can distort comparisons within and between countries. Salaries are linked to average national income and can thus affect costs. HIV prevalence does not affect the cost-effectiveness of every intervention but, where costs are incurred in diagnosis of a case of HIV (such as with VCT), the lower the prevalence, the higher the cost per HIV-positive case detected. For example, all studies of prevention of mother-to-child transmission show a relation between prevalence rates and cost-effectiveness.

Our review includes data from low-income countries in each intervention group, typically with high HIV prevalence. For mother-to-child transmission and tuberculosis treatment, we included data from both low-income and middle-income countries with a wide range in HIV prevalence, and the rankings of the types of intervention were consistent. For some other interventions, such as tuberculosis prevention, costs are likely to be higher in wealthier countries with lower rates of HIV infection. Two possible exceptions are blood safety and VCT. For blood safety, the major costs are probably supplies and equipment, which are likely to be similar across countries. For VCT, the estimated cost was similar to other estimates that have been made for Africa. A drawback to the VCT data is that the study used an index for HIV transmission efficiency that was ten times that typically used by the UN programme on HIV/AIDS. Together with very high rates of reported behaviour change, we might have overstated the effectiveness of this intervention elsewhere.

The evidence base could be improved by more cost-effectiveness studies that included all economic costs and used standard methods. Guidelines for cost-effectiveness analysis, including those for HIV/AIDS prevention, should be more widely and rigorously used. Ideally, analyses for several interventions in a single setting should be undertaken. In view of the powerful advocacy for access to antiretroviral therapy for HIV-infected adults, and the poor evidence currently available, work on the cost and effectiveness of such treatment in African settings is a priority. But in five other intervention areas—peer education for prostitutes, diagnosis and treatment of STDs, VCT, prevention therapy for tuberculosis, and co-trimoxazole prophylaxis for HIV-positive patients with tuberculosis—we depend on the results of only one study. Moreover, apart from tuberculosis, there are no data for treatment of opportunistic infections. New analysis could initially focus on interventions for which we have effectiveness data, but for which costs are not documented, and vice versa.

How can the existing data be used to inform policy? Cost-effectiveness rankings do not, on their own, indicate which health interventions are priorities for public funding. A recent framework based on seven questions has proposed that an intervention should be publicly funded if it is cost effective and it is (1) a public good; or (2) associated with important externalities and demand is inadequate; or (3) represents a catastrophic cost and insurance is not available; or (4) beneficiaries are poor.

Table 2 shows how this economic framework supplements the cost-effectiveness data we have collated. The use of a more comprehensive framework makes little difference. No intervention is ruled out with the first six questions, and the determining factor for public finance is cost-effectiveness.

Despite the limitations of our review and difficulties with generalisation, cost-effectiveness can be used for some broad prioritisation among interventions. The World Development Report of 1993 suggested that any intervention achieving a DALY gain for $50 or less ($62 in year 2000 prices) was highly cost effective in the context of the poorest countries. The general inference was that these interventions should be made available to all those in need before less cost-effective options are provided to a few. On this basis, several preventive interventions (targeted condom distribution, blood screening, nevirapine for the prevention of mother-to-child transmission and STD treatment), and two treatment interventions (co-trimoxazole prophylaxis for patients with HIV and tuberculosis) and tuberculosis treatment should have first call on new funds for HIV/AIDS in Africa. Within intervention categories first priority should be given to the intervention that is a clear best buy—for example, short-course nevirapine treatment for mothers and babies, and targeted condom distribution.

Table 2

Economic Factors Affecting Priority of Health Interventions for Public Funding

	Public good?	Important externalities	Adequate demand?	Catastrophic cost?	Voluntary insurance available for catastrophic cost?	Benefit group poor?	Cost effective (US$ cost per DALY)
Condom distribution	No	Yes	No	No	N/a	Yes	1–99
Blood safety	No	Yes	Yes	No	N/a	Yes	1–43
Peer education for prostitutes	No	Yes	No	No	N/a	Yes	4–7
MTCT	No	No	?	No	N/a	Yes	1–731
STDs	No	Yes	No	No	N/a	Yes	12
VCT	No	Yes?	No	No	N/a	Yes	18–22
TB short course	No	Yes	Yes	Yes	N/a	Yes	2–68
Co-trimoxazole prophylaxis	No	No	?	No	N/a	Yes	6
Home care	No	No	?	Yes	No	Yes	77–1230
TB preventive therapy	No	Yes	No	?	No	Yes	169–288
ARV therapy	No	?	Yes	Yes	No	Yes	1100–1800

Reading from left to right, answers to the seven questions included in the framework are suggested, in the order in which they should be asked. ARV=antiretroviral therapy, MTCT=mother-to-child transmission, TB=tuberculosis, VCT=voluntary counselling and testing. N/a=not applicable.

In practice, cost-effectiveness will need to be balanced with several other considerations. Affordability is one important issue; in the context of health budgets, a cost-effective intervention is not necessarily affordable when it is relevant to many people, and public funding will result in high demand. In Africa, this concern is most likely to apply to interventions to prevent mother-to-child transmission. Even with only restricted provision of antiretroviral treatment to HIV-positive adults, it could also become relevant for VCT services. Antiretroviral treatment for HIV-positive adults may not be as cost-effective as some other interventions, but the overwhelming pressure being placed on governments to provide such care is impossible to ignore. Recent estimates are that 20%, 40%, and 50% of health resources are already being consumed by HIV infected persons in Malawi, Zambia, and Zimbabwe, respectively.

In addition, HIV-infected people and the non-governmental organisations assisting them represent an increasingly important political force. Therefore, provision of care and support is more politically attractive, at least in the short term. Furthermore, care and support are essential parts of an enabling environment (in which people are empowered to address their difficulties) that is required to reduce discrimination and stigmatisation. By contrast, people at risk of becoming infected, the young in particular, are a more disparate and less easily organised group, with no clear cut or well articulated interests, and weak advocates. Prioritisation of care can be reinforced by difficulties in implementing or expanding the more cost-effective preventive interventions. Effective prevention strategies for the most vulnerable populations are still not scaled up to levels that could have a major impact on the HIV epidemic, even where funds are available.

The kind of cost-effectiveness evidence presented here can, however, help to inform policy decisions on resource allocation between prevention and care. For example, the results and estimates of the reachable population (i.e., the population size that is feasible to cover with each intervention) have been used to explore the consequences of alternative ways to use an additional $400 million per year. At a WHO workshop (HSI/WHO/HQ, WHO/AFRO, UNAIDS. Costing and prioritisation of WHO's contribution to the International Partnership Against AIDS in Africa. Geneva, Sept 4–5, 2000), participants estimated that with this increase in funding, about 750 000 more people with HIV/AIDS in Africa could be treated every year, and almost one million infections prevented (17.9 million DALYs gained). A 10% spending reallocation from treatment towards more prevention (defined as management of STDs, blood safety, VCT, prevention of mother-to-child transmission, and preventive programmes among prostitutes) would increase the total DALYs gained by over 15%.

Allocation of new funds for HIV/AIDS requires more than rankings of cost-effectiveness. Nevertheless, value for money is important, especially in African countries, where resources are particularly scarce and needs are so great. Existing cost-effectiveness data are few, and much more high quality research is needed for detailed planning and programming. Yet even the available data make it clear that a spending programme for HIV/AIDS relief in Africa that neglects to bring cost-effectiveness evidence into the consultation process risks unnecessary sacrifice of hundreds of thousands of prevention opportunities, treatment opportunities, and lives.

Changing Minds:
Botswana Beats Back AIDS

The greatest public health challenge in our time is AIDS, and the place it has wreaked the most havoc is in Africa. Sub-Saharan Africa has about 10 percent of the world's people, but about 60 percent of the people infected with HIV; in more than a dozen countries, it has become common practice to hire two or even three people to fill each job because of the certainty that workers will die within a short time after training. About 26 million people in Southern Africa are infected now, and that means, even with future successes, the epidemic will be a major presence for many years to come. If a new global health campaign is to be waged, the defeat of AIDS in Africa must be a priority. The myths about AIDS in Africa persist—that money to fight AIDS is lost to corruption, that Africans have resisted condoms and behavior change more than others. The truth is progress has been made in several places in Africa, and much more progress is expected as sophisticated HIV treatments become more accessible to Africans. The most remarkable case of progress against HIV in Africa has been in Botswana, which is now leading what has been called the "decade of treatment" in Africa.

Whoever called Africa the dark continent must have been blind, because there is very little of Africa that could be called dark. Here in the great southern expanses of the continent, day after day, the sky is empty of clouds; intense light fills the whole blue bowl.

The morning is cool and promising; then the day opens itself up and heat and light pour in like a hot liquid. The searing brightness drives cows, goats, people, and cars under trees for shade. But the evening forgives sins and cools again, relenting in a mild night amid the perfume of wood smoke and flowering trees.

Tucked away hundreds of kilometers from the coasts, Botswana is said to be the last remnant of ancient Africa, the last open wilderness on the continent. Despite the depredations of the times, forty-five thousand elephants still roam in herds in this country. The lions, zebras, and giraffes move freely on great expanses of savannah. This place was largely unreached by the colonial thrust 120 years ago from all the coasts inward. Botswana writer Bessie Head said that Botswana has always been a bit apart from the fray. "A bit of ancient Africa was left almost intact to dream along its own way."

From *Rx for Survival: Why We Must Rise to the Global Challenge* by Philip J. Hilts (Penguin, 2005), pp. 130–164. Copyright © 2005 by Philip J. Hilts. Reprinted by permission of Penguin Group (USA).

When Botswana was declared independent in 1966 its prospects were grim. It was one of the two or three poorest nations in the world, with its citizens earning fewer than two dollars per day, and nothing to give hope for more.

It sits up on high tableland, grass and bush savannah mostly, with a great desert—the Kalahari—in the middle and west of the country. There is too little rain for agriculture in most of the country, just enough to grow grasses for cattle. At the time of independence, the nation, the size of Texas, had a total of four miles of road, essentially no electricity, and no telephones. The airport was a dirt strip with a shack sometimes housing an attendant. From the shack it was a hefty hike to the new capital city, Gaborone.

But from that low state, everything changed. Diamonds were unearthed soon after; the best and purest cache in the world lay beneath the country's red and ochre soil. Many nations have rich resources and [they] squander them; not so in Botswana. The government planned wisely and built a parliamentary democracy strong on merit and clean of corruption. With the diamond money the government built schools, roads, clinics, and wells across the dry nation.

Botswana soon was the star of Africa—with the highest growth rate in the world, 9.2 percent—between 1966 and 1996. Average life spans shot up over a similar period, from 46 to 67.5 years expectancy in 1999. School enrollment went up from about 40 percent of children to 98.4 percent. Literacy went up to over 80 percent, among men and women equally. Ninety-seven percent of the Batswana have safe drinking water within 2.5 kilometers of home. Child immunization for five diseases rose to up over 95 percent.

There were problems amid the prosperity, of course. The wealth was not evenly divided. Wealth is measured in cattle in Botswana, and a little over 2.5 percent of Batswana own 40 percent of the nation's cattle. Two-thirds of the women and one-third of the men have no cattle at all. In a country with no industry and very tough farming, that is serious. Even in the midst of prosperity, unemployment remained high—it's estimated to be 40 percent. Half the people still earn less than a dollar a day, though because land is communally held and essentially free for citizens, most everyone has a place to go to lay her head, even if it's only a hut on parched ground.

At first, it looked like the horseman of pestilence would pass Botswana by. The epidemic of AIDS flared up in the United States, then in the Caribbean, then Europe, and finally it became its most intense across a belt in the middle of Africa, from Kenya to Uganda and down to Zambia. Until about 1993, Botswana was spared.

Then the epidemic struck hard, like a silent storm sweeping down the eastern boundary of the nation, dropping an unwelcome viral rain. By 2000, Botswana found itself with almost 40 percent of its adult population infected with HIV, one of the highest rates of infection ever recorded for any disease. Some towns had even higher rates—up to 52 percent in one town that serves as a way station on the truck route from Zimbabwe.

This small, successful country thus became the worst spot in what is arguably the worst epidemic in human history. The people, led by President Festus Mogae, an Oxford-trained development economist and a former official of the World Bank, began to worry about the future of the nation. As Mogae himself said, this is the worst thing that had ever happened there, bar none.

Just as matters got to their worst, a plan emerged.

The bold project in Botswana started with an idea inside the executive offices at Merck & Co. in Whitehouse Station, New Jersey, in talks with experts at the Harvard AIDS Institute and Harvard University.

The plan was to tackle an entire nation at once. Too many AIDS projects were short or limited to one aspect of the problem, such as counseling for those infected, or providing condoms, or writing educational materials. Why not take one nation and try to provide a complete set of services, from education in schools to prevention messages in the media, to condom distribution, to testing for HIV, to counseling, to full AIDS treatment in hospitals and clinics? It would be most unusual particularly because it would offer sophisticated triple-drug AIDS treatments to all the people of the nation, free of charge, permanently. Could such a comprehensive effort turn back the epidemic? That was the bet.

Merck was at the top of the pharmaceutical industry, hugely profitable. But there was trouble ahead; the entire pharmaceutical industry was beginning a terrible slide in the eyes of the public.

The drugs beginning to come on the market in 1996 were the greatest breakthrough in AIDS treatment since the beginning of the epidemic. They were a new category of drugs called "protease inhibitors." These, combined with earlier drugs, made a potent antiviral "cocktail."

The excitement over the drugs was uncontainable. Even before they were released patients were clamoring for them. People who had three hundred thousand to a million virus particles per deciliter of their blood found that the drugs knocked those counts to zero. Some small numbers of viruses undoubtedly remained, but they could not be detected.

The physical consequences of it was dramatic. Lazarus stories began to be reported of patients who were within days or hours of death when they got the drugs. Two weeks later they were up and about, gaining weight and getting ready to go back to work. Over time, the miracle was confirmed. As many as 90 percent of those infected were able to knock out the virus and restart their lives; it is now estimated that the drugs can add ten, and possibly many more, healthy years onto the life of an infected person.

There were fifty-one thousand deaths from AIDS in the United States in 1995. After the drugs appeared in the United States the death rate dropped immediately and soon was down to fifteen thousand per year, and is still dropping in 2005. Doctors are cautious about pronouncements like this, but they began to say that near normal life spans might be ahead for many of those infected with HIV.

The industry had avoided getting into the business of researching and producing AIDS drugs. But when they did, the companies made a strategic blunder. They decided to set prices for AIDS drugs at extremely high levels, enough to produce gasps even in the business press.

These drugs were true lifesavers. But the companies set their prices between ten and fifteen thousand dollars per person for a year's supply. That did not take into account the other thousands of dollars needed for medicines and doctors by people infected with HIV, and it was a price that would have to be paid every year into the future.

Worse, when the companies were asked to send the drugs to Africa and Asia for cheaper prices, they flatly refused. They feared that if there were two prices their arguments that high prices were justified would be undermined. Soon, generic manufacturers were defying Western patents and making the drug cocktails for less than $150 for a year's supply *and still making a profit.* Thus it was clear just how inflated the Western drug company prices were.

The companies made further blunders, arguing that a good price wouldn't matter in Africa because those countries didn't have the infrastructure—doctors and clinics and refrigerators—to handle proper distribution of the drugs anyway, so high prices weren't important.

Famously, Andrew Natsios, the head of the U.S. Agency for International Development, testified in Congress that Africans shouldn't get the drugs.

> If we had [the drugs] today, we could not distribute them. We could not administer the program because we do not have the doctors, we do not have the roads, we do not have the cold chain. . . . If you have traveled to rural Africa you know this, this is not a criticism, just a different world. People do not know what watches and clocks are. They do not use western means for telling the time. They use the sun. These drugs have to be administered during a certain sequence of time during the day and when you say take it at 10:00, people will say, what do you mean by 10:00?

The argument was made that if the drugs were not taken properly, and doses were missed, then the virus could quickly become resistant. That could render them useless for all.

The reaction to all this was outrage. A Harris poll showed that the drug companies were popular and respected up until 1997, when their rating by the public was 79 percent favorable. By 2000, it had dropped to 59 percent, and did not stop there. By 2004 it had sunk to 44 percent favorable and 48 percent unfavorable. People were now saying the companies were greedy and drug prices were too high. No industry in the poll's history had ever fallen so far and so fast in the public esteem.

Roy Vagelos, by then former chairman of Merck, began to condemn the companies outright. "This industry delivered miracles, and now they're throwing it all away," he said. "They just don't get it." As if to reinforce Vagelos's comments, Pat Kelly, president of Pfizer's American drug division, said when told the drug companies were now esteemed as low as cigarette companies, "We find it incredible that we could be equated to an industry that kills people as opposed to cures them."

Poorer nations said that they could not afford the drugs, but they had real AIDS emergencies and they had no choice but to seize the formulas for the drugs and make cheap copies themselves to give to their people. Brazil and India started making and using emergency "generic" AIDS drugs. In South Africa the law allowed the government to force pharmaceutical companies to grant licenses for making the drugs during an emergency. The drug companies took the South African government to court to maintain control over the drugs and their prices; and the companies lost.

Merck officials were feeling the pinch of the criticism. This was particularly difficult for Merck, which had a reputation for charitable instincts, and its executives routinely quoted their founder's remarks about how health was first and profits second.

The company had just succeeded in a philanthropic venture that was the most remarkable in the history of the pharmaceutical industry. The company has a drug that is a very powerful antiparasite medicine; it was the top seller among veterinary drugs. Scientists inside the company knew that it would also be useful against parasites in humans, but unfortunately the humans infected with parasites were not in major markets. The patients who needed the drug couldn't pay the price, so Merck at first did not make a human version of it.

In the end, Merck decided to make the drug and donate it to the people who needed it. Further, they offered to create a delivery system. By the late 1990s 25 million people per year were getting the drug. Some areas, where half the people had in the past been infected with river blindness now saw essentially no cases. (River blindness is a parasitic disease passed by the bites of black flies: the flies thrive near rivers which are also the most fertile farmland, but which must be abandoned for higher ground when too many people go blind from the bites.) The gains turned out to be more than health. Farm output jumped up, as people were able to work more and in areas previously off-limits because of infestations of worm-carrying black flies.

Another Merck project done in combination with the Harvard AIDS Institute was the "enhancing care initiative," a project to understand how best to deliver AIDS care in developing countries. It soon became clear that piecemeal projects by universities, NGOs, and governments, with poor communication among the groups, would be ineffective in addressing the epidemic.

Gradually, Merck executives and Harvard public health doctors were learning that even if the price problem could be resolved, the harder issue would be finding ways of delivering the drugs. "Improving infrastructure was key to making a difference in HIV," said Guy Macdonald, then Merck vice president. "Just donating drugs is not enough." He said it was frustrating that the public debate was about unconscionable prices, when he knew that even if the prices drop to zero, the drugs still would not get to the people who need them.

Macdonald and other executives had become so isolated from mainstream opinion that they had lost sight of the main point—whether AIDS was a tough problem in other ways wasn't the issue. Setting high prices was an act that on its own was unconscionable.

In the middle of the public relations crisis over AIDS drug prices, in August 1999, Macdonald went to a Merck management committee saying that Merck should go beyond the previous "enhancing care" project and do something bolder on AIDS. The Gates Foundation surprised Merck by offering to put in $50 million if Merck would match it. So, the scale of the project was set at $100 million, to be spent over five years in some nation that needed an anti-HIV program.

The five countries that on paper appeared to be the best candidates were those in Africa hard-hit by HIV—Uganda, Malawi, Senegal, Rwanda, and Botswana. The country chosen had to have both an HIV problem and, far more important, a government willing to tackle it. That left out many countries immediately, most especially South Africa, where the government refused to build an AIDS treatment program and rejected Western drugs.

Botswana was the smallest and likely the most manageable spot for a project. What decided the matter finally was President Festus Mogae, who told Merck and Gates at their first meeting that the project should have started yesterday; a sense of urgency was needed.

Then CEO Ray Gilmartin and Vice President Guy Mcdonald made the announcement on July 10, 2000.

What began in Botswana in 2000 soon grew into the most important experiment on AIDS ever done. No one had tackled an entire nation before with a comprehensive program including the most sophisticated AIDS drugs, especially not in Africa. It became the largest and longest running public health project on the African continent.

It was also in a new and growing category of aid, called "public-private partnerships," in which international private groups become partners with governments to carry out large projects.

The plan raised startlingly clear questions: It was said the price of the drugs was killing tens of thousands—what if the drugs were free? It was said that even if the drugs were shipped to developing nations, they could not get them to patients because governments were ineffective. It was said that even if the drugs were offered on all street corners, people wouldn't take them because they did not even want to know if they were infected, and besides they didn't believe in Western treatments. This bold project would answer those questions and more.

Why Botswana? It seems strange that such a small, peaceful country should be the center of a firestorm, the nation to reach the highest level of infection anywhere on earth. People puzzled over it for a few years, from a dozen different perspectives.

It was an unusual conjunction of three dark stars over the time and the place—unusual happenings in love, in the virus itself, and in the movement of the people along the roads that made the difference.

First, one of the worst pieces of biological news in recent decades has been the finding that HIV changes routinely. Gene trading with other bugs, it has developed a new, more troublesome strain in southern Africa. The viral type that began the HIV epidemic in America, Africa, and Europe in the early 1980s was one that came to be called HIV-lb. The kinds most common in Central and East Africa have been HIV-la and HIV-Id. Only a decade later did a new one, HIV-lc, spread in southern Africa. This one spreads and multiplies faster than any previous type. Upon genetic examination, it has been found to have at least two features that make it more dangerous than its predecessors. First, it has an ability to latch onto the surface of vaginal and penile skin cells more readily than other strains. Second, it multiplies more readily than other strains. More particles more able to attach to skin cells may make the C strain more adept at spreading.

At the peak of the epidemic in Uganda, 15 to 20 percent of pregnant women, who are the sentinel group, were infected at the peak of the epidemic in Uganda with A and D types. But in the countries infected with the C type the rates are double and triple that.

A second feature of the epidemic alighting in southern Africa is that the people in Botswana and neighboring areas are extremely mobile for people of a traditional culture. They move constantly, from farm to cattle post to city and back again, thus assuring that couples are constantly apart and vulnerable to second loves.

This mobility is at least partly a legacy of colonialism. When the British were making colonies all around Botswana, they needed labor. The solution was to impose a "hut tax" on the Tswana tribes, payable only in British money. People who would normally pay in cattle or goats were thus forced to get "paying" work. That paying work, of course, was not in their local villages. It was in the mines of South Africa, or on distant estates of white settlers. So the Batswana got used to husbands on the move and family members in general traveling to get work.

The impact of this high mobility is that there is no settled family life. Typically, mothers and fathers do not raise their children themselves; grandmothers and aunts raise the children. Young couples are often away from each other, and during these times they develop other relationships.

This leads to the most important reason why AIDS in southern Africa runs at a higher rate: It is fractured relationships. Not sex, but relationships.

Despite the mythology to the contrary, Africans are not more promiscuous than other people. As researchers have discovered in recent years, the term "promiscuity" has little meaning in the real world, as patterns of behavior are mixed everywhere.

Relationships and sex are different. To start with the sexual habits of two rather different groups—Americans and Batswana—it is clear that the Batswana are in general *not* more promiscuous. Studies show that the peoples with the greatest number of sexual partners are the Americans, French, and Germans, while Asians and Africans lag in both the number of partners and the frequency of sex. The Americans have more sexual partners than the Africans do, and their sexual adventures are more abrupt—more one-night stands and sudden runs of activity. The number of sex partners is not the essential problem in southern Africa. It is something more fundamental.

In traditional Africa the basic standard is, in fact, the standard of the Old Testament. Faithfulness is valued, but male authority is paramount. A man who is prosperous enough traditionally may have a mistress or a second wife, as in the Bible.

In Tswana tradition both man and woman have a responsibility to be faithful. It was assumed that men tended toward infidelity and wives were counseled in the marriage ceremony itself to accept their husband's straying. The old saying was "A man, like a bull, cannot be confined in a *kraal* [corral]." The modern version now heard in discussions of faithfulness uses the two local staples of the diet—cornmeal and rice—to make the metaphor: "A man cannot live on *pap* all his life, but must sometimes have rice."

The infidelities of men and women were not treated as equally serious in the old culture, but a man did have a duty to care for his wife, including faithfulness in most situations. He could be punished for affairs.

What has gotten lost in the mix of Western and African culture is the punishment for unfaithfulness for husbands. Families are now fractured in time and space, and tribal authority to rein in the behavior of men has dissipated. The logic and cohesiveness of the culture has developed fissures. Now, this ancient version of "faithfulness" has become dangerous.

Roads in Africa are the streams that carry HIV—where the roads go, the virus can follow. When HIV first began to move in Africa, it was the trans-African highway from Kenya, through Uganda, to the Congo that carried this live cargo. It was hauled in the cabs of the trucks, inside the bodies of the drivers. They made stops and deposited swarms of particles along the way, from whence the virus spread to cities and rural areas surrounding the truck routes.

The image of the road describes not only the travel of the virus, but in another way describes its entrance into human society. Imagine rain falling on pavement. The water pools and does not break through, except where there are cracks in the surface. At those breaks the rain can penetrate, moving down and then spreading beneath the pavement. Once it is under the surface water compacts the grains of sand and dirt, eliminating pockets. The settling in some spots and not others means the roadway begins to waffle and crack. Potholes appear and edges of the roadway begin to shear off. The breakup of the entire road is underway.

So it is with HIV: As this particle rains down on humans, it seeks out the small breaks in culture. As families break and partners turn to others, a seam is opened. As people move from place to place, seeking comfort from other partners, the fissure opens. The undermining of society begins.

In science the couplings of love are sometimes expressed in mathematical models. They are like fluid flows, or combustion spreading.

When two people have a relationship, and one or the other occasionally strays, thereby introducing a third element to the movement of biological fluids, the risk begins. But if the unfaithfulness is not sustained over many encounters, the virus may remain trapped within the more or less faithful pair.

Add the extra partner on a regular basis, though, and the relationship basically becomes no longer two-way, but three- or four-way. These openings among others in the "couples" become stable passages through which the virus can move.

In each individual sexual encounter the odds of passing the virus vary—from one time in fifty to one in ten thousand, depending on other things. (For example, for a period of three to six weeks after a person is first infected, the virus multiplies rapidly in the body, so tens to hundreds of thousands of viral particles are present in every milliliter of blood. Later in the infection the immune system is at work and the number of virus particles drops ten- to a hundredfold. Or, in another example, if a person has another sexually-transmitted disease, the sores from it greatly increase the chance of passing the virus.)

All this means that a steady diet of sex, say, three times per week, can make passing the virus a near certainty between two people over a year or two. Add the third party, and a great amplification of mathematical risk occurs.

Unfaithfulness exists in all cultures, but there are several different patterns. In some places, such as the United States and Europe, multiple sex partners are common, but mostly the partnerships are serial—people don't usually go with more than one partner at a time. This is referred to as "serial monogamy." In Asia a somewhat similar pattern exists, but with an emphasis on flings. Dalliances with other partners occur, but partners return to their main squeeze after short affairs. In these situations, HIV can be passed easily, but not so quickly.

Unfortunately, add a dollop of faith*ful*ness to the mix, and you have the pattern of stable but overlapping relationships common in southern Africa. In Botswana, polygamy is outlawed and infidelity is frowned upon. But at the same time, the traditional marriage ceremony still carries this counseling for the bride: Wives are advised that for a happy marriage it is important not to ask a husband where he has been at night. Loyalty is more important, lest a woman create mistrust. The same advice is not passed on to the groom.

Thus, Africans are caught in transition between two African values—faithfulness and male authority.

What has been learned from this collision of values has meaning for those who want to stop the HIV epidemic. It suggests that from a purely practical point of view the two warring factions who emphasize abstinence and faithfulness or condoms and knowledge have both been right in some ways.

Faithfulness is vital. But also, condoms can be very effective. And in both cases, knowledge and testing are indispensable.

In the three successful cases of prevention documented around the world—in Senegal, Thailand, and Uganda—there were two key factors. People reduced the number of partners they had, and increased their condom use. There are fancier ways of saying these things: "Be faithful" and "change your behavior" emphasize a different approach to loving, while "condomize" and "safe sex" emphasize practical precautions.

When the international team arrived to begin their challenge to AIDS in Botswana, they started with a plan that read precisely like something from a corporate office. What's needed? Computers, IT plans, counselors, doctors, trucks, refrigerators. How long it would take to bring them in? How long would the training take? How much it would cost?

It said that in Botswana, a nation of about 1.5 million people, something more than 300,000 adults were infected with HIV. They had calculated that not all were equally sick, but that about 110,000 of them were already in need of triple-drug treatment. The project would treat 19,000 the first year, and then ramp up to the full 110,000 over the next four years.

Some 108 new doctors, nurses, and pharmacists would be brought in. An array of new prefab clinic buildings would be put up. Training by experts would be provided for all the key people in Botswana's health system, from social workers to counselors and MDs. Condoms would be provided in the millions, both free and through discounted "social marketing" plans. Education would be brought in to the public schools from grade school to university. Market research, including focus groups and other polling methods, would be used to create the right messages and deliver them to the right citizens. The center of the system would be a computer network that would keep track of all patients, from the time they came forward for tests, all over the country on an instantaneous basis. That way, when a patient showed up at a clinic, his or her updated record would be in hand at any place.

The system was built to address the problems they expected to see: difficulty ensuring that patients took their medicines; effort needed to visit patients in their homes as well as waiting for them in the clinic; full counseling for the difficult problems of stigma; and effective public messages about HIV and the project itself.

The experts who worked in Botswana were completely unprepared for what they actually faced. If they had known, they probably would have first hired some experts on Botswana culture and begun with visits to tribal chiefs in the villages and bureaucrats in the city. They learned, of course, but it was a painful process with long, long delays.

While it was tough going, the project was something never really tried before, and it led the way in Africa. It did prove that critics were wrong about bringing sophisticated medicines to Africa; Africans take their medicine as well as or better than Westerners, and the drugs are just as effective as anywhere else, even considering that the patients coming in are sicker in Africa.

The obstacles were not the expected logistical difficulties—from watches to strict routines of medicine taking—but cultural, and the full array of expensive clinics, doctors, and nurses once thought to be essential could actually be a hindrance to delivering lifesaving treatment. The Botswana experiment lit up the path that other countries in Africa and Asia have now started following.

Donald de Korte, a Dutch doctor who had been Merck's managing director in South Africa for some years, was brought on to lead the grand experiment. He is a tall man, with sandy, gray-streaked hair and a ruddy complexion.

He became the director of what was named ACHAP, for the African Comprehensive HIV/AIDS Partnership. He is blunt, impatient with those not working to speed, but has a sense of humor about life and the difficulties ahead.

He looks like a man comfortable outdoors, and in one of his early adventures in Botswana he did head out into one of the wilderness parks in a four-wheel safari vehicle. He was soon halted by the tough driving—deep mud, to be exact—and it took two days for rescuers to reach him and haul him out.

He was confident, a game fellow.

De Korte was aware that the project was a risk, for Botswana, for Merck, and for himself personally because of the chance of failure. And after laying the ground for the plan for months, he had already encountered unexpected trouble.

De Korte found that there were widespread vacancies in Botswana's health staff—perhaps as many as 25 percent of the positions were unfilled. But the ministries that had failed to fill the jobs, the government now insisted, would also be in charge of recruiting the 108 new doctors, pharmacists, and nurses needed in the new program. De Korte asked to hire the 108 separately, fresh staff to work alongside current staff, rather than filling current holes in the system.

But as time went on, intransigence increased. The parallel staff was not only disapproved, but after de Korte had recruited dozens of doctors and nurses from other African countries, the usual process took over and destroyed recruitment.

The usual procedures—two interviews spaced months apart, approval by several ministries, finding funds from the treasury—would normally take a year and a half before a new hire could come to work. Needless to say, doctors in Africa, much in demand, cannot keep their lives and families on hold while waiting for consultations to be followed in careful order in Botswana. So in the second year of the program, more than forty prospective doctors had been recruited—and then were lost to the system because of delay.

It took two and a half years after the announcement of the program in Botswana to begin bringing doctors in. And when they arrived there were still disputes among government officials, and the doctors went for months without pay.

When it came to building a little prefabricated clinic on the grounds of the main hospital, Princess Marina in the capital, de Korte was told to allow the government to build it through the usual system. Eighteen months later nothing had been done. De Korte pleaded with the president himself, and was finally given power to build the clinic; it was up in less than three months.

The normal operation of the government was set by old cultural habits. The tribes had an unusually democratic way of governing themselves. They were led by chiefs, but all major decisions traditionally have been submitted to the whole population of the village. That meant meeting at the *kgotla*—a public space under a large tree where all villagers could assemble. At such meetings, which could last for days or weeks, each citizen was expected to speak his piece and debate the matter at hand. Then the chief would make a decision, usually based on what he felt the consensus was. This type of consultation before decisions is a conservative mechanism that has served the tribes well for eight hundred years, keeping them stable and peaceful.

The principle is one of consultation (*therisanyo* in Setswana, one language used in Botswana). The dignity of each person in society is maintained by recognizing his position, as shown by the gesture to get his consent when a matter is at issue.

Translated into modern government, that means that many officials have the right to be heard from, and to give or withhold support for the measure going forward. The effect is of a system that is careful to follow rules and works well to weed out corruption and renegade decision makers.

The doctor hired to lead the toughest challenge, building the drug treatment program for the nation, was Ernest Darkoh. His family is from Ghana, but he was raised in Kenya and Tanzania, and then went to school in the United States. He earned his MD from Harvard, but felt that ordinary medicine really didn't address the larger problems of the world, so he decided to study public health. After gaining his MPH, he realized that the big problems in health all have major parts that are about management and money. So he took an MBA from Oxford. Then, to get real world experience, he took a job with the McKinsey consulting firm, well known for its work in health systems.

Eventually, he was asked to take on the challenge of Botswana.

When he arrived one of his first encounters was with the board appointed by the government to oversee the big project. He jumped in enthusiastically, talking about rolling out plans.

But he found that the government-appointed board had not been authorized to make any decisions. It was a board built, in the midst of a raging epidemic, to discuss matters. He was told the board had no mandate yet, could not hire or fire or make any lasting decisions.

"The sole reason for creating the team was to be empowered and to make decisions," Darkoh said, but in fact what was created was another layer of bureaucracy even before the work started. "It was literally very scary," he said. "This was not the way to run a national emergency. From the beginning it was ten steps forward, twelve steps back. You have major victories, and then the next day you're ready to pack up and go."

A central problem was also encouraging people to get HIV tests. Under the elaborate system developed on the American model, testing involved multiple trips and permissions and counseling, raising great barriers to getting lifesaving drugs to the people. The solution proposed was to change the whole approach to testing, removing the elaborate counseling sessions and visits, and creating rather a routine medical procedure done whenever a patient shows up at the hospital. Patients had the right to refuse, but if they didn't object, they would get HIV tests.

Darkoh said that the policy was controversial because it went against policies established over years in the West built to guarantee consent and confidentiality. It was built in the years when disclosure of infection would bring down stigma, alienation, and potential job loss. Because there had been no good treatments, and the government had not offered care to the sick, it had been clear that forced testing could have led to harm, and would have done little good for the infected.

But by the time the epidemic had swept Botswana, the situation had changed greatly. The stigma was still present, but the government was offering lifesaving drugs and full care, at no cost. And with more than a third of the whole nation infected, it was a vital state interest to find out who was infected and to try to treat them and counsel them not to spread the disease further. It was President Mogae who broke the jam over testing, and he pressed through the new "routine testing" policy.

Two years into the project the infrastructure was still being built, and the 19,000 patients treated in the first year turned out to be much less—3,200 after eighteen months, and most of that was done by a handful of doctors working absurd hours because manpower recruitment was still caught in the hands of multiple, slow decision makers.

By 2003, de Korte was frustrated and under great stress. He had become a customer for pharmaceuticals himself: He started on Prozac. Darkoh had been ready to resign more than once.

Eventually the friction got to be too much. The board relieved de Korte of command in the fall of 2003, in hopes that the conflicts had to do with de Korte's tough personality and demands for performance. Few of those watching thought much would have been done at all without de Korte's pressure. But to prevent a complete and public breakdown of the project, Merck and Gates took the prudent step of bringing in a manager from within Botswana; Tsetsele Fantan, who had been the manager of the HIV program at the nation's largest mine, [and appointing him] director of the Merck-Gates project.

Harriet would have appeared stately if she were able to stand to her full height, but she was still a little weak. She has been on the new drugs only a short time. Less than four weeks before, she was on a pallet in a dark room at the back of her mother's house. She could not walk.

Most healthy people have immune system cell counts of about 1,000; hers were down to 35 and diving to zero. By contrast, her blood was swarming with more than 300,000 HIV particles per deciliter of blood, and that count was rising. Her medical counts, in any clinic around the world at the time, were the numbers of death.

She was ready for it. She said she could do nothing more for her children. She had stopped weeping over her youngest daughter who had HIV, as she didn't have even the moisture in her eyes to make tears.

But one afternoon a friend who worked at the hospital came by to get her. She brooked no arguments, but simply lifted Harriet up and with some help carried her to a car, and then into the hospital. She was given a round of tests, though she was semiconscious and barely cared.

Within weeks she regained her appetite both for food and life. She started to walk again. Reporters in the capital of Botswana were keeping count, and they said that Harriet was the fifth person in the entire nation to openly admit being infected.

On the other side of the coin was a schoolteacher from a village south of the capital who asked that reporters not use his name. He understood about the disease and how it was passed. He knew in looking over his last few years that he had taken risks with girlfriends. He suspected he might well have the virus. But when he faced the terrible moment of going to get tested he realized with crystal clarity what it was all about.

He knew that he would likely walk into the testing center as one person, relatively healthy and well liked, a young man with friends, a good job, and a family, and walk out of the center as a ghost, a man whose friends did not want to touch him, whose neighbors would turn away in the street. His job would be in danger. It was as if he were being asked to end his life as a human and become a monster, some kind of shamed creature, all voluntarily. Death didn't seem so serious a consequence compared to a life of being humiliated daily.

After many attempts and great pressure from his wife, he finally did get tested. But now that he knows he has the virus, he still keeps it a secret from his friends and the school where he works.

Botswana has no medical school, so all the Batswana who are doctors must go abroad for training. Not all come back, and among those who do, few ever went to work for the government. The pay is low and there has been little effort to recruit them. It is a significant loss for the country.

Ndwapi Ndwapi, born in a village in northern Botswana and schooled in the capital, spent ten years being trained in medicine in the United States. When he came back his training was perfect—infectious diseases—for him to be grabbed up by his nation's emergency AIDS program.

He is a tall young man with round cheeks and a small mustache. His gaze is intense. When he speaks he sounds like a man in the trenches, under constant fire. He is now head of the largest HIV clinic in Botswana, in the hospital in the capital. He had had to fight his own government to get there.

When he arrived back from his training in the United States, he says, "They were very glad to see me" at the ministry of health. But they took some months before giving him an assignment; they wanted him to take his place at the bottom of the ladder in Botswana—a posting in a low-ranking job two hundred miles from the capital at a salary of about $2,500 per year. He could have earned about $120,000 to start in the United States and could have moved quickly into his specialty; accounting for cost of living differences, a $30,000 salary would be expected in Botswana.

"It's completely nuts," he says, "and it's symptomatic of the system."

Having left and come back, he can see Botswana in the light of other ways of doing things. So his first response to his treatment when he returned was anger and frustration. He was in some ways in the same position as the outsiders from ACHAP who were fighting to make things happen on an emergency basis. He had to battle for his job in an AIDS program even though it was desperately short of doctors, especially doctors who understood Batswana culture and language.

He says, frustrated as he was, that he understood where the slowness came from. Now that he was back he felt it was his duty to help change the dysfunctional things about the slow system in Botswana.

It is a cultural inheritance with a distinguished past now caught in an emergency that makes it break down.

In emergencies, it fails, says Ndwapi. "The money's there, the will is there, the know-how is there, but somehow, you can't hire the people you want to hire. We know for a fact that there are doctors throughout Africa and other places in the world that would love to do this job. And yet, somehow, it just doesn't happen," he says.

To protest, to fight the dignified system of respect and consultation is considered shameful, putting your personal interests above the good of the group. But Ndwapi began to argue loudly for quick decision making, and for bypassing some usual procedures. "I was more than willing, and I still am more than willing, to take responsibility for whatever happens," he says, "for whatever comes down on my head."

"If I make decisions on my own sometimes, I can sleep at night because I know that the alternative is a hundred times worse than anything I can do. The alternative is doing nothing."

As we sat in a tiny cubicle at the Princess Marina Hospital, near a long corridor lined on both sides by patients who have been waiting for several hours, he says that the war on HIV must come down to this, in some ways, everywhere: fighting to make a system not set up to handle health emergencies take them on aggressively.

He mentions the prefab clinic that had been the subject of fights between ACHAP and the government. It was finally built, he says, "but you know it's not being used yet."

"Why?"

"Because there are no phones. Because there is no paper. In Botswana, we run out of paper." There is money to buy paper, but it doesn't get bought and delivered. The supply office refers to the ministry that controls procurement, and that ministry refers to another ministry, which says it has not sent supply money because it's been asked to close its books to end the fiscal period by another ministry. When you get to the ministry of finance, they say the folks are wrong, they are still supposed to be buying paper.

"You see what I mean?"

But on a day in May 2003, the prefab clinic does, finally, open for business. Gradually, changes happen. Slowly, new ways of doing things are installed.

Ndwapi is a man struggling to push his culture forward much faster than it wants to go. In the face of years of success that suggests the culture has done well, he must argue that it is now stumbling. It is a struggle not just for the lives of those in the hallway, but for the future of the country, and for a way nations can work together.

It's tough, there are fights and suspicions, but it is the best way to do it, Ndwapi says. "The best way is to bring in an organization like ACHAP. Charge it with the responsibility of doing this, okay? And just give them the mission of setting it up and then integrating it into this nation." They must start it, hire and train local people, and then turn it over, he says. "Any time in the past two years we got anything significant done, it was because ACHAP has been involved. They took charge."

By 2004, the original leader, Donald de Korte was gone, and Darkoh and Ndwapi were beleaguered, but the fights had begun to pay off. All of the clinics envisioned in the original plan were finally up and running, albeit three years late. After the decision to make testing easier, those infected began coming forward in larger and larger numbers.

"We're going to take off," Ndwapi says. "This is beginning to work. You know, this is a country that is good with the big things. The little things on the ground are not so good, but the big things, knowing which way to go, we get those right."

The program did take off, finally. The number of people who were tested, enrolled in regular AIDS care, and getting the HIV drugs rose from 3,000 in early 2003, to 20,000 in mid 2004, to 43,000 by early 2005, and it is still rising.

The project has now surpassed all other African and Asian countries in its delivery of HIV care and drugs. Following Botswana's lead, twelve other African nations have now started bringing drug treatment to their HIV-infected citizens, and this time in Africa has been dubbed "the decade of treatment" for

Africa, when the world's drugs and expertise finally began to turn to those who need them most.

A United Nations project called "3 by 5," has begun working toward the goal of treating three million people in the poor countries with top-end AIDS drugs by the end of 2005. The goal will not be met, but the pressure is on, and it is estimated that one million or more HIV-infected people in poor countries will be under treatment by that date.

Ernest Darkoh, after three years before the mast in Botswana, now says that there are strong lessons for others to follow when delivering AIDS drugs to poor countries. First, don't expect to see results quickly; development is not linear, it takes time to build up and learn what the real problems are before it takes off. Second, though it is natural to think of bringing on a treatment program slowly, beginning with central clinics in the capital, it is a mistake. The sickest patients come forward first, and because they need ten times more help than others, they overwhelm the system. There is also little to be gained from a slow scale-up—the problems at each new village clinic are the same as the previous ones, and must be learned anew at each place. So, he says, go for it. Scale up quickly, go to the rural areas nearest the patients' homes all at once. Then, bring in experts to work in each place for six months to train those who will have to take over.

Another lesson: Make testing as easy as possible, because you need to get to patients before they are at their sickest, and they can be better treated and returned to work the earlier they come in.

The final message is that "patients will spend ninety-nine percent of their lives in their communities, not at a hospital or a health facility." Therefore, it is possibly wasteful and dangerous to emphasize the building of a "brick-and-mortar health care infrastructure." In fact, drugs can be delivered from the back of a truck if need be. But the vital job is to track the patients and their conditions.

"Therefore," says Darkoh, "for any new program, the highest priority and the bulk of the initial effort should go toward establishing a robust and reliable patient-tracking, monitoring, and evaluation system," with or without walls.

Along the way, the data from Botswana have proved skeptics wrong—Africans actually take their medicines much better than Westerners. On average, Africans take 90 percent of all their pills, compared to Americans, who take about 70 percent of their prescribed dose. In addition, in testing verbal reports, researchers found that Africans are more truthful when telling how many pills they missed compared to Americans. The Africans overestimate by 3 percent, whereas Americans overestimate their compliancy by 20 percent.

The cost of treating the Batswana, even though it was the first program and emphasized the need for high-level training and fully qualified doctors and nurses, cost about one-tenth or less than treatment in the United States. In Botswana treatment cost has so far ranged from $580 to $1,580 per person for a year. The cost is expected to drop considerably in coming years.

To come back to the patients themselves: Harriet, one of the first to get the drugs, within weeks of beginning treatment went to work as a volunteer, counseling other infected women. Eventually she got a paid job and became one of the top counselors. By the end of 2003 she was engaged to be married. Her life was beginning again.

And in the village of Molepolole, I visited Edwin Moses. He works as a consultant on a radio drama that tells stories of village life, including about AIDS.

On the road out of Gaborone, you turn at the filling station and then veer right toward the Kalahari Desert. There is a track that threads little fenced-in plots with their characteristic yards—no grass, but the reddish dirt has been carefully tidied by raking.

Two of Edwin's neighbors died of HIV recently, one from an infection triggered by the virus and another from suicide.

Edwin comes out of the concrete-block house, greeting me with a shy smile and a charming giggle. He is fundamentally a happy guy—never mind his circumstances.

He is infected with HIV. But his beaming face is also the new face of the epidemic in southern Africa. Edwin and people like him represent the beginning of a tide moving the other way. After he became infected and his infant son died, he began to change his life. He stayed away from women for a time, then hitched up with a woman from his village who was also infected, and also wary of love.

But together the two thought it through, and decided to marry. They are both healthy now, but they are tested and monitored regularly, and are ready to take the drugs when it's best for them to start. They even got a little plot of land, and are building a house.

Then they went the final step. They had a baby, and because medicine can now keep babies free of virus, she is negative. She has a future.

Edwin says he looks forward to telling little Virginia (named after the friend who brought him and his wife, Mariah, together) their story. "It will be a treasure to her," he said. The baby, ten months old, is making squawking noises and cooing sounds as we speak. She is nonchalantly standing with one hand on her mother's knee.

"She will know all about how it happened, how she was born, and knowing everything, it will make her free. It will be a great treasure. We are living for her now."

Edwin goes to funerals every weekend, but now he asks to speak for a few minutes, to suggest that people get themselves tested for the virus, so they can get help and treatment when the time comes. AIDS no longer means death, he says, we can live with it.

The feeling among those who have worked in Botswana on AIDS is a double sense that it is the hardest and most patience-consuming work they have ever done. They also feel that building programs here, with tens of thousands now getting regular high-level care, is an extraordinary achievement. The work is pioneering and should make the way somewhat easier, at least, as new programs are now built across Africa.

From Botswana and from each of the other successes I have described, there are fundamentals to be learned about the new "smart aid" approach.

First, in each case, there was strong leadership present—Fazle Abed and his coworkers at BRAC in Bangladesh, Ram Shrestha and his colleagues in Nepal,

David Heymann and the polio team in India and elsewhere, and the combination of President Festus Mogae and the team at the African Comprehensive HIV/AIDS Partnership in Botswana. This does not exhaust the list of leaders on these projects, for good leaders tend to attract others who are effective. In each of these stories, one of the key issues leaders emphasize is a commitment to get results no matter what the obstacles or number of years it takes. These leaders believe there is simply nothing more important for their nations than to deliver on this work. There can be no higher ambition than to make them succeed even if the work is hard and requires many tries to get each part of the project right. Blunders were made in each case, as I hope I have spelled out.

A further feature in common is that in three cases, the work was accomplished with inexpensive medicine and local volunteers. The fourth case, Botswana, used medicine that is not so cheap but still is a bargain relative to its cost in the developed nations. In each case, there is a high health impact for the money spent.

One thing that is vital is that scientific research was used effectively, coupled with good management, in each instance. None of these results would have been possible without scientific experiments showing what was effective, because people in the past have often made reasonable but quite incorrect guesses about what works in health. Scientific backup is indispensable.

Some of this is not obvious. We thought we knew a great deal about vitamin A, how the body uses it, what its chemistry is, and so forth. When scientists did the research, they discovered that we knew nothing about the single most vital fact about this body chemical. In Bangladesh, Fazle Abed, after three decades of work in the villages, has found that no matter how well he thought he knew the culture and how to get people organized, he didn't really know until he tried. In that regard, development is like science—it requires actual experimentation, data collection, and analysis to get things done, even when it's assumed things work a specific way. A good case in point is the topic of maternal mortality and infant mortality during the day of birth; in Western countries we think we know what's important for health at birth. But we have been wrong. We are just now discovering how to deliver babies with high success rates. For example, we know that germs can cause infection for both mother and baby at birth, but our Western solution has been a kind of blanket hygiene of the entire birth environment. However, in most places that is not possible. So now we must find out what the important parts of hygiene at birth are. Is it clean hands for the birth attendant? In many countries, probably not, because the midwives do not examine and handle mothers and babies in the intrusive way we do in most developed nations; the chances of infection are smaller from that route. But it turns out that how and when the cord is cut might be vital. Or putting the baby to the mother's breast immediately—not common in medical settings—may save many infants from dying due to lack of warmth or early nutrition. In new settings, we need new science.

The good management part of any project should go without saying, as successful work in any field requires it. In delivering aid it is important to keep in mind that culture is critical, and so local workers and a sense of local values must be present.

Another point worth making is that each project went forward with explicit plans to build up local infrastructure and systems for the long haul. Good programs don't come and go; they come and stay. This is one reason why they are so cost-effective.

The best practitioners say the essence of successful development is a humility before challenging work and a commitment to learn what is the right thing to do—not in general, but in this intervention, in this week or month. The other important factors—science, monitoring, management, cautious spending—all follow from this.

POSTSCRIPT

Is the International Community Focusing on HIV/AIDS Treatment at the Expense of Prevention in Africa?

There are a number of interesting points that arise from these selections. First, one of early concerns about widespread HIV/AIDS treatment programs in Africa was related to the belief that African patients would not take their medicine regularly, or that limited health infrastructure would lead to lax surveillance of patients. Siddhartha Mukherjee, for example, made this argument in a 2000 article in *The New Republic* entitled "Take Your Medicine." As the Hilts study of the Botswana case showed, Africans took 90 percent of all their pills as compared to 70 percent for Americans.

Second, the problems of AIDS orphans and mother-to-child transmission clearly seem relevant to this debate as well. Sub-Saharan Africa has the largest population of AIDS orphans in the world, that is, children for whom both parents have died of AIDS. Clearly, drugs that prolong the life of a parent also enhance the life of a child. Drugs taken shortly before the birth of a child by an HIV-infected mother greatly reduce the chances of transmission. For more on the issue of mother-to-child transmission, see a 2001 World Health Organization technical consultation entitled "New Data on the Prevention of Mother-to-Child Transmission of HIV and Their Policy Implications." A good book on the AIDS orphan problem in Uganda, Zambia, and South Africa is Emma Guest's *Children of AIDS: Africa's Orphan Crisis* (Pluto Press, 2001).

Finally, it is sobering to think that hundreds of HIV/AIDS cases could be prevented for the cost of treating one person with the disease (even as the cost of treatment falls). This is a terrible choice for a government to have to make, and ideally government budgets would be large enough to do both. Many African government health department budgets were reduced drastically in the 1980s and 1990s as the result of IMF and World Bank imposed structural adjustment policies. Debt service payments (with several African countries ranking amongst the most indebted in the world) also reduce the amount of current spending on social services. In the face of such a horrendous epidemic, some have argued for debt forgiveness if it is tied to increases in public health spending (see, e.g., a 2005 United Nations University Research Paper #2005/9, by Addison, Mavrotas, and McGillivray, entitled "Aid, Debt Relief and New Sources of Finance for Meeting the Millennium Development Goals"). For additional general information on the HIV/AIDS issue see, UNAIDS' "2007 AIDS Epidemic Update" (http://data.unaids.org/pub/EPISlides/2007/2007_epiupdate_en.pdf).

Internet References . . .

AllAfrica Global Media

AllAfrica Global Media is the largest provider of African news online, with offices in Johannesburg, Dakar, Abuja, and Washington, DC. Over 700 stories are posted daily in French and English, in addition to multimedia content.

http://allafrica.com

African Governments on the WWW

African Governments on the WWW catalogues Internet links to many government organizations in Africa.

http://www.gksoft.com/govt/en/africa.html

Africa Action

Africa Action provides information on political movements related to a wide variety of African issues. It hosts its own information and contains links to other Internet sites.

http://www.africaaction.org/index.php

Centre for Democracy and Development

The Centre for Democracy and Development is a U.K.-based nongovernmental organization promoting democracy, peace, and human rights in Africa.

http://www.cdd.org.uk/

African Union

The Web site of Africa's premier pan-African organization, the African Union (formerly the Organization of African States).

http://www.africa-union.org

UNIT 5

Politics, Governance, and Conflict Resolution

*T*he terrain of politics, governance, and conflict resolution is simultaneously one of the most hopeful and distressing realms in contemporary African studies. Although the formal role of women in African politics has increased in recent years, and popular uprisings in Egypt and Tunisia have led to the downfall of dictators, the African continent suffers from more instances of civil strife than other regions of the world. Scholars and commentators intensely debate the connections between contemporary political developments; historical patterns of governance; global geopolitics; and local traditions of decision making, public discourse, and conflict resolution.

- Is Multi-Party Democracy Taking Hold in Africa?
- Does Increased Female Participation Substantially Change African Politics?
- Is Corruption the Result of Poor African Leadership?
- Are African-Led Peacekeeping Missions More Effective Than International Peacekeeping Efforts in Africa?

ISSUE 16

Is Multi-Party Democracy Taking Hold in Africa?

YES: Michael Bratton and Robert Mattes, from "Support for Democracy in Africa: Intrinsic or Instrumental?" *British Journal of Political Science* (July 2001)

NO: Joel D. Barkan, from "The Many Faces of Africa: Democracy Across a Varied Continent," *Harvard International Review* (Summer 2002)

ISSUE SUMMARY

YES: Michael Bratton, professor of political science at Michigan State University, and Robert Mattes, associate professor of political studies and director of the Democracy in Africa Research Unit at the University of Cape Town, find as much popular support for democracy in Zambia, South Africa, and Ghana as in other regions of the developing world, despite the fact that the citizens of these countries tend to be less satisfied with the economic performance of their elected governments.

NO: Joel D. Barkan, professor of political science at the University of Iowa and senior consultant on governance at the World Bank, takes a less sanguine view of the situation in Africa. He suggests that one can be cautiously optimistic about the situation in roughly one-third of the states on the African continent, nations he classifies as consolidated democracies and as aspiring democracies. He asserts that one must be realistic about the possibilities for the remainder of African nations, countries he classifies into three groups: stalled democracies, those that are not free, and those that are mired in civil war.

There was a great deal of enthusiasm among Africanists in the early 1960s when more than 40 African nations gained independence and formed popularly elected governments. This enthusiasm was tempered when a large proportion of these countries succumbed to one-party rule or military regimes by the end of the decade. The 1970s and 1980s were largely characterized by the persistence of undemocratic forms of governance. Lacking popular support, undemocratic regimes and guerilla movements often sought Soviet or American patronage within the context of the cold war. The United States, in the name of anti-communism, financially and militarily backed a number of unsavory political leaders and guerilla insurgents during this period. These ranged from Mabuto Sese Seko in former Zaire to UNITA rebels in Angola. Seko, perhaps one of the most corrupt of African dictators, plundered his country for over 20 years during the cold war with the full support of the United States. In Angola, the U.S. and then-white-ruled South Africa sustained a bloody civil war by supporting UNITA rebels in the late 1980s.

The end of the cold war largely led to the end of perverse outside intervention in African affairs. Combined with this change in the external environment was a groundswell of internal support for political reform, which some commentators attribute to the democratic changes occurring in Eastern Europe in the late 1980s that many Africans observed through the international media. The result has often been referred to as Africa's "second wave" of democratization in which, between 1991 and 2000, multiparty elections were held in all but 5 of Africa's 47 states.

In this issue, Michael Bratton and Robert Mattes find as much popular support for democracy in Zambia, South Africa, and Ghana as in other regions of the developing world. However, citizens of these countries tend to be less satisfied with the performance of their elected governments than those in comparable non-African nations. The authors interpret these results to mean that support for democracy in Africa is more intrinsic (an end in itself) than instrumental (a means to an end—such as improving material standards of living). This finding highlights the importance that Africans attach to the basic political rights afforded by democracy. It also contradicts other research indicating that governments in new democracies mainly legitimate themselves through economic performance.

In contrast, Joel D. Barkan takes a more sober view of the situation in Africa. He states that those assessing the political situation in Africa roughly break down into two camps, the optimists and the realists. These two groups tend to draw very different conclusions because they focus their attention on different countries in Africa. Barkan contends that one can be cautiously optimistic about the situation in roughly one-third of the states on the African continent, nations he classifies as consolidated democracies (Botswana, Mauritius, and South Africa) and as aspiring democracies (15 countries). The prospects for the remainder of African nations are much more uncertain. Barkan classifies these states into three groups, the stalled democracies (13 countries), those that are not free (10 countries), and those that are mired in civil war (roughly 6 countries). As a realist, he believes that there is a tendency to over celebrate progress in the first two groups and to retreat from the challenges in the third, fourth, and fifth groups.

YES

Michael Bratton and Robert Mattes

Support for Democracy in Africa: Intrinsic or Instrumental?

Popular support for a political regime is the essence of its consolidation. By voluntarily endorsing the rules that govern them, citizens endow a regime with an elusive but indispensable quality: political legitimacy. The most widely accepted definition of the consolidation of democracy equates it squarely with legitimation. In a memorable turn of phrase, Linz and Stepan speak of democratic consolidation as a process by which all political actors come to regard democracy as 'the only game in town.' In other words, democracy is consolidated when citizens and leaders alike conclude that no alternative form of regime has any greater subjective validity or stronger objective claim to their allegiance.

This article explores how the general public in new multiparty political regimes in sub-Saharan Africa is oriented towards democracy. What, if anything, do Africans understand by the concept? Do they resemble citizens in new democracies elsewhere in the world in their willingness to support a regime based on human rights, competing parties and open elections? And beyond democracy as a model set of rights and institutions, are citizens in Africa satisfied with the way that elected regimes operate in practice? All of these questions are coloured by the fact that many of Africa's democratic experiments are taking place in countries with agrarian economies, low per capita incomes and minuscule middle classes. Under such unpropitious conditions, observers have every reason to wonder whether elected governments have the capacity to meet citizen expectations and, if they cannot, whether citizens may therefore quickly lose faith in democracy.

We assume that citizens will extend tentative support to neo-democracies, if only because they promise change from failed authoritarian formulae of the past. But what is the nature of any such support? Is it *intrinsic,* based on an appreciation of the political freedoms and equal rights that democracy embodies when valued as an end in itself? Or does support reflect a more *instrumental* calculation in which regime change is a means to other ends, most commonly the alleviation of poverty and the improvement of living standards?

The resolution of this issue has direct implications for regime consolidation. Intrinsic support is a commitment to democracy 'for better or worse;'

From *British Journal of Political Science,* July 2001, pp. 447–450, 453–460, 469–473. Copyright © 2001 by Michael Bratton and Robert Mattes. Reprinted by permission of the authors.

as such, it has the potential to sustain a fragile political regime even in the face of economic downturn or social upheaval. By contrast, instrumental support is conditional. It is granted, and may be easily withdrawn, according to the temper of the times. If citizens evaluate regimes mainly in terms of their capacity to deliver consumable benefits or to rectify material inequalities, then they may also succumb to the siren song of populist leaders who argue that economic development requires the sacrifice of political liberties.

Let us be clear. We do not dispute that evaluations of democracy in new multiparty regimes are likely to be based in good part on the performance of the government of the day. After all, it is very unlikely that citizens in neo-democracies would possess a reservoir of favourable affective dispositions arising from a lifetime of exposure to democratic norms. If democracy is a novel experience, how could such socialization have taken place? Instead of bestowing 'diffuse support,' citizens fall back on performance-based judgements of what democracy actually does for them.

We wish to divide regime performance, however, into distinct baskets of goods: an *economic* basket (that includes economic assets, jobs and an array of basic social services) and a *political* basket (that contains peace, civil liberties, political rights, human dignity and equality before the law). The African cases provide a critical test of the importance of political goods to evaluations of democracy. If the denizens of the world's poorest continent make 'separate and correct' distinctions between 'a basket of economic goods (which may be deteriorating) and a basket of political goods (which may be improving),' then citizens everywhere are likely to do so. And if political goods seem to matter more than economic goods in judging democracy, then we can cast light on the 'intrinsic v. instrumental' debate. If democracy is valued by citizens as an end in itself in Africa, then this generalization probably holds good universally.

In this study we find that citizen orientations to democracy in Africa are most fully explained with reference to both baskets of goods. With one interesting country exception, satisfaction with democracy (the way elected governments actually work) is driven just as much by guarantees of political rights as by the quest for material benefit. Support for democracy (as a preferred form of government) is rooted even more deeply in an appreciation of new-found political freedoms, a finding that runs counter to the conventional view that the continent's deep economic crisis precludes regime consolidation. At least so far, new democratic regimes in Africa have been able to legitimate themselves by delivering political goods.

Scope of the Study

Our substantive focus is intentionally restricted—to attitudes to democracy, among masses rather than elites—because our geographical coverage is broader than most studies in Africa. This article uses standard survey items to compare political attitudes in Ghana, Zambia and South Africa, thus bridging the major regions of the sub-Saharan sub-continent and situating public opinion in Africa in relation to other new democracies in the world.

All three countries underwent an electoral transition to multi-party democracy during the last decade but their political trajectories have since diverged. Both of South Africa's competitive polls (in April 1994 and June 1999) were ruled substantially free and fair by independent observers. By contrast, Zambia's founding elections of October 1991 were far more credible than its dubious second contest of November 1996. For its part, Ghana experienced improved electoral quality, with flawed elections in November 1992 being followed by a December 1996 poll that drew almost universal praise. Thus, with reference to the institution of elections alone, South Africa's democracy has stabilized, Ghana's is gradually consolidating, and Zambia's is slowly dying.

In reality, democracy is a fragile species throughout Africa. It is far from clear that a pervasive political culture exists to promote and defend open elections, let alone any other democratic institution. Regime transitions in Africa commonly resulted from intense struggles between incumbent and opposition elites, whose interest in self-enrichment was sometimes more palpable than their commitment to democracy. Even elected leaders have tampered with constitutional rules in order to prolong a term of office or to sideline rivals. And the armed forces continue to lurk threateningly in the wings: about half a dozen of Africa's new democracies succumbed to military intervention within five years of transition. Only in places like South Africa in 1994 (and possibly Nigeria in 1999), where transitions were lubricated by pacts among powerful insiders, are there signs that a culture of compromise and accommodation has penetrated the ranks of the political elite.

The extent to which a commitment to democracy has radiated through the populace is also open to question. After all, regime transitions in Africa were sparked by popular protests that were rooted in economic and political grievances. While the protesters had clear ideas about what they were *against* (the repressions and predations of big-man rule) they did not articulate an elaborate or coherent vision of what they were *for*. Judging by the issues raised in the streets, people seemed to want accountability of leaders and to eliminate the inequities arising from official corruption. To be sure, these preferences loosely embodied core democratic principles. And multiparty elections quickly became a useful rallying cry for would-be political leaders. But, during the tumult of transition, relatively little attention was paid to the institutional design of the polity. Emerging from life under military and one-party rule, citizens could hardly be expected to have in mind a full set of democratic rules or to evince a deep attachment to them.

This article takes stock of what has been learned from the first generation of research on political attitudes in new African democracies in the 1990s. . . .

The Meaning of Democracy in Africa

In considering the meaning of 'democracy' in Africa, the first possibility is that the term has not entered popular discourse, especially where indigenous languages contain no direct semantic equivalent. Some cultural interpretations emphasize that the word changes its meaning in translation, sometimes even signifying consensual constructs like community or unity. Or, because African languages borrow new terminology from others, a phonetic adaptation from a European language (like 'demokrasi') may have become common currency.

In one form or another, democracy seems to have entered the vocabulary of most African citizens. When the 1997 Ghanaian survey asked respondents 'What is the first thing that comes to mind . . . when you think of living in a democracy?,' 61.5 per cent were able to provide a meaningful response, rising to 75 per cent in 1999. Interestingly, even more respondents felt that Ghana was a democracy in 1997, implying that some people who could not specify a meaning for democracy could nevertheless recognize one if they saw one. In both countries, the salience of the concept was a function of education, with democracy having meaning in direct proportion to a respondent's years of schooling.

Contrary to cultural interpretations, we contend that standard liberal ideas of civil and political rights lie at the core of African understandings of democracy. In Zambia in 1993 and 1994, participants in two rounds of focus groups were asked 'What does democracy mean to you?' In the ensuing discussions, democracy was most commonly decoded in terms of the political procedure of competitive elections in which 'people are free to vote if they want to' and 'have a right to choose their own leaders.' Informants described how they resented having been forced to vote for the former ruling United National Independence Party (UNIP) and decried the political intimidation exerted by the party's youth wing. They favourably compared a choice of candidates under a multi-party regime with the system of 'appointed representatives' under a one-party state.

An open-ended question in the 1999 survey in Ghana about 'the first thing that comes to your mind . . . when you hear the word "democracy"' elicited the following responses, in frequency order: civil liberties and personal freedoms (28 per cent of all respondents), 'government by the people' (22 per cent), and voting rights (9.2 per cent). The only other major response was 'Don't know' (24.8 per cent) and very few respondents offered a materialistic interpretation (2.5 percent). These findings seem to suggest that Ghanaians view democracy almost exclusively in political terms, with an emphasis on selected civil liberties (especially free speech), collective decision-making and political representation.

Survey findings point to a much more materialistic world view in South Africa. In 1995, South Africans were asked to choose from a list of diverse meanings (both political and economic) that are sometimes attached to democracy. At the top of the popular rankings, 91.3 per cent of respondents equated democracy with 'equal access to houses, jobs and a decent income' (with 48.3 per cent seeing these goods as 'essential' to democracy). This earthy image of democracy far outstripped all other representations: for example, regular elections (67.7 per cent), at least two strong parties (59.4 per cent), and minority rights (54.5 per cent). To be sure, a majority of South Africans did associate democracy with procedures to guarantee political competition and political participation, but their endorsement of these political goods was far less ringing than the almost unanimous association of democracy with improved material welfare. Tellingly, only small minorities found it 'essential' to democracy to hold regular elections (26.5 per cent) or guarantee minority rights (20.6 per cent).

Because South Africa is a deeply divided society with mutually reinforcing fault lines of race and class, one would expect that various social groups would hold disparate views of democracy. We have noted elsewhere 'massive racial differences in agreement with regime norms.' Whites are much more likely than blacks to agree that regular elections, free speech, party competition and minority rights are essential to democracy. This procedural interpretation of democracy most likely reflects their own minority status and their reliance for protection on constitutional and legal rules. South African blacks, for their part, attach just as much or more importance to narrowing the gap between rich and poor. And while many South Africans of all races say they accept the necessity of redistributing jobs, houses and incomes, blacks seem to focus more on 'equality of results' while whites stress 'equality of opportunity.'

We reach four working conclusions based on recent research on citizen conceptions of democracy in three African countries. First, Africans here are more likely to associate democracy with individual liberties than with communal solidarity, especially if they live in urban areas. Secondly, popular conceptions of democracy have *both* procedural *and* substantive dimensions, though the former conception is more common than the latter. Thirdly, citizens rank procedural and substantive attributes in different order across countries. Zambians place political rules at the top of the list of democratic attributes, whereas South Africans relegate such guarantees behind improvements in material living standards. Finally, rankings differ even within the category of political goods: whereas Zambians (and to a lesser extent South Africans) grant primacy to elections, Ghanaians elevate freedom of speech to the top of their own bill of democratic rights.

These cross-national differences can be interpreted in terms of the life experiences of citizens under each country's old regime. Zambians may regard democracy mainly in terms of competitive multi-party elections because of their disappointing experiences with the ritual of 'elections without choice' under Kenneth Kaunda's one-party state. Ghanaians, for their part, emphasize freedom of speech as a reaction against the tight controls over communication imposed by the previous military regime, whose populist ideology was the only approved form of political discourse. Finally, South Africans place socio-economic considerations at the heart of their notion of democracy because of the integrated structure of oppression experienced under apartheid. Impoverished under the old regime, they see the attainment of political freedom as only the first step in rectifying manifold inequalities in society. In this conception, democracy has an inclusive meaning; it is as much a means to social transformation as a politically desirable end in itself.

Support for Democracy in Africa

The best way to ask questions about popular support for democracy is in concrete terms and in the form of comparisons with plausible alternatives. Since democracy has motley meanings, it is not useful to ask whether people support it in the abstract. It is far better to elicit opinions about a real regime with distinctive institutional attributes, such as a 'system of governing with free elections and many parties.' And if citizens support democracy as the 'least worst' system (the so-called 'Churchill hypothesis,') it is worth testing their levels of commitment against other regime forms that they have recently experienced or could conceive of encountering in the future.

Table 1 reports results of survey questions of this sort from various world regions, with sub-Saharan Africa represented by Ghana, Zambia and South Africa. In so far as these countries are representative of the region, Table 1 shows that the level of public commitment to democracy is much the same in Africa as in other regions of the world that have recently undergone regime

Table 1

Public Attitudes to Democracy: Preliminary Cross-National Comparisons

	Support democracy	Satisfied with democracy	Supportive and satisfied	Supportive but not satisfied
European Union	78	53	—	—
Southern Europe	84	57	79	11
Greece	90	52	84	11
Portugal	83	60	77	9
Spain	78	60	75	12
East and Central Europe	65	60	72	6
Czech	77	56	70	8
Poland	76	61	70	4
Romania	61	77	68	4
Bulgaria	66	61	75	2
Slovakia	61	49	62	14
Hungary	50	53	79	4
South America	63	50	45	22
Uruguay	80	54	57	29
Argentina	77	53	55	28
Chile	53	48	38	17
Brazil	41	46	32	16
Sub-Saharan Africa	64	48	41	18
Ghana (1997)	74	53	46	13
Zambia (1996)	63	53	49	14
South Africa (1997)	56	38	29	13
South Africa (blacks)	61	45	35	11
South Africa (whites)	39	7	5	18

Note: Regional means are raw estimates, uncorrected for proportional population size of countries. Further fnn. to Table 1 can be found in the electronic version of the journal available at www.cup.cam.ac.uk.

change. Excluding Southern Europe, almost two out of three citizens in new democracies extend legitimacy to elected government as their preferred political regime: the relevant mean figures are 65 per cent for East and Central Europe, 63 per cent for South America, and 64 per cent for the three countries of sub-Saharan Africa. Indeed, the average level of support in Africa (64.3 per cent) is virtually identical to the combined mean for Latin America and post-Communist Europe (64.2 per cent).

Moreover, deviation in support for democracy around the regional mean is lower for the three African countries than for other parts of the world. The countries with the lowest and highest levels of support for democracy are separated by just 18 percentage points in the African cases, but by 27 points for Eastern Europe and 39 points in South America. We interpret this to mean that authoritarian regimes have been widely discredited across the continent. Although the citizens of Ghana and Zambia may not have committed themselves to democracy as firmly as the citizens in Uruguay and the Czech Republic, they evince less nostalgia for hardline rule than citizens in Hungary and Brazil. Once again, though, South Africa is an exception. And we would need many more confirming cases before we could be sure that legitimating sentiments are evenly spread across all African countries.

Indeed, variations are evident within Africa in the extent to which citizens support new regimes. Of the three cases under review, Ghana displays the highest levels of citizen commitment to democracy. In 1997, fully 73.5 per cent of citizens thought it somewhat or very important for Ghana to 'have at least two political parties competing in an election.' The intensity of this support appears to be strong, as reflected by the 55.9 per cent of respondents who thought these institutions 'very important.' And the quality and depth of this support is underlined by the even higher proportions who granted importance to the right of citizens to form parties representing diverse viewpoints (82.5 per cent), to the openness of the mass media to political debate (89.3 per cent), and to the regular conduct of honest elections (92.7 per cent). While there is some possibility that respondents are acquiescing here to non-controversial 'motherhood' questions, Ghanaians nonetheless appear to consistently favour a full basket of liberal political rights.

Among the countries considered, legitimation of the new regime was lowest in South Africa, where citizens do not yet feel a widespread attitudinal commitment to democracy. A 1997 survey asked respondents to choose between the following statements: '[When] democracy does not work . . . some say you need a strong leader who does not have to worry with elections. Others say democracy is always best.' Since only a bare majority chose the democratic option (56.3 per cent, up from 47 per cent in 1995, but dropping back again below 50 per cent in 1998), support for democracy appeared to be weaker there than in the other African countries. Other responses underscore the shallowness of democratic legitimacy and the appeal of authoritarian alternatives in South Africa. In 1997, about one-third of the population thought that, under democracy, 'the economic system runs badly' (29 per cent), order is poorly maintained (30.2 per cent), and leaders are 'indecisive and have too much squabbling' (35.1 per cent). And more than half of all South Africans

(53.8 per cent) stated that they would be 'willing to give up regular elections if a non-elected government or leader could impose law and order and deliver jobs and houses.'

Thus, the potential constituency for forceful rule appears to be larger in South Africa than in South America, where an average of just 15 per cent of citizens considers that 'in some circumstances an authoritarian government can be preferable to a democratic [one].' Sentiments for a strong man were higher in South Africa (30.8 per cent) than in Chile (19 per cent) and Brazil (21 per cent), where authoritarian nostalgia is usually considered to be high. Question wording may have had a significant effect, with the cue of higher material living standards ('jobs and houses') inducing even some of democracy's supporters to abandon it. But, at minimum, this finding draws attention to the role of instrumental calculations in the assessments of democracy by many South African citizens.

South Africa's deviance is explicable again, however, in terms of its cultural diversity. White South Africans were much less likely to judge that 'democracy is always best' (39 per cent) than the country's African citizens (61 per cent). And, while 'coloureds' situated themselves between blacks and whites when granting such support to democracy (53 per cent), Asian South Africans were the least supportive of all (27 per cent). Thus the cautious, even retrogressive, attitudes of ethnic minorities tended to depress overall levels of commitment to democracy in South Africa. Examined alone, African citizens can be seen to support this form of regime at the highest level of any ethnic group in South Africa (61 per cent), a level not too different from citizens in Zambia (63 per cent) and the sub-Sahara region as a whole (64 per cent).

In Zambia, the question on support for democracy differed slightly, while still focusing on a political system featuring elections and posing a comparison with a realistic alternative regime. Respondents were asked to choose: Is 'the best form of government . . . a government elected by its people' or 'a government that gets things done?' On the assumption that support erodes as regimes mature, especially if citizens' expectations are not fully realized, we thought that support for 'elected government' would decline over time. To date this has not happened in Zambia. Public support for democracy held steady, at 63.4 per cent in 1993 and 62.9 per cent in 1996. As in Ghana, other related items bespoke an electorate with a relatively firm syndrome of democratic commitments. In 1996, 73 per cent preferred 'a choice of political parties and candidates' to 'a return to a system of single-party rule.'

Satisfaction with Democracy in Africa

Democracy looks better in theory than in practice. In elected regimes worldwide, more citizens support democracy as their preferred form of government than express satisfaction with the way that it actually works. This generalization holds true not only for Third Wave neo-democracies but, even more so, for the established regimes of Western Europe. . . .

Unlike support for democracy, satisfaction with democracy is not as widespread in the three African countries as it is in South America and Eastern

Europe. Satisfaction lags support by a wider margin in the sub-Saharan region (16 per centage points) than in the other two world regions (13 and 5 per centage points respectively). Substantively, fewer than half (48 per cent) of the citizens in these new African democracies report satisfaction with key aspects of the performance of elected regimes. Once more, the African average is pulled down by South Africa, with Ghana and Zambia displaying popular approval of regime performance at levels similar to consolidating democracies like Uruguay and Argentina. Although different racial groups in South Africa again evince distinct levels of satisfaction (45 per cent for blacks versus just 7 per cent for whites), black South Africans in this instance trail their fellow citizens elsewhere on the continent in their contentment with democracy in practice (39 per cent). Instead, they tend to more closely resemble the citizens of Brazil (41 per cent), more of whom are unhappy with democracy than are satisfied with it. . . .

Explaining Satisfaction with Democracy

. . . Satisfaction with democracy among African citizens appears to depend upon their assessment of the performance of government, particularly its performance at delivering *both* economic *and* political goods. Taken together, these factors explain between a quarter and two-fifths of the variance in expressed satisfaction in three African countries. Apart from social background, no set of factors—whether general performance, economic goods or political goods—can be discarded without a significant loss of explanatory power. Any ecumenical explanation of satisfaction with democracy in Africa must make reference to government performance in *both* its political *and* economic dimensions.

But what about the relative weight of economic and political explanations? We note that the delivery of economic goods sometimes has large independent effects on satisfaction with democracy in individual countries. Cross-nationally, however, such effects are rather inconsistent. We therefore conclude that economic effects are subject to the exigencies of time and place, such as gradual economic recovery in Ghana and persistent economic crisis in Zambia. We therefore doubt that a general explanation of satisfaction with democracy can be constructed from economic data alone. At the same time, we note that the effects of political factors, while occasionally weaker than those of economic factors, prevail more consistently across countries. This observation suggests that the delivery of political goods is a more reliable general predictor of satisfaction with democracy and a more promising foundation on which to construct a theory of democratic consolidation. This line of argument is explored further in the next section.

Explaining Support for Democracy

We turn, finally, to explain support for democracy as a preferred regime type. As Table 2 shows, our analysis accounts for 12 to 17 per cent of the variance in popular support (see adjusted R^2 statistics). Our explanation was less complete in this case perhaps because of the impalpability of the issue at hand: citizens

Table 2

Multiple Regression Estimates of Support for Democracy

	S. Africa			Ghana			Zambia		
	B	(s.e.)	Beta	B	(s.e.)	Beta	B	(s.e.)	Beta
Social background factors									
Gender			0.028			-0.014			0.048
Age			-0.019			0.000			0.042
Education			-0.009			0.001	0.064	(0.010)	0.200***
General performance factors									
Approval of government performance	0.084	(0.014)	0.145***	0.368	(0.064)	0.169***	0.122	(0.030)	0.128***
Satisfaction with democracy	0.082	(0.011)	0.196***	0.393	(0.052)	0.195***			-0.010
Economic factors									
Assessment of current economic conditions			-0.013			0.038			0.001
Assessment of current personal QOL			-0.017			0.023			-0.021
Assessment of future personal QOL	0.035	(0.011)	0.077***			-0.023			0.067
Support for market reforms			0.038			-0.006			0.017
Delivery of economic goods			-0.002	0.066	(0.031)	0.062***			0.059
Political factors									
Interest in politics	0.054	(0.015)	0.077***			0.015	0.115	(0.025)	0.144***
Trust in government institutions			0.032			-0.003			0.018
Perception of government responsiveness			0.031			-0.008			0.012
Perception of official corruption			-0.003	-0.140	(0.072)	-0.045*			-0.020
Delivery of political goods			0.028	0.158	(0.025)	0.153***	0.461	(0.080)	0.181***
N	3,500			2,005			1,182		
R	0.349			0.417			0.382		
R^2	0.122			0.174			0.146		
Adjusted R^2	0.120			0.171			0.142		

(*)Significant at 0.05
(**)Significant at 0.01
(***)Significant at 0.001

may find it more difficult to assess the qualities of abstract constitutional rules than the concrete performance of actual governments. In any event, public opinion in Africa seems to be less fully formed, and more contradictory, when it comes to support for democracy.

Nevertheless, Table 2 does reveal interesting findings. First, it reconfirms that attitudes to democracy cannot be inferred from standard social background characteristics. Again, gender and age were irrelevant to the legitimation of democracy in all countries studied and education had a positive impact only in Zambia. These findings are consistent with the observations that 'the more education a person has, the more likely he or she is to reject undemocratic alternatives' but that, overall, 'social structure [has] little influence . . . on attitudes towards the new regime.' If African societies do not contain entrenched pockets of generational or gender-based resistance to democratization, then the prospects for the consolidation of democratic regimes would seem to be slightly brighter than is sometimes thought.

Secondly, regime legitimacy in Africa depends upon popular appraisals of government performance. Consistently, in all three countries, support for democracy was strongest among citizens who felt that elected governments were generally doing a good job. But approval of government performance was closely connected to party identification, with supporters of the ruling party in each country being much more approving. Thus we must investigate further whether citizens are accrediting government performance—and thereby supporting democracy—out of 'knee-jerk' loyalty to a ruling party rather than a rational calculation that democratic governments deserve legitimation because they are more effective.

In Ghana and South Africa, support for democracy also was accompanied by expressions of satisfaction with democracy. We take this as further evidence that regime legitimation in Africa rests squarely upon performance considerations. On the up-side, popular demand for government performance increases the likelihood that citizens will make use of the rules of democratic governance to hold their leaders accountable. On the down-side, it also raises the possibility that citizens may conflate the performance of governments (that is, the achievements of incumbent groups of elected officials) with the performance of regimes (that is, the rules by which governments are constituted). The risk thus arises that, faced with continued mismanagement by ineffective governments, Africans may throw the baby out with the bathwater. By punishing government under-performance, they may inadvertently dismiss democracy.

Thirdly, and notwithstanding what has just been said about performance, we find little systematic evidence from Africa that citizens predicate support for democracy on the delivery of economic goods. Generally speaking, the legitimation of democratic regimes does not depend on citizen assessments of personal or national economic conditions, either now or for the future. Only in South Africa are assessments of future personal conditions linked positively to support for democracy. Strikingly too, when other relevant factors are controlled for, citizen perceptions of economic delivery have no discernible effects on the endorsement of democracy in either Zambia or South Africa.

The delivery of economic goods only seems to matter in Ghana, though the influence of this instrumental consideration is far from the strongest in the Ghana model.

Instead, we are led back again to the impact of political factors. For the first time, we find that citizen interest in politics had a positive effect on attitudes to democracy in two out of the three countries (Zambia and South Africa). It stands to reason that democracy will not consolidate where citizens remain disinterested in, and detached from, the political process; before people can actively become democracy's champions, they must orient themselves towards involvement in political life. One wonders why Ghanaians, who display the highest levels of interest in politics among the Africans surveyed, do not automatically support democracy. The answer appears to lie, at least in part, in the popular perception of rampant official corruption in that country. Many persons who are predisposed by their interest in politics to become active citizens are 'turned off' from democracy by what they see as the illicit machinations of civilian politicians. As one would expect, perceptions of official corruption are negatively associated with support for democracy in all three African countries; only in Ghana, however, is this relationship statistically significant.

Finally, and most importantly, the delivery of political goods bears a strong and significant relationship to the popular legitimation of democracy. In judging democracy, the Africans that we surveyed think of government performance first and foremost in political terms. Unlike the delivery of economic goods, a factor that is relevant in only one country, this relationship holds in at least two country cases. . . .

Conclusion

In this [selection], we have established that levels of popular support for democracy are roughly similar in three neo-democracies of sub-Saharan Africa as in other Third Wave countries. Almost two-thirds of eligible voters in these African countries say that they feel some measure of attachment to democratic rules and values. Under these circumstances, the popular consolidation of democracy in at least some African countries does not seem an entirely far-fetched prospect.

Yet the African cases stand apart from other new democracies in terms of lower levels of mass satisfaction with actual regime performance. The fact that African survey respondents support democracy while being far from content with its concrete achievements suggests a measure of intrinsic support for the democratic regime form that supersedes instrumental considerations. But, although support for democracy may be quite broad, we cannot confirm that it is deep. We do not yet know if citizens will vigorously defend the political regime if economic conditions take a decisive turn for the worse or if rulers begin to backtrack on hard-won freedoms.

Joel D. Barkan  **NO**

The Many Faces of Africa: Democracy Across a Varied Continent

$\mathbf{A}$ decade ago, seasoned observers of African politics including Larry Diamond and Richard Joseph argued that the continent was on the cusp of its "second liberation." Rising popular demand for political reform across Africa, multiparty elections, transitions of power in several countries, and negotiations toward a new political framework in South Africa led these experts to conclude that the prospects for democratization were good. Today, these same observers are not so sure. They describe Africa's current experience with democratization in terms of "electoral democracy," "virtual democracy," or "illiberal democracy," and are far more cautious about predicting what is to come. What is the true state of African democracy? And what is its future?

Governance Before the 1990s

Africa's first liberation was precipitated by the transition from colonial to independent rule that swept much of the continent, except the south, between 1957 and 1964. The West hoped that the transition would be to democratic rule, and more than 40 new states with democratic constitutions emerged following multiparty elections that brought new African-led governments to power. The regimes established by this process, however, soon collapsed or reverted to authoritarian rule—what Samuel Huntington has termed a "reverse wave" of democratization. By the mid-1960s, roughly half of all African countries had seen their elected governments toppled by military coups.

In the other half, elected regimes degenerated into one-party rule. In what was to become a familiar scenario, nationalist political parties formed the first governments. The leaders of these parties then destroyed or marginalized the opposition through a combination of carrot-and-stick policies. The result was a series of clientelist regimes that served as instruments for neo-patrimonial or personal rule by the likes of Mobutu Sese Seko in Zaire or Daniel Arap Moi in Kenya—regimes built around a political boss, rather than founded in a strong party apparatus and the realization of a coherent program or ideology.

From *Harvard International Review*, Summer 2002, pp. 72–77. Copyright © 2002 by the President of Harvard College. Reprinted by permission via Sheridan Reprints.

This pattern, and its military variant (as with Sani Abacha in Nigeria), became the modal type of African governance from the mid-1960s until the early 1990s. These regimes depended on a continuous and increasing flow of patronage and slush money for survival; there was little else binding them together. Inflationary patronage led to unprecedented levels of corruption, unsustainable macroeconomic policies that caused persistent budget and current account deficits, and state decay, including the decline of the civil service. Most African governments still struggle with this structural and normative legacy, which has obstructed the process of building democracy.

Decade of Democratization?

Africa's second liberation began with the historic 1991 multiparty election in Benin that resulted in the defeat of the incumbent president, an outcome that was replicated in Malawi and Zambia in the same year. The results of these elections raised expectations and created hopes for the restoration of democracy and improved governance across the continent. By the end of 2000, multiparty elections had been held in all but five of Africa's 47 states—Comoros, the Democratic Republic of Congo, Equatorial Guinea, Rwanda, and Somalia.

Along with the new states of the former Soviet Union, Africa was the last region to be swept by the so-called "third wave" of democratization, and as with many of the successor states of the former Soviet Union, the record since has been mixed. In stark contrast to the democratic transitions that occurred in the 1970s and 1980s in Southern and Eastern Europe and Latin America (excluding Mexico), most African transitions have not been marked by a breakthrough election that definitively ended an authoritarian regime by bringing a group of political reformers to power. While this type of transition has occurred in a small number of states, most notably Benin and South Africa, the more typical pattern has been a process of protracted transition: a mix of electoral democracy and political liberalization combined with elements of authoritarian rule and, more fundamentally, the perpetuation of clientelist rule. In this context, politics is a three-cornered struggle between authoritarians, patronage-seekers, and reformers. Authoritarians attempt to retain power by permitting greater liberalization and elections while selectively allocating patronage to those who remain loyal. Meanwhile, patronage-seekers attempt to obtain the spoils of office via electoral means, as reformers pursue the establishment of democratic rule. The boundaries between the first and second of these groups, and sometimes between the second and third, can be blurred because political alignments are very fluid. Liberal democracy is unlikely to be consolidated until reformers ascend to power.

The result is what Thomas Carothers has termed a "gray zone" of polities, describing countries where continued progress toward democracy beyond elections is limited and where the consolidation of democracy, if it does occur, will unfold over a long period, perhaps decades. This characterization does not necessarily mean that the third wave of democratization is over in Africa. Rather, we should expect Africa's democratic transitions to be similar to those of India or Mexico. In the former, the party that led the country to independence did not lose an election for three decades, and periodic alternation of power between parties did not occur until after 40 years. In the latter, the end of one-party rule and its replacement by an opposition committed to democratic principles played out over five elections spanning 13 years rather than a single founding election. Such appears to be the pattern in Africa, where two-thirds of founding and second elections have returned incumbent authoritarians to power, but where each iteration of the electoral process has usually resulted in a significant incremental advance in the development of civil society, electoral fairness, and the overall political process.

That many African polities fall into the gray zone is confirmed by the most recent annual *Freedom in the World* survey conducted by Freedom House. Of the 47 states that comprise sub-Saharan Africa, 23 were classified by the survey as "partly free" based on the extent of their political freedoms and civil liberties. Only eight (Benin, Botswana, Cape Verde, Ghana, Mali, Mauritius, Namibia, and South Africa) were classified as "free" while 16, including eight war-torn societies (Angola, Burundi, the Democratic Republic of Congo, Ethiopia, Eritrea, Liberia, Rwanda, and Sudan) were deemed "not free."

The overall picture revealed by these numbers is sobering. Less than one-fifth of all African countries were classified as free, and of these, only two or three (Botswana, Mauritius, and perhaps South Africa) can be termed consolidated democracies. On the other hand, if one excludes states in the midst of civil war, one-fifth of Africa's countries are free, one-fifth not free, and three-fifths fall in-between. That is to say, four-fifths of those not enmeshed in civil war are partly free or free, a significant advance over the continent's condition a decade ago. Only a handful are consolidated democracies, but few are harsh dictatorships of the type that dominated Africa from the mid-1960s to the beginning of the 1990s. As noted by Ghana's E. Gyimah-Boadi, "Illiberalism has persisted, but is not on the rise. Authoritarianism is alive in Africa today, but is not well."

Optimists and Realists

The current status of democracy in Africa varies greatly from one country to the next, and one should resist generalizations that apply to all 47 of the continent's states; one size does not fit all. Notwithstanding this reality, those who track events in Africa have divided themselves into two distinct camps: optimists and realists. Those in the United States who take an optimistic view—mainly government officials involved in efforts to promote democratization abroad, former members of President Bill Clinton's administration responsible for Africa, members of the Congressional Black Caucus, and the staff of some Africa-oriented nongovernmental organizations—trumpet the fact that multiparty elections have been held in nearly 90 percent of all African states. They note that most African countries have now held competitive elections twice and that some, including Benin, Ghana, and Senegal, have held genuine elections three times, at least one of which has resulted in a change of government. The optimists further note that the quality of these elections has improved in some countries, both in terms of efficiency and of fairness.

Electoral commissions seem to have been more independent, even-handed, and professional in recent elections than in the early 1990s. Opposition candidates and parties have greater freedom to campaign and have faced less harassment from incumbent governments. The presence of election observers, both foreign and domestic, is now widely accepted as part of the process. Perhaps most significant, citizen participation in elections has been fairly high, averaging just under two-thirds of all registered voters.

Recognizing that elections are a necessary but insufficient condition for the consolidation of democracy, the optimists also point to advances in several areas, listed below in their approximate order of accomplishment. First, there has been a re-emergence and proliferation of civil society organizations after their systematic suppression during the era of single-party and military rule. Second, an independent and free press has also re-emerged, spurred on by the privatization of broadcast media in several countries. Third, members of the legislature have increasingly asserted themselves in policymaking and overseeing the executive branch. Fourth, the judiciary and the rule of law have been strengthened in countries such as Tanzania, and human rights abuses have also declined. Fifth, there have been new experiments with federalism—the delegation or devolution of authority from the central government to local authorities—to enhance governmental accountability to the public and defuse the potential for ethnic conflict, most notably in Nigeria but also in Ethiopia, Ghana, Tanzania, Uganda, and South Africa. One or more of these trends, especially the first two and perhaps the third, can be found in most African countries that are not trapped in civil war.

Optimists also point to less exclusive membership in the governing elite, which has expanded into the upper-middle sector of society far more than during the era of authoritarian rule. In country after country, repeated multiparty elections have resulted in significant turnover in the national legislatures and local government bodies, sometimes as high as 40 percent per cycle. While the quality of elected officials at the local level remains poor, members of national legislatures are younger, better educated, and more independent in their political approach than the older generation they have displaced. Although further research is needed to confirm any major change in the composition of these bodies, new politicians and legislators also appear more likely than their predecessors to be democrats and to focus on issues of public policy and less likely to be patronage seekers.

Finally, public opinion across Africa appears to prefer democracy over any authoritarian alternatives. Surveys undertaken for the Afrobarometer project in 12 African countries between 1999 and 2001 found that a mean of 69 percent of all respondents regarded democracy as "preferable to any other kind of government," while only 12 percent agreed with the proposition that "in certain situations, a non-democratic government can be preferable." Moreover, 58 percent of all respondents stated that they were "fairly satisfied" or "very satisfied" with the "way democracy works" in their country.

Realists—who criticize what they contend was a moralistic approach to US foreign policy by the Clinton administration and disparage the use of democratization as a foreign policy goal—take a far more cautious view of what is occurring on the continent. Considering the same six developments that optimists cite as examples of democratic advances, realists note that all six are present in fewer than six countries. They also see much less progress than the optimists when nations are considered one by one. First, regular multiparty elections across the continent have resulted in an alternation of government in only one-third of the countries that have held votes. Moreover, only about one-half of these elections have been regarded as free and fair, with results accepted by those who have lost. It is also debatable in most of these countries whether recent elections have been of higher quality than those held in the early 1990s.

Second, although the re-emergence of civil society and the free press is a significant advance from the era of authoritarian rule when both were barely tolerated or systematically suppressed, civil society remains very weak in Africa compared to other regions and is concentrated in urban areas. Political parties are especially weak and rarely differentiate themselves from one another on the basis of policy. Apart from the church, farmers' organizations, or community self-help groups in a smattering of countries (such as Kenya, Côte d'Ivoire, and Nigeria), civil society barely exists in rural areas where most of the population resides. The press, especially the print media, is similarly concentrated in urban areas and thus reaches a relatively small proportion of the entire population. Only the broadcast media penetrates the countryside, but it is largely state-owned. Although private broadcasting has grown in recent years, especially in television and FM radio, stations cater almost exclusively to urban audiences. With a few exceptions, AM and short- and medium-wave radio—the chief sources of information for the rural population—remain state monopolies.

Third, while the legislature holds out the promise of becoming an institution of countervailing power in some countries, it remains weak and has rarely managed to effectively check executive power. Fourth, the judicial system in most countries is ineffectual, either because its members are corrupt or because it has too few magistrates and too poor an infrastructure to keep pace with the number of cases. Human rights abuses also continue, though less frequently and with less intensity than a decade ago. Fifth, Africa's experiments with federalism, though apparently successful, are confined to six states. Finally, the extent to which Africans have internalized democratic values is hard to judge. Although the Afrobarometer surveys indicate broad support for democracy, the results also suggest that such support is "a mile wide and an inch deep." An average of only 23 percent of respondents in each country described their country as "completely democratic."

Both optimists and realists are correct in their assessments of what is occurring in Africa. But how can both views be valid? The answer is that each presents only one side of the story. On a continent where the record of democratization is one of partial advance in over one-half of the cases, those assessing progress toward democratization, or lack of it, tend to dwell either on what has been accomplished or on what has yet to be achieved. These divergent assessments are proverbial examples of those who view the glass as either half-full or half-empty. Optimists and realists also draw their conclusions from slightly different samples. Whereas optimists focus mainly on states that are partly free or free, realists concentrate on states that are partly free or not free.

Optimists and realists are also both right because there are several Africas rather than one. In fact, at least five Africas cut across the three broad categories of the Freedom House survey. First are the consolidated and semi-consolidated democracies—a much smaller group of counties than those classified as free. This category presently consists of only two or three cases, such as Botswana, Mauritius, and perhaps South Africa. The second group consists of approximately 15 aspiring democracies, including the remaining five classified by Freedom House as free but not yet consolidated democracies plus roughly 10 classified as partly free where the transition to democracy has not stalled. All these states have exhibited slow but continuous progress toward a more liberal and institutionalized form of democratic politics. In this group are Benin, Ghana, Madagascar, Mali, Senegal, and possibly Kenya, Malawi, Tanzania, and Zambia. Third are semi-authoritarian states, countries classified as partly free where the transition to democracy has stalled. This category consists of approximately 13 countries including Uganda, the Central African Republic, and possibly Zimbabwe. Fourth are countries that are not free, with little or no prospect for a democratic transition in the near future. About 10 countries make up this group, including Cameroon, Chad, Eritrea, Ethiopia, Rwanda, and Togo. Finally, there are the states mired in civil war, such as Angola, Congo, Liberia, and Sudan. Each of these five Africas presents a different context for the pursuit of democracy.

Inhibiting Democracy

Several conditions peculiar to the continent make Africa a difficult place to sustain democratic practice. They explain why Africa lags behind other regions in its extent of democratic advance, why political party organizations are weak, and why the ties between leaders and followers are usually based on clientelist relationships. These conditions in turn create pressures for more and more patronage, a situation that undermines electoral accountability and leads to corruption.

Africa is the poorest of the world's principal regions: per capita income averages US$490 per year. This condition does not affect the emergence of democracy but does impact its sustainability. On average, democracies with per capita incomes of less than US$1,000 last 8.5 years while those with per capita incomes of over US$6,000 endure for 99. The reasons for this are straightforward. Relatively wealthy countries are better able to allocate their resources to most or all groups making claims on the state, while poor countries are not. The result is that politics in a poor country is likely to be a zero-sum game, a reality that does not foster bargaining and compromise between competing interests or a willingness to play by democratic rules.

Almost all African countries remain agrarian societies. With few exceptions like South Africa, Gabon, and Nigeria, 65 percent to 90 percent of the national populations reside in rural areas where most people are peasant farmers. Consequently, most Africans maintain strong attachments to their places of residence and to fellow citizens within their communities. Norms of reciprocity also shape social relations to a much greater degree than in urban industrial societies. In this context, Africans usually define their political interests—that is to say, their interests as citizens vis-à-vis the state—in terms of where they live and their affective ties to neighbors, rather than on the basis of occupation or socio-economic class.

With the exceptions of Botswana and Somalia, all African countries are multi-ethnic societies where each group inhabits a distinct territorial home-land. Africans' tendency to define their political interests in terms of where they live is thus accentuated by the fact that residents of different areas are often members of different ethnic groups or sub-groups.

Finally, African states provide much larger proportions of wage employ-ment, particularly middle-class employment, than states in other regions do. African states have also historically been large mobilizers of capital, though to a lesser extent recently. Few countries have given rise to a middle class that does not depend on the state for its own employment and reproduction. In this context, people seek political office for the resources it confers, for their clients' benefit, and for the chance to enhance their own status. In the words of a well known Nigerian party slogan, "I chop, you chop," literally, "I eat, you eat."

POSTSCRIPT

Is Multi-Party Democracy Taking Hold in Africa?

In addition to the factors raised by Barkan as inhibiting democracy in Africa, which some would dispute, there are other conditions frequently evoked to foreground a discussion of multiparty democracy in Africa. First, prior to independence, most African nations had little to no experience with democracy (as it is conceived in the West) at the national scale. If anything, the colonial period served to reinforce undemocratic tendencies, as the main purpose of unelected colonial administrations was to extract resources. Second, most African nations inherited national borders from the colonial era that cut across ethnic boundaries. The "unnaturalness" of these boundaries has made African states more difficult to govern. Third, and a point alluded to in the introduction to this issue, external powers have meddled in African affairs, often to the detriment of more representative government. The French, in particular, have intervened on a number of occasions in their former colonies when the leadership was not supportive of French interests. Finally, the role of the military is very poorly defined in many African contexts. In the absence of strong civilian rule, there is a tendency for this institution to assert control when there are economic or political problems. Here again, the way in which the military was used in the colonial era probably has contributed to this ill-defined role. While some scholars are highly cognizant of the aforementioned factors when assessing political change in Africa, others assert that attempts to put the "blame" for Africa's nontransition to democracy on outsiders or former colonial powers is simply a convenient excuse for the corruption and mismanagement of African leaders.

For those interested in further reading, Claude Ake offers a different perspective on African democracy than Bratton and Mattes, arguing that "the democracy movement in Africa will emphasize concrete economic and social rights rather than abstract political rights." See *Democracy and Development in Africa* (The Brookings Institution, 1997). A good example of a more thoroughly pessimistic view (or realistic depending on your perspective) of the prospects for democracy in Africa is George Ayittey's *Africa in Chaos* (St. Martin's Press, 1998).

ISSUE 17

Does Increased Female Participation Substantially Change African Politics?

YES: Elizabeth Powley, from "Rwanda: Women Hold Up Half the Parliament," in Julie Ballington and Azza Karam, eds., *Women in Parliament: Beyond Numbers* (International Institute for Democracy and Electoral Assistance, 2009)

NO: Carey Leigh Hogg, from "Women's Political Representation in Post-Conflict Rwanda: A Politics of *Inclusion* or *Exclusion*?" *Journal of International Women's Studies* (November, 2009)

ISSUE SUMMARY

YES: Elizabeth Powley, a specialist on gender and postconflict reconstruction, argues that Rwandan women are beginning to consolidate their dramatic gains that came with a gender-sensitive constitution in 2003 and parliamentary elections, which saw females win 48.8 percent of seats in the Chamber of Deputies. These successes were built on the specific circumstances of the Rwandan genocide, a quota system, and a sustained campaign by the women's movement in Rwanda. The women's caucus in the Rwandan parliament reviews existing laws and introduces amendments to discriminatory legislation, analyzes proposed laws with an eye to gender sensitivity, and works closely with women's organizations.

NO: Carey Leigh Hogg, program officer for Vital Voices, describes how the ruling political party in Rwanda has advocated for greater inclusion of women under the premise that this will improve the political climate, yet this same party also suppresses political dissent and ethnic identification. She argues that the Rwandan case shows how women's political identities can be dangerously frozen in a situation where the ruling party is intent on building national unity by quieting dissent.

Ellen Johnson-Sirleaf, a former World Bank official and grandmother of six, became the world's first black female head of state in 2005 when Liberia's voters elected her president by a margin of 19 percent. She and other female politicians in Africa have appealed to voters by arguing that they are better able to clean up corruption and end violent confrontation than their male counterparts. Women hold, or have recently held, high executive offices in other African countries as well, including the vice presidencies of Mozambique and Zimbabwe, as well as the deputy presidency in South Africa. On the parliamentary front, women hold an average of 17.5 percent of legislative seats in sub-Saharan Africa (compared with 18 percent worldwide). Rwanda has the world's highest ratio of women in parliament, at 49 percent. Also, of the 50 legislatures with the most female members in the world, 11 are found in Africa.

Are we in the midst of a female political renaissance in Africa? When a wave of democracy broke out across Africa in the 1990s, women were not prominently represented in leadership positions. Some say that it took the African electorate time to recognize that little was changing at the helm with male politicians. Now female politicians are successfully joining the fray under the assertion that they have the temperament and ethical bearing needed to clean up the system. Critics argue that such views of women in the political sphere are overly essentialized and that solely putting women into political office will not change the system. Worse yet may be political parties that simply use female politicians to maintain or expand their grip on power. Other scholars maintain that female politicians really do govern in a different manner. Although some may maintain that this has to do with biology, others suggest that it has more to do with the traditional roles that women have performed in African societies as caregivers, providers, and peace makers.

YES

Elizabeth Powley[1]

Rwanda: Women Hold Up Half the Parliament

In October 2003, women won 48.8 percent of seats in Rwanda's lower house of Parliament.[2] Having achieved near-parity in the representation of men and women [in] its legislature, this small African country now ranks first among all countries of the world in terms of the number of women elected to parliament.

The percentage of women's participation is all the more noteworthy in the context of Rwanda's recent history. Rwandan women were fully enfranchised and granted the right to stand for election in 1961, with independence from Belgium. The first female parliamentarian began serving in 1965.[3] However, before its civil war in the early 1990s and the genocide in 1994, Rwandan women never held more than 18 percent of seats in the country's Parliament.[4]

The 1994 genocide in Rwanda, perpetrated by Hutu extremists against the Tutsi minority and Hutu moderates, killed an estimated 800,000 people (one-tenth of the population), traumatized survivors, and destroyed the country's infrastructure, including the Parliament building. Lasting approximately 100 days, the slaughter ended in July 1994 when the Tutsi-dominated Rwandan Patriotic Front (RPF), which had been engaged in a four-year civil war with the Hutu-dominated regime of President Juvenal Habyarimana, secured military victory. Once an opposition movement and guerilla army, the RPF is now a predominately (but not exclusively) Tutsi political party. It is in power in Rwanda today.

During the nine-year period of post-genocide transitional government, from 1994 to 2003, women's representation in Parliament (by appointment) reached 25.7 percent and a new gender-sensitive constitution was adopted. But it was the first post-genocide parliamentary elections of October 2003 that saw women achieve nearly 50 percent representation.

The dramatic gains for women are a result of specific mechanisms used to increase women's political participation, among them a constitutional guarantee, a quota system, and innovative electoral structures. This case study will describe those mechanisms and attempt to explain their origins, focusing in particular on the relationship between women's political representation and the organized women's movement, significant changes in gender roles in post-genocide Rwanda, and the commitment of Rwanda's ruling party, the RPF, to

Reproduced by permission of International IDEA from *Women in Parliament: Beyond Numbers*. © 2005 International Institute for Democracy and Electoral Assistance.

gender issues. It will also briefly introduce some of the achievements and challenges ahead for women in Rwanda's Parliament.

The Constitutional Framework

In 2000, nearing the end of its post-genocide transitional period, Rwanda undertook the drafting of a new constitution and established a 12-member Constitutional Commission. Three members of the commission were women, including one, Judith Kanakuze, who was also the only representative of civil society on the commission. She played an important role both as a "gender expert" within the commission ranks and as a liaison to her primary constituency, the women's movement in Rwanda.[5]

The commission was charged with drafting the constitution and with taking the draft to the population in a series of consultations designed to both solicit input and sensitize the population as to the significance and principal ideas of the document.[6] Although political elites controlled both the content and the process of the consultations with Rwanda's largely illiterate population, it was—at least on the face of it—a participatory process, and its participatory nature allowed for significant input by women and women's organizations.[7]

The women's movement mobilized actively around the drafting of the constitution to ensure that equality became a cornerstone of the new document. The umbrella organization, Collectifs Pro-Femmes/Twese Hamwe (Pro-Femmes) and its member NGOs brought pressure to bear on the process and carefully coordinated efforts with women parliamentarians and the Ministry of Gender and Women in Development.

Rwanda's new constitution was formally adopted in May 2003.[8] It enshrines a commitment to gender equality. The preamble, for instance, cites various international human rights instruments and conventions to which Rwanda is a signatory, including specific reference to the 1979 Convention on the Elimination of All Forms of Discrimination Against Women (CEDAW). It also states a commitment to "ensuring equal rights between Rwandans and between women and men without prejudice to the principles of gender equality and complementarity in national development." Title One of the constitution also establishes, as one of its "fundamental principles," the equality of Rwandans. This respect for equality is to be ensured in part by granting women "at least" 30 percent of posts "in all decision-making organs."

It is important to note, however, that, although Rwanda's constitution is progressive in terms of equal rights, gender equality and women's representation, it is limiting in other important ways; specific concerns have been raised about restrictions on freedom of speech around issues of ethnicity.

The Quota System and Innovative Electoral Structures

Since the genocide, several innovative electoral structures have been introduced to increase the numbers of women in elected office.[9] Towards the end of its transition period, Rwanda experimented with the representation of women in the Parliament. Two women were elected to the then unicameral legislature on the basis of descriptive representation, with a mandate to act on behalf of women's concerns. Those two women came not from political parties but from a parallel system of women's councils (described in more detail below) that had been established at the grass-roots level throughout the country.

The 2003 constitution increased exponentially the number of seats to be held by women in all structures of government.

The Senate

In the upper house of Rwanda's (now) bicameral legislature, the Senate, 26 members are elected or appointed for eight-year terms. Some members of the Senate are elected by provincial and sectoral councils, others are appointed by the president and other organs (e.g., the national university). Women, as mandated in the constitution, hold 30 percent of seats in the Senate.

The Chamber of Deputies

The lower house of the Rwandan Parliament is the Chambre des Députés (Chamber of Deputies). There are 80 members serving five-year terms, 53 of whom are directly elected by a proportional representation (PR) system. The additional seats are contested as follows: 24 deputies (30 percent) are elected by women from each province and the capital city, Kigali; two are elected by the National Youth Council; and one is elected by the Federation of the Associations of the Disabled.

The 24 seats that are reserved for women are contested in women-only elections, that is, only women can stand for election and only women can vote. The election for the women's seats was coordinated by the national system of women's councils and took place in the same week as the general election in September 2003. Notably, in addition to the 24 reserved seats in the Chamber of Deputies, the elections saw an additional 15 women elected in openly competed seats. Women thus had in total 39 out of 80 seats, or 48.8 percent.

The Women's Councils

The Ministry of Gender and Women in Development first established a national system of women's councils shortly after the genocide, and their role has since been expanded. The women's councils are grass-roots structures elected at the cell level (the smallest administrative unit) by women only, and then through indirect election at each successive administrative levels (sector, district, province). They operate in parallel to general local councils and represent women's concerns. The ten-member councils are involved in skills training at the local level and in awareness-raising about women's rights. The head of the women's council holds a reserved seat on the general local council, ensuring official representation of women's concerns and providing links between the two systems.

Berthe Mukamusoni, a parliamentarian elected through the women's councils, explains the importance of this system as follows:

> In the history of our country and society, women could not go in public with men. Where men were, women were not supposed to talk, to show their needs. Men were to talk and think for them. So with [the women's councils], it has been a mobilization tool, it has mobilized them, it has educated [women] . . . It has brought them to some [level of] self-confidence, such that when the general elections are approaching, it becomes a topic in the women's [councils]. 'Women as citizens, you are supposed to stand, to campaign, give candidates, support other women.' They have acquired a confidence of leadership.[10]

While the women's councils are important in terms of decentralization and grassroots engagement, lack of resources prevents them from maximizing their impact and they are not consistently active throughout the country. Members of local women's councils are not paid, and because they have to volunteer in addition to performing their paid work and family responsibilities the councils are less effective than they could be. Nevertheless, women in these grass-roots councils have been successful in carving out new political space. And the 2003 constitution increased their importance by drawing on these structures to fill reserved seats for women in the Chamber of Deputies.

The Factors Giving Rise to Women's Increased Parliamentary Presence

The Women's Movement and Civil Society Mobilization

Immediately after the genocide, while society and government were in disarray, women's NGOs stepped in to fill the vacuum, providing a variety of much-needed services to the traumatized population. Women came together on a multi-ethnic basis to reconstitute the umbrella organization Pro-Femmes, which had been established in 1992. Pro-Femmes, which coordinated the activities of 13 women's NGOs in 1992, now coordinates more than 40 such organizations.[11] It has been particularly effective in organizing the activities of women, advising the government on issues of women's political participation, and promoting reconciliation.

Women in Rwanda's civil society have developed a three-pronged mechanism for coordinating their advocacy among civil society (represented by Pro-Femmes), the executive branch (Ministry of Gender and Women in Development), and the legislative branch (Forum of Women Parliamentarians).

An example of the effectiveness of this mechanism is the process the Rwandan women's movement initiated around the ratification of the new constitution. To elicit concerns, interests, and suggestions regarding a new constitution, Pro-Femmes held consultations with its member NGOs and women at the grassroots level. They then met with representatives of the Ministry of Gender and Women in Development and the Forum of Women

Parliamentarians to report members' concerns. Together the three sectors contributed to a policy paper that recommended specific actions to make the constitution gender-sensitive and increase women's representation in government, which was submitted to the Constitutional Commission. Once the draft constitution sufficiently reflected their interests, Pro-Femmes engaged in a mobilization campaign encouraging women to support the adoption of the document in the countrywide referendum.

Through the coordination mechanism that Pro-Femmes has forged with women in the executive and legislative branches of government, the women's movement has an increasingly powerful voice. A 2002 report commissioned by the US Agency for International Development (USAID) recognized the significant challenges faced by Rwandan civil society, including limited capacity, problems of coordination, and excessive control by the government,[12] but commended the significant role Pro-Femmes plays in shaping public policy. The study concluded that women's NGOs are the "most vibrant sector" of civil society in Rwanda and that "Pro-Femmes is one of the few organizations in Rwandan civil society that have taken an effective public advocacy role."[13] Its effectiveness is a result of a highly cooperative and collaborative relationship forged with women in government. Unfortunately, the close relationship has also compromised Pro-Femmes' independence and ability to criticize the government.

Changing Gender Roles

In addition to an effective women's movement, the dramatic gains for women in Parliament can also be traced to the significant changes in gender roles in post-genocide Rwanda. Women were targeted during the genocide on the basis not only of their ethnicity, but also of their gender: they were subjected to sexual assault and torture, including rape, forced incest and breast oblation. Women who survived the genocide witnessed unspeakable cruelty and lost husbands, children, relatives and communities. In addition to this violence, women lost their livelihoods and property, were displaced from their homes, and saw their families separated. In the immediate aftermath, the population was 70 percent female (women and girls).[14] Given this demographic imbalance, women immediately assumed roles as heads of household, community leaders and financial providers, meeting the needs of devastated families and communities. The genocide forced women to think of themselves differently and in many cases develop skills they would not otherwise have acquired. Today, women remain a demographic majority in Rwanda, comprising 54 percent of the population and contributing significantly to the productive capacity of the nation.

The overwhelming burdens on women and their extraordinary contributions are very much part of the public discourse in Rwanda. In April 2003, speaking about the parliamentary elections, President Paul Kagame said, "We shall continue to appeal to women to offer themselves as candidates and also to vote for gender sensitive men who will defend and protect their interests." He continued, "Women's under-representation distances elected representatives from a part of their constituency and, as such, affects the legitimacy of political decisions. . . . Increased participation of women in politics is, therefore, necessary for improved social, economic and political conditions of their families and the entire country."[15]

The Commitment of the Rwandan Patriotic Front

The Rwandan government, specifically the ruling RPF, has made women's inclusion a hallmark of its programme for post-genocide recovery and reconstruction.[16] This approach is novel in both intent and scope; it deserves further study in part because it contradicts the notion that the inclusion of women is solely a "Western" value imposed upon developing countries.

The government's decision to include women in the governance of the country is based on a number of factors. The policy of inclusion owes much to the RPF's exposure to gender equality issues in Uganda, where many of its members spent years in exile. Uganda uses a system of reserved seats to guarantee women 20 percent of the seats in Parliament: one seat from each of the 56 electoral districts is reserved for a woman. Men and women in the RPF were familiar with this system, as they were with the contributions and successes of women in South Africa's African National Congress (ANC). Within its own ranks, too, women played a significant role in the success of the movement. They played critical roles from the RPF's early days as an exile movement through the years of armed struggle. Such involvement provided them with a platform from which to advocate for women's inclusion during the transitional phase and consolidate their gains in the new constitution.

The RPF's liberation rhetoric was embraced by its own members and was applied to the historic exclusion of women as well as the Tutsi minority; this gender sensitivity is now government policy. As John Mutamba, an official at the Ministry of Gender and Women in Development explains, "Men who grew up in exile know the experience of discrimination. . . . Gender is now part of our political thinking. We appreciate all components of our population across all the social divides, because our country . . . [has] seen what it means to exclude a group."[17] RPF members who embraced notions of gender equality have informed the development of gender-sensitive governance structures in post-genocide Rwanda.

During the transitional period, before quotas were established in Rwanda, the RPF consistently appointed women to nearly 50 percent of the seats it controlled in Parliament. Other political parties lagged behind in their appointment of women, and therefore women never made up more than 25.7 percent of the Parliament during the transitional period.[18]

The RPF dominated the transitional government and consolidated its grip on power in the August 2003 post-transition election of President Paul Kagame and the installation of a new Parliament in October 2003. The RPF, together with its coalition, controls 73.8 percent of the openly contested seats in the Chamber of Deputies. The women's seats were not contested by political parties, but observers charge that a majority of the women in the reserved seats are also sympathetic to the RPF. Freedom House, in its most recent survey of nations, ranked Rwanda as "not free," with concern about political rights and civil liberties.[19] This puts Rwandan women and the women's movement in a precarious position, as they owe their ability to participate in democratic institutions to a political party that is less than fully democratic, and cannot be truly independent of the state.

Achievements and Challenges Ahead

In addition to performing all the functions their male counterparts do, women in Rwanda's Parliament have formed a caucus, the Forum of Women Parliamentarians, with international funding and support. This is the first such caucus in Rwanda, where members work together on a set of issues across party lines. Member of Parliament (MP) Connie Bwiza Sekamana explains, "When it comes to the Forum, we [unite] as women, irrespective of political parties. So we don't think of our parties, [we think of] the challenges that surround us as women."[20] The Forum has several roles: it reviews existing laws and introduces amendments to discriminatory legislation, examines proposed laws with an eye to gender sensitivity, liaises with the women's movement, and conducts meetings and training with women's organizations to sensitize the population to and advise about legal issues.

A key legislative achievement was the revoking of laws that prohibited women from inheriting land in 1999. Rwandan women parliamentarians, particularly the 24 who specifically represent the women's movement but also those who contested open seats and represent political parties, feel that it is their responsibility to bring a gender-sensitive perspective to legislating.

As elsewhere in the world, there are challenges related to descriptive representation. Many of the new parliamentarians are inexperienced legislators and have to overcome stereotypes about their (lack of) competence as leaders and their supposed naiveté, as well as some resistance to the fact that they owe their positions to the new quotas. There is an obvious status difference between those seats that are reserved for women and those that are gained in open competition with men, at both the local and the national levels.

It is also problematic, in the long term, to consider all Rwandan women a single constituency. Currently, the women's movement is represented most effectively by one organization, Pro-Femmes, and there is a great deal of consensus among women parliamentarians about the needs and priorities of women. In a mature democracy, however, women disagree on policies and desired political outcomes, even those, such as the use of quotas, which directly affect women's access to power. Perhaps because the quotas are so new and because the dominant voices in the women's movement supported their introduction so vigorously, there has not been public dissent within the movement about their utility.

There is, however, a sense, as in many other parts of the world, that quotas, reserved seats and descriptive representation are only a first step. Aloisea Inyumba, former women's minister, explains that at this point in Rwanda's development the new electoral mechanisms in Rwanda are needed to compensate for women's historic exclusion: "If you have a child who has been malnourished, you can't compare her to your other children. You have to give her a special feeding."[21]

It also remains to be seen what impact women will have, particularly on those issues that are not traditionally "women's issues." These women carry a double burden, as they must find ways to insert a gender perspective into a new range of issues—foreign affairs, for example—and yet remain loyal to their constituency of women in a country where the basic development needs are so great and women still lag behind men in terms of rights, status, and access to resources and education.

Conclusion

The representation of women in Rwanda's Parliament can be seen in the larger context of two trends: the use of quotas in Africa and the post-conflict situation. The rate of increase of numbers of women in Parliament has been faster in sub-Saharan Africa in the last 40 years than in any other region of the world, primarily through the use of quotas.[22] And, according to the Inter-Parliamentary Union (IPU), in the last five years post-conflict countries have "featured prominently in the top 30 of the IPU's world ranking of women in national parliaments," and these countries have been effective at using quotas and reserved seats to "ensure the presence and participation of women in [their] newly-created institutions."[23]

The ten years since the Rwandan genocide have been ones of enormous change for all Rwandans, but most dramatically for women. Rwanda is still vastly underdeveloped and the great majority of Rwandan women are disadvantaged vis-à-vis men with regard to education, legal rights, health and access to resources. Furthermore, the nearly equal representation of men and women in Rwanda's Parliament has been achieved in a country that is less than democratic and where a single political party dominates the political landscape.

Despite these challenges, women are beginning to consolidate their dramatic gains, with the new gender-sensitive constitution of 2003 and parliamentary elections that saw them earn 48.8 percent of seats in the Chamber of Deputies. These successes were the result of the specific circumstances of Rwanda's genocide, the quota system, and a sustained campaign by the women's movement in Rwanda, in collaboration with women in government and with the explicit support of the Rwandan Patriotic Front. The Rwandan case provides us with examples of gender-sensitive policy making and innovative electoral mechanisms that could be models for other parts of the world.

References

1. This case study draws on and excerpts previously published material by the same author. Powley, Elizabeth, 2003. *Strengthening Governance: The Role of Women in Rwanda's Transition*. Washington, DC: Women Waging Peace; and Powley, Elizabeth, 2005. "Rwanda: La moitié des sièges pour les femmes au Parlement" [Rwanda: half the seats for women in Parliament], in Manon Tremblay (ed.). *Femmes et parlements: un regard international* [Women and parliaments: an international view]. Montreal: Remueménage.

2. Inter-Parliamentary Union (IPU), 2003. "Rwanda Leads World Ranking of Women in Parliament," 23 October. See <http://www.ipu.org/press-e/gen176.htm>.

3. "Africa: Rwanda: Government." Nationmaster, <http://www.nationmaster.com/country/rw/Government>.

4. Inter-Parliamentary Union (IPU), 1995. *Women in Parliaments 1945–1995: A World Statistical Survey.* Geneva: IPU.

5. Judith Kanakuze, personal interview, July 2003.

6. "Legal and Constitutional Commission," <http://www.cjcr.gov.rw/eng/index.htm>.

7. Hart, Vivien, 2003. "Democratic Constitution Making." United States Institute of Peace, Special Report 107, <http://www.usip.org/pubs/specialreports/sr107.html>.

8. Constitution of the Republic of Rwanda. <http://www.cjcr.gov.rw/eng/constitution_eng.doc>.

9. For a more complete description of electoral mechanisms designed to increase women's participation, including triple balloting in the 2001 district-level elections, see Powley 2003, op. cit.

10. Berte Mukamusoni, personal interview, translated in part by Connie Bwiza Sekamana, July 2002.

11. For more information on women's NGOs in Rwanda, see Newbury, Catharine, and Hannah Baldwin, 2001. "Confronting the Aftermath of Conflict: Women's Organizations in Postgenocide Rwanda," in Krishna Kumar (ed.). *Women and Civil War: Impact, Organizations, and Action.* Boulder, CO: Lynne Rienner Publishers, pp. 97–128.

12. "Rwanda Democracy and Governance Assessment," produced for USAID by Management Systems International, November 2002, p. 35.

13. Ibid., p. 37.

14. Women's Commission for Refugee Women and Children, 1997. *Rwanda's Women and Children: The Long Road to Reconciliation.* New York: Women's Commission, p. 6.

15. "Rwandan President Urges Women to Stand for Public Office." Xinhua News Agency, 23 April 2003, <http://www.xinhua.org/english/>.

16. Rwandan Government, "Good Governance Strategy Paper (2001)," <http://www.rwandal.com/government/president/speeches/2001/strategygov.htm>.

17. John Mutamba, personal interview, July 2003.

18. Powley 2003, op. cit.

19. Freedom House, *Freedom in the World 2004,* <http://www.freedomhouse.org/research/freeworld/2004/table2004.pdf>.

20. Connie Bwiza Sekamana, personal interview, July 2002.

21. Aloisea Inyumba, personal interview, July 2002.

22. Tripp, Aili Mari, 2004. "Quotas in Africa," in Julia Ballington (ed.). *The Implementation of Quotas: Africa Experiences,* Stockholm: International IDEA.

23. Inter-Parliamentary Union (IPU), 2004. "Women in Parliaments 2003: Nordic and Post-Conflict Countries in the Lead," <http://www.ipu.org/press-e/gen183.htm> (accessed 8 September 2004).

Carey Leigh Hogg  NO

Women's Political Representation in Post-Conflict Rwanda: A Politics of *Inclusion* or *Exclusion?*

Introduction

One need not look further than recent news headlines regarding the status of women in Rwanda to note the international community's proclamation of the nation as a 'beacon of hope' for gender equality in Sub-Saharan Africa. Such reports range from claims that women in post-conflict Rwanda are now the most politically represented women on the planet, holding the world's highest percentage of female parliamentarians at 56%, to assertions that Rwandan women are now leading the rehabilitation of a nation left in tatters after 1994's horrific genocide. What is rarely addressed, however, is the strange confluence of two opposing trends in Rwanda's post-conflict environment: that the Rwandan Patriotic Front (RPF)-led government has advocated for women's greater political *inclusion* under the premise that women will 'better' the political climate, while simultaneously *excluding* any form of political dissent or ethnic identification.

This article therefore ventures into uncharted territory by asking two questions: First, does the discourse surrounding the Government of National Unity's (GNU) campaign to increase women's participation in formal politics uncritically assume that women parliamentarians will have a *different* relationship to politics? If so, to what extent does the Rwandan government pare women representatives' identities down to non-ethnic female subjects, seen only as promoting peaceful reconciliation? Secondly, given what external actors increasingly term an authoritarian state, does the notion that women will change the political climate have any substantive *meaning* in post-genocidal Rwanda? If there indeed *is* a lack of political space in Rwanda today, how does one examine the claim made by one unnamed Rwandese civil servant that *"[the RPF] puts women in the National Assembly because they know they [the women] will not challenge them?"*

. . . Viewing the Rwandan case with a critical and gendered lens generates deeper meaning for how women political representatives' identities can be dangerously frozen and 'subjectified' in post-conflict contexts, especially those intent on building 'national unity' by way of quieting dissent.

From *Journal of International Women's Studies*, November 2009, pp. 34–55 excerpted. Copyright © 2009 by Journal of International Women's Studies—JIWS. Reprinted by permission.

Part I: The 'New Politically Represented Woman' as 'Subject' in Post-Genocide Rwanda

. . . In 1998, Francis Fukuyama wrote that increased women's political leadership would foster a more cooperative and less conflict-prone world. While those in the poststructuralist feminist camp might cringe at the essentialising tone of this statement, it seems as if his assertion is a moot point—'women' have yet to test Fukuyama's supposition, for women still remain grossly underrepresented, constituting only 18.6% of parliaments worldwide.

Yet the RPF-led government of this small central African nation has put Fukuyama's call to task, implementing a series of mechanisms to increase women's political participation so that women now hold the majority of seats in the Rwandan Parliament. Much of the GNU's discourse that led to this dramatic rise in women's representation depends upon the concept that 'women as a group' have a *different* relationship to politics and that increasing their representation will lead to better governance and a more egalitarian society. This can be seen in the ways the rhetoric surrounding the RPF's campaign for greater women's representation pares women's identities down to 'nonethnic' subjects who are more prone to 'build peace.'

Difference Feminism

In the discourse surrounding the GNU's campaign to increase women's political representation, there exists an uncritical assumption that Rwandese women representatives have a different relationship to politics and that their increased parliamentary presence will automatically improve the political climate. While this belief proceeds without sufficient interrogation in Rwanda, it has been one of the most disrupted concepts within Western feminist political theory, reaching back to the division between the 'difference' and 'equality' schools of feminist thought. Those in the first camp have taken hold of the notion that women, by virtue of their capacity for motherhood, hold a sense of connectedness with others and therefore are more prone to transcend a *"political life dominated by a self-interested, predatory, individualism."* Those who utilise this 'difference' feminist argument to advocate for increased numbers of women in formal political life typically argue that the world would be more peaceful if women took the lead in creating policy, or as McAllister states, in 'reweaving the web of life.'

The notion that women representatives are less self-interested and less prone to factionalisation becomes further heightened when it is contextualised in the post-conflict context of Rwanda. Therefore, the same 'difference' feminists who depict women politicians as being more 'peaceful' argue that women are inherently less warlike than men; that their maternalist pacifism engenders an aversion to war.

On the other side of the feminist spectrum are those in the 'equality' school of thought, who believe that women who *do* become involved in formal politics will not put their 'femininity' above their particular political ideology; the implication being that women representatives will be just as prone to a self-interested and factionalised politics as 'political men.' Hunt notes that the most visible female political leaders, including Indira Gandhi, Margaret Thatcher, and Golda Meir showed little interest in promoting issues of common concern to women. One of the reasons 'equality' feminists have hesitated to utilise 'difference feminism' arguments is because such arguments often falsely assume that individuals can be grouped under the heading 'women' simply due to a shared set of anatomical attributes.

As Dahlerup and Friedenvall have put it, *"One may argue that women as a group are both the Achilles' heel of the feminist movement and its raison d'être."* The concept of 'woman' then becomes particularly paradoxical for those lobbying for women's political participation, for in advocating women's representation *by women,* referred to as 'descriptive' representation, proponents tend to elide other vectors of women's 'intersectional' or multi-axial identities.

Intersectionality in Rwanda

The theory of 'intersectionality' argues that those individuals dubbed 'women' may be more inclined to identify with other socially-attributed aspects of their persona, such as class, race, and ethnicity, to name a few. Intersectionality is especially applicable to the Rwandan case, for as Sharlach writes, "In 1994's Rwanda, a woman's loyalty to her ethnic group almost always overrode any sense of sisterhood to women of the other major ethnic group." Yet, in the dense web of power relations that comprises contemporary Rwandan society, women's *de facto* ethnic identification is shrouded under what Baines terms a 'dark veil of silence.' . . .

In all fairness, it is hardly surprising that the discourse surrounding women's increased parliamentary participation in Rwanda fails to address such ethnic classification, as ethnicity is far beyond a 'sensitive' issue in the post-conflict state. Baines, who has written extensively on the intersection of ethnicity and gender in post-conflict Rwanda, writes that when she first embarked upon her fieldwork in the country, she was instructed by UN officials never to refer to Rwandese ethnic groups by name; only 'old caseload returnees' (Tutsi refugees in exile since 1959), 'new caseload returnees' (Hutu refugees who fled in 1994), or 'rescapes' (Tutsi genocide survivors). Thus, in the void created by a deafening silence obscuring women representative's ethnic identities, the 'politically represented Rwandan woman' is created as a new 'subject' to fill the non-ethnic space. . . .

RPF's Arguments for Increasing Women's Representation

Women's Shared Experience of the Genocide

Powley, a scholar who has written extensively on women's role in 'strengthening governance' in Rwanda, observes that the RPF often argues that women will have a different relationship to politics simply because they were disproportionately affected by the genocide itself. . . .

Evidence of the gendered nature of the genocide can be found in much of the Hutu extremist ideology used to promote the ethnic cleansing, portraying Tutsi women as seductresses who would use their sexuality to trick and entrap Hutu men, identifying even Hutus who married or associated with Tutsi women as traitors to their 'kind.' Such discourse continued throughout the slaughter with the systemic use of rape as a weapon; as in many ethnic cleansings, rape was utilized as a means to humiliate and control the entire Tutsi population.

The horrific killings of the genocide were carried out at an unparalleled pace, as more than 800,000 individuals were murdered in the span of only 100 days. Though women did not participate in the genocide to the same degree as men, they were certainly not without blame, as women were both objects and agents of the genocide. . . .

Yet, with the genocide's end in July of 1994, women were disproportionately burdened with rehabilitating what was left of the country. Women and girls—who at that point constituted 70 percent of the population—faced an almost insurmountable task in having to reweave decimated social structures and severed traditional networks. Women were left to assume multiple roles as heads of households, community leaders, and financial providers. They were quite literally left to pick up the pieces of their destroyed society, building shelters, adopting almost 50,000 orphans, and burying a near 1 million dead bodies.

Thus, most Rwandans today acknowledge that women bore the brunt of the genocide, both in their sexual victimization and in their spearheading of Rwanda's social and economic reconstruction. Furthermore, the lasting effects of the genocide on gender roles in contemporary Rwanda do not go unnoticed; today, women constitute a demographic majority in Rwanda at 54% of the population. Women head 35% of households, are primarily responsible for rearing the next generation of Rwandans, and produce the majority of the nation's agricultural output. In sum, both in the immediate wake of the Rwandan genocide and today, women are seen as taking the lead in healing a nation permeated by loss and despair.

While the common experience of the genocide suggests that Rwandan women *will* have a different relationship to politics, the RPF has pared women's identities down to *women* representing *all Rwandan women* through its commitment to increasing women's political representation. This essentialisation dangerously elides women's *de facto* identifications as Hutu or Tutsi, as well as some of the more extremist women's roles in perpetrating genocidal violence.

Constitution and 'Elections'

Such essentialisation can be seen in the discourse surrounding the RPF's commitment to 'gender equality' as enshrined in 2003's constitution. The RPF first came to power in July of 1994, ending the genocide with its takeover of Kigali. By November of that year it had installed a 'Transitional National Parliament' that included ten women. Over the next nine years, the number of women holding political office steadily increased to 25.7 percent just before the 2003 elections. Throughout the entirety of the transitional period, the GNU expressed a strong commitment to women's inclusion in politics, organizing 'women's councils' at the cell, sector, district, and provincial levels, establishing a Ministry of Gender, and implementing electoral gender quotas for the national parliament. The latter has become the most visible of such efforts, recently garnering an unprecedented 56% of seats for Rwandese women in the lower house of Parliament.

Women representatives' demographic stronghold of the Rwandan parliament is partly the result of the GNU's commitment to women's political inclusion, as enshrined in Article 76 of the new Rwandan Constitution:

> The Chamber of Deputies is composed of eighty (80) members consisting of: ". . . . twenty-four (24) members of the *female sex* with two per Province and the City of Kigali elected by the Councils of Districts, of Cities, and of the City of Kigali, to which are added the Executive Committees of the women's organizations at the level of the Province, City of Kigali, Districts, and Sectors."

Yet the constitutional guarantee of gender equality still raises questions as to the substantive equality actually afforded through such mechanisms. Powley notes that:

> ". . . although Rwanda's constitution is progressive in terms of equal rights, gender equality and women's representation, it is limiting in other important ways; specific concerns have been raised about restrictions on freedom of speech around issues of ethnicity."

While Powley cautions against the constitutional restrictions on ethnic identifications and political dissent, one might question her assertion that the constitution supports 'gender equality'; that is, if 'gender' is taken as an analytical tool to examine all aspects of an individual's socially constructed identity, including her ethnic identification and even political leanings, and not as a mere synonym for 'woman.' However, as scholars Devlin and Elgie found in a series of interviews with Rwandan MPs, though the interviewees often referred conscientiously to 'gender,' most seemed to use the term interchangeably with 'women.' As one MP put it explicitly, *"we say gender but we mean women."*

'Women's Solidarity' Amongst Parliamentarians

Another issue that emerged in Devlin and Elgie's study was the focus on 'female solidarity' between women parliamentarians. These scholars repeatedly received unsolicited remarks referring to women's 'solidarity,' even though their study was crafted to examine the extent to which women's increased representation in Rwanda had affected any substantive change in policy outcome. . . .

Devlin and Elgie found that the maintenance of 'female solidarity' within the Rwandan parliament emerged as one of the female deputies' utmost priorities. In fact, some of the women MPs noted that they had tried to recruit women substitutes so as not to suffer a drop in 'group' numbers.

Perhaps Rwandese women parliamentarians have most visibly demonstrated their strong commitment to 'women's solidarity' through the Forum of Rwandan Women Parliamentarians (FFRP), a women's caucus formed in the mid-1990s. The FFRP has worked closely with women's civil society organizations such as Pro-Femmes on a number of different issues, including the revoking of pre-genocidal laws that had prohibited women from inheriting land. Powley and Pearson demonstrate the role of the FFRP in bringing women deputies together to draft the 2006 bill combating gender-based violence. They note that the drafting process was 'highly participatory' in that women parliamentarians capitalized upon a strong relationship with their female constituents to solicit input and sensitize citizens to the bill's content.

Certain advocates of special group representation for women might interject here to note that such a communicative relationship between women deputies and their constituents can actually function to *circumvent* charges that the identities of women representatives are being essentialised. Briefly, feminist political theorists such as Young and Mansbridge theorise that historically marginalised groups such as 'women' are best represented by 'women' *only if* 'representation' is viewed as a processual relationship between a woman representative and her female constituency, rather than a relationship of mere substitution. . . .

One could construe Rwandan women's common experience of the genocide as an instance in which such a processual definition of 'representation' between a woman representative and her female constituency would circumvent Mansbridge's 'historical communicative mistrust'; particularly in light of the sexual violence women suffered from at the hands of men. Yet the question 'mistrust of whom' still remains in the post-genocidal Rwandan case; if Mansbridge bases her promotion of women's descriptive representation on women's historical oppression by, and subsequent mistrust of men, can this theory apply in a post-genocidal context in which Hutu extremist women were also perpetrators of genocide launched against Tutsi and moderate Hutu women?

Part II. Utilisation of the 'Politically Participatory Rwandan Female Subject' in an Increasingly Authoritarian State

Now fourteen years after the genocide, Rwanda's 'Government of National Unity' disseminates rhetoric abundant with references to 'democracy' and 'reconciliation' whilst embarking upon policies pointing to a dictatorial regime intent on the exclusion of political dissent and the consolidation of power. Paradoxically, the RPF's 'democratisation' discourse has relied heavily upon the increased *inclusion* of women in its parliamentary ranks alongside the systemic *exclusion* of ethno-political dissent. As aforementioned, the RPF's promotion of women's greater representation has been grounded in the assumption that the 'nonethnic woman representative' as *subject* will have a different relationship to politics, and therefore that women's greater inclusion will 'better' the post-conflict political climate. While the regime's cursory usage of the 'difference' feminist argument is problematic in that it falsely freezes women's intersectional identities, this utilisation is further compounded with the RPF's implicit and uncritical coupling of women's increased parliamentary presence with 'the guise of democratic transition.'

The GNU has silenced both political dissent and ethnic identification, as demonstrated by the RPF's numerous human rights abuses and systemic lack of political space. Furthermore, the RPF has tethered women's greater political representation with the process of 'democratisation,' which is problematic for two reasons. First, women's much-touted attainment of more than fifty percent of the Rwandan parliament has in effect been the result of a non-democratic regime's promotion of 'gender equality,' as will be discussed below. Secondly, though it is too soon to decipher whether women's demographic stronghold of the Rwandan parliament will lead to an opening up of political space, it is safe to say that even the 2003 elections' ushering in of the world's only functional gender parity has had little impact on the RPF's *exclusion* of ethnopolitical dissent. This is evidenced by the fact that even though women have constituted a critical mass within the lower house of parliament since 2003, there have been no substantial changes in policy outcomes that move beyond 'women's issues,' or that disrupt the RPF's agenda. These argumentative strands can be woven together to seriously question RPF assertions that women's greater presence has led to a greater tolerance of difference, a higher propensity to build and maintain peace, and an automatic translation to the advancement of democratic ideals.

The Rise of a Single-Party State

In the early 1980s, a group of exiled Tutsis residing in Uganda joined hands to form the Rwandan Patriotic Front (RPF), the political arm of the Rwandan Patriotic Army (RPA), which in 1990 would fall under the leadership of General Paul Kagame, now the President of Rwanda. When the RPF took over Kigali in early July of 1994 to end the genocide, it initially showed a great commitment to power-sharing between Hutus and Tutsis, creating a multiparty transitional government entitled the 'Government of National Unity.' This first government included both Hutu and Tutsis in high-ranking positions; the president, prime minister and ministers of justice, interior, and foreign affairs were Hutu, while the speaker of the national assembly was Tutsi. Yet this diverse composition never made it past the infancy stages of the transitional government; in 1995, five of the most illustrious Hutu in the government resigned in protest due to what they claimed was a lack of substantive power.

Even if one were to apply a skeptical lens to such claims, it would be hard to argue against overwhelming evidence that power has increasingly been concentrated in the hands of the RPF. A number of amendments unilaterally made by the RPF have introduced a strong executive headed by President Paul Kagame and have redrawn the composition of parliament so that the RPF now dominates the government. For example, a study conducted by Gakusi and Mouzer finds that the RPF now occupies a disproportionately large portion of governmental posts. . . .

In all fairness, one must concede that the RPF inherited a devastated country rife with security concerns, in which Reyntjens argues necessitated a trade-off between control and freedom:

> . . . the RPF initially seemed to waver between, on the one hand, political openness and inclusiveness (witness the setting up of a government of national union and the return to Rwanda of a number of non-RPF civilian and military officeholders) and, on the other, a violent mode of management and discriminatory practices (witness the large number of civilians killed by the RPF. . . .

Yet since the initial stages of the transitional government, the RPF-dominated 'Government of National Unity' has moved away from ethnic inclusion and political openness, leaning increasingly towards a policy of discrimination rife with human rights abuses. . . .

Rwandese Women Representatives as Promoting 'Democratisation?'

Most external observers note that the Rwandan government's claims of successfully moving the nation forward along the path towards democratic governance are based on Huntington's 'transition paradigm' of 'democratisation.' The 'transition paradigm,' canonical for political scientists and international development practioners alike, assumes that a nation undergoing political transition will proceed towards democratic governance when democratic rule, freedom, good governance, and the rule of law are able to trump authoritarianism, oppression, human rights abuses, and corruption. Yet in light of the RPF's oppressive practices, *inter alia* human rights abuses and quieting of political dissent, the current authoritarian nature of the regime hardly falls within the parameters of democratic governance.

Multiple scholars see women's greater representation in both civil society and democratic decision-making bodies as harbingers of democratic governance. Jaquette notes that multilateral and bilateral donors are exerting an increasing amount of pressure on developing countries to increase women's political participation, particularly on nations emerging from conflict such as Rwanda. Women's political participation is seen as a means of promoting this 'transition paradigm' and subsequently for securing its status as a 'donor darling' for procuring international aid. It can be said that the RPF has utilised women's greater degree of political representation in part as a guise for other aspects of the 'transition paradigm' in which it has failed. These failures can be exemplified in three particular instances: the 2003 elections, or the very way in which women first came to occupy half of Rwanda's parliament; women's complicity in the fusion of civil society with the government; and women parliamentarians' inability to promote tolerance for ethnic and political dissent.

The Parliamentary 'Elections' of 2003

. . . International observers reported that the parliamentary elections were marred by the manipulation of ballot-box stuffing, lack of secrecy of the vote, and a lack of transparency in the counting procedure. . . . With the RPF winning two-thirds of parliamentary seats, the majority of remaining posts went to the Liberal Party (PL-10%) and the Social Democratic Party (PSD-12%). Multiple sources have confirmed that these are the only other significant parties outside of the RPF coalition, both of which are rarely critical of the RPF and typically work in tandem with them. . . .

The systemic lack of political space and diversity in Rwanda can also be found in the fusion of grassroots organizations with the Rwandan government itself. For example, the amalgamation of women's civil society organizations (CSOs) with the Rwandan government cements women representatives' *de facto* complicity with a regime that has painted a veneer of 'democracy' over distinctly undemocratic processes. . . .

In concluding that women did not attain nearly fifty percent of the Rwandan parliament via a fully democratic process, nor that there exists room within women's civil society for the promotion of ethnic tolerance or political freedom in Rwanda, how must one assess the RPF's claim that women's greater political representation automatically furthers the country along a path towards 'democratisation?' The only other way in which this claim could be validated would be if one were to prove that women representatives have even minutely destabilized the authoritarian status quo within parliament itself.

Substantive Change in Policy Outcomes?

In a study examining the effect of women's increased representation in the Rwandan parliament, Devlin and Elgie found that Rwandese women MPs had a different political 'style' than men—working more closely with grassroots women and placing more of an emphasis on 'female solidarity' when working to pass particular pieces of legislation. Women MPs have worked closely with women's CSOs to add new 'women's issues' to the policy agenda, such as HIV/AIDS, a Gender-Based Violence amendment, and increased property rights for Rwandese women. Yet Devlin and Elgie found little evidence that women's increased representation had augmented policy outcomes to any significant extent, e.g. to any level that moves outside of the RPF's policy parameters.

This is not to say that women's increased representation in Rwanda has been without value. The specific pieces of legislation women MPs *have* spearheaded are of great consequence to the women of the country, as these legal achievements include important milestones such as the 2003 Inheritance Act that eradicated pre-genocidal legal restrictions on women's rights to inherit land outside of their fathers' or husbands' approval. In addition, Devlin and Elgie found that some of the more experienced women MPs have actually spent *less* time working on 'women's issues,' so that ". . . some women are moving to become 'parliamentarians' rather than constituency workers." While the legislative attainments of Rwandese women MPs should not be discounted, it is important to note that their successes have primarily focused upon reforms of benefit to *women*. This is hardly surprising in a system dominated by the RPF, as the lack of political freedom severely limits women's ability to influence policy seen as running counter to the RPF's agenda.

To give a contrasting argument due consideration, some scholars have defended the RPF's efforts for greater inclusion of women in Rwandese political life. For example, Powley attempts to debunk the notion that the RPF could be including women *". . . as a means of diverting attention from the absence of more ethnically plural and representative government."* She argues that if the decentralisation programme that has already been initiated is allowed time to flourish, there is no way that the government could remain ethnically exclusive. Yet, folding back to the previous argument, this ethnic exclusion is exactly what *is* occurring in Rwanda, through both a 'Tutsification' and 'RPF-ization' of political power. If the population of Rwanda today is 85% Hutu, how can a government that is dominated by a party ruled by Tutsi elite be considered 'representative?'. . . Aren't women representatives not only 'women,' but also 'Hutu,' also 'Tutsi,' also invested in protecting human rights and the freedom to express independent political belief?

Conclusion: Equality vs. Difference; Inclusion vs. Exclusion

The RPF's contrasting drives for 'women's *inclusion*' and 'politico-ethnic *exclusion*' strike a familiar chord within feminist political theory's long-running, love–hate relationship with 'equality' and 'difference.' On the one hand, feminists *should* be at the ready with praise for Rwanda's attainment of the world's only functional parliamentary gender parity, as it seemingly overnight has accomplished the 'gender equality' feminists in Western democracies have been working towards for decades. On the other hand, some feminists might frown upon the relatively undemocratic means by which Rwandan women have ascended into the lofty halls of formal politics, as well as the way in which the RPF has pared women down to nonethnic subjects who are fundamentally driven to create peace and reconciliation.

. . . There is merit in highlighting the ways that the Rwandan government has boosted the identity-axis of 'woman' up, while sweeping that of 'Hutu' or 'Tutsi' under the proverbial rug. Furthermore, the applicability of what Western feminist political theory has developed regarding the discourses of 'difference' and 'equality' *vis-à-vis* women's increased political representation remains questionable when used as a lens to examine the Rwandan case. One must keep in mind that the canonical discourses on 'equality' and 'difference' have been theorised in stable democratic contexts; e.g. not in a fledgling democracy such as Rwanda whose government must strike a balance between tolerance of ethnopolitical *dissent* and a disavowal of ethnic hatred and discrimination. The Rwandan post-conflict environment, therefore, leads to new paths of investigation in terms of how such theories can be destabilised and regenerated to further an understanding of women's political representation in non-Western contexts.

In querying the extent to which the RPF supports women's increased political representation because it assumes women representatives will be 'malleable' enough not to destabilize the authoritarian status quo, this discussion has come full circle back to the discourses of 'equality' and 'difference.'

The RPF has been mistaken in appealing to a fixed, biologically determinist conception of 'woman' in its campaign to increase women's political representation. This is not only because the GNU has pared women representatives' multi-axial identities down to unilateral 'peaceful' and 'nonethnic' female subjects, it has also banked upon women's 'different' relationship to politics in order to render their supplication to authoritarian rule.

To argue against the RPF's utilisation of the 'difference' feminist school of thought to boost women's greater political participation is not intended to lessen the importance of Rwandan women's shared experience of a *gendered* genocide. Often those who take the opposing stance to 'difference feminist' arguments fall prey to the false pretence of arguing for women's equal political participation under a banner of 'gender-neutrality'; an abstract notion of citizenship that many feminist political theorists view as the root cause of the continuing andocentrism of modern day political institutions. The reality of the Rwandan case is that women *have* suffered the horrors of the genocide 'as women'; but they have also suffered through the systematic butchering of their people as 'Hutus,' as 'Tutsis'; as victims *and* perpetrators of the killings; a fact that is all too forgotten in Rwanda's contemporary landscape of a 'unified sameness.'

What is more troubling than the 'equality' and 'difference' debate within feminist circles, however, is the realisation that the flag of 'gender equality' in Rwanda is waved high with flying colours—while calls to 'ethnic tolerance' and 'political freedom' are simultaneously shrouded under a dark veil of silence. Brown in particular has wondered why the 'woman question' is so frequently addressed alongside references to 'equality' while discourses of ethnicity are relegated to the sphere of 'tolerance.' She concludes that while the discourse of 'equality' most often assumes an umbrella of 'sameness' under which sexual difference can be subsumed, the rhetoric of 'tolerance' is employed to control an ethnic 'difference' that is seen as more threatening to a unified nation. Brown's observation has an *almost* perfect fit with the Rwandan case— Rwandese women representatives have been essentialised under a 'unified sameness,' as the epitome of the female *Banyarwandan* subject seen as promoting peace, yet have not been constructed as 'different' enough to destabilise the RPF's tight grip over state power. In contrast, ethnic identification has been discursively constructed as a force so dangerous that it threatens the very security of the 'new' and 'unified' Rwanda—here misaligning with Brown in that ethnicity is not only not 'tolerated'; it is *disavowed*.

In a society such as post-conflict Rwanda, in which the discursive construction of gender is inextricably linked with ethnopolitical cleavages, one should take pause to think on whether the goal of 'gender equality' has been *substantively* achieved in the world's only democratic-decision making body in which women outnumber men. For aren't 'women' also 'Hutu,' also 'Tutsi?' Don't they have the 'equal' right to express political dissent, even if it goes against the RPF's agenda? If 'gender equality' *vis-à-vis* women's political representation is to be attained in Rwanda, women need to be afforded the political space to transform the oppressive policies of the single-party regime. 'Women' can unite together to transform the Rwandese political climate; not as 'women,' not as 'nonethnic' subjects who express a maternal altruism towards all, but as individuals coming together to further the ideals of democratic freedom. Veneranda Nzambazamariya, the Rwandan winner of the Millennium Peace Prize for Women, spoke these words to women survivors in the immediate aftermath of the genocide, though they might be more applicable today: *"Let yourselves be consoled, you have been sacrificed by systems it is necessary to change. Unite so as to transform problems into opportunities for action."*

POSTSCRIPT

Does Increased Female Participation Substantially Change African Politics?

One of the more interesting facets of the high level of female representation in Rwanda's parliament is the fact that their presence is, in part, supported by a quota system. Affirmative action at the level of elected officials is a growing trend (with at least 11 other African countries using quotas as well). Although it is difficult to determine if the actual practice of governance is changing in Rwanda with larger numbers of female parliamentarians, it is hoped that laws are now being scrutinized by law makers who are more sensitive to gender issues. Over the long run, this is likely to change the legal landscape in Rwanda, which has implications for women being able to own property, become educated, and run for office (irrespective of a quota system). Obviously, and as discussed in the NO article, what makes this case difficult to evaluate is the oppressive nature of the ruling regime and the degree to which it may be using the female quota system to its own ends.

In other African country settings, it is not clear if female politicians will always seek to shape laws and programs in a way that will be beneficial to women. Some have suggested that female politicians may fail to act on behalf of women if there is not a strong women's movement in the country. As such, electing women to parliament may not be as meaningful if this representation at higher levels is not backed up by grassroots organization (Aili Mari Tripp et al., *African Women's Movements: Changing Political Landscapes* [Cambridge University Press, 2009]).

It is important to note that the introduction of electoral quota systems has been met by stiff resistance in some African countries (as it has in other parts of the world). Although some constituents oppose such quotas because they are against increased female representation, others are allies who share a concern about male domination in politics, yet are concerned about a lack of effectiveness of female politicians elected through quotas, sometimes labeled "quota women." According to Tripp et al. ("Sub-Saharan Africa: On the Fast Track to Women's Political Representation," in Drude Dahlerup, ed., *Women, Quotas and Politics* [Routledge, 2006]), "many African critics of quotas, even among feminists, believe that quotas will lead to tokenism and might become yet another mechanism in the service of patronage politics."

Another good general reference on recent trends of female participation in parliamentary politics is Mona Krook, "Women's Representation in Parliament: A Qualitative Comparative Analysis," *Political Studies* (2010).

ISSUE 18

Is Corruption the Result of Poor African Leadership?

YES: Robert I. Rotberg, from "The Roots of Africa's Leadership Deficit," *Compass* (2003)

NO: Arthur A. Goldsmith, from "Risk, Rule, and Reason: Leadership in Africa," *Public Administration and Development* (2001)

ISSUE SUMMARY

YES: Robert I. Rotberg, director of the Program on Intrastate Conflict and Conflict Resolution at Harvard University's John F. Kennedy School of Government, holds African leaders responsible for the plight of their continent. He laments the large number of corrupt African leaders, seeing South Africa's Mandela and Botswana's Khama as notable exceptions. According to Rotberg, the problem is that "African leaders and their followers largely believe that the people are there to serve their rulers, rather than the other way around."

NO: Arthur A. Goldsmith, professor of management at the University of Massachusetts–Boston, suggests that African leaders are not innately corrupt but are responding rationally to incentives created by their environment. He argues that high levels of risk encourage leaders to pursue short-term, economically destructive policies. In countries where leaders face less risk, there is less perceived political corruption.

Rightly or wrongly, corruption is perceived to be a major obstacle to development in Africa. Transparency International's annual corruptions perceptions index (a survey of surveys on this issue) ranked Chad, Nigeria, Equatorial Guinea, Ivory Coast, Angola, and the Democratic Republic of Congo among the top ten countries perceived to be the most corrupt in the world in 2005. Chad tied for first with Bangladesh as the worst in the world of the 159 countries surveyed. Among those countries perceived to be the least corrupt in Africa were Botswana, Tunisia, South Africa, and Namibia.

Arguments about the genesis of corruption in Africa typically break into those that emphasize internal factors and those that stress structural or political economic conditions (often termed externalist explanations). The first type of argument typically looks to aspects of African culture, society, and tradition to explain the presence and persistence of corruption. For example, regional ties, ethnic allegiances, obligations to the extended family, patron-client relationships, and (occasionally) moralistic explanations are used to elucidate why corruption takes place in Africa. In contrast, structural arguments begin with the premise that a proclivity for corruption is universal, but that history and macro-level political economic conditions have created or reinforced opportunities for corruption in Africa. Examples of such factors include a colonial experience that reinforced anti-democratic traditions, cash-strapped governments that cannot afford to pay civil servants a living wage, or predatory international corporations that seek to access markets and resources through bribery rather than normal government channels. The two points of view presented in this issue represent the two sides of this debate, yet they approach a middle ground of explanation that is slightly more nuanced than the typical arguments heard from either side.

In this issue, Robert I. Rotberg, director of the Program on Intrastate Conflict and Conflict Resolution at Harvard University's John F. Kennedy School of Government, holds Africa's kleptocratic, patrimonial leaders, like Robert Mugabe of Zimbabwe, responsible for giving Africa a bad name, creating poverty and despair, and inciting civil wars and ethnic conflict. He attributes this problem to several factors, including the tendency of the African electorate to "acquiesce for long periods to the autocratic actions of their leaders," the lack of a hegemonic bourgeoisie that is independent of government, a weak civil society, little expectation for political leaders to be fair, and presidents who believe they are the embodiment of the state. He praises the ethical leadership of Botswana's Khama, South Africa's Mandela, and Mauritius' Ramgoolam.

Arthur A. Goldsmith, professor of management at the University of Massachusetts–Boston, suggests that African leaders are not innately corrupt, but are responding rationally to incentives created by their environment. In finding that there is a correlation between political risk and corruption, and between low political risk and liberal economic reform, he argues that high levels of risk encourage leaders to pursue short-term, economically destructive policies. He suggests that the risks of governing may be reduced by the spread of multi-party democracy, a form of governance that will make transitions in power more orderly and reduce the chances of execution or imprisonment for leaders upon departure from office.

YES

Robert I. Rotberg

The Roots of Africa's Leadership Deficit

Leadership in Africa is typified more by disfiguring examples—the Idi Amins and Robert Mugabes—than by positive role models such as Nelson Mandela and Seretse Khama. Other clusters of developing nations, such as Southeast Asia or Latin America, exhibit wide variations in leadership quality, but none is so extreme in its range. During the past three decades roughly 90 percent of sub-Saharan Africa's leaders have behaved despotically, governed poorly, eliminated their people's human and civil rights, initiated or exacerbated existing civil conflicts, decelerated per capita economic growth, and proved corrupt.

Why should sub-Saharan Africa show such an extensive disparity between the many nation-states that have been and are poorly led and those few that consistently have been led well? Are the distinctions particularly African? Are they a product of colonial misrule? Do they reflect a common problem of transition from dependency to independence? Do they emanate from deep-rooted poverty and a lack of economic growth? Is sub-Saharan Africa's lamentable leadership record, in other words, attributable to exogenous variables beyond its control, or does Africa respond less favorably to a leadership challenge of the same order as every other region's?

The positive examples of African leadership stand out because of their clear-minded strength of character, their adherence to participatory democratic principles, and their rarity. In contrast, the negative examples include so many varieties—predatory kleptocrats; autocrats, whether democratically elected or militarily installed; simple-minded looters; economic illiterates; and puffed-up posturers—that caricaturing or merely dismissing them would mislead. These single-minded, often narcissistic leaders are many and share common characteristics: they are focused on power itself, not on the uses of power for good; they are indifferent to the well-being of their citizens but anxious to receive their adulation; they are frequently destructive to and within their own countries, home regions excepted; unreachable by reason, they are quick to exploit social or racial ideologies for political and personal purposes; and they are partial to scapegoating, blame-shifting, and hypocrisy.

Good leaders globally, not only in sub-Saharan Africa, guide governments of nation-states to perform effectively for their citizens. They deliver high security for the state and the person; a functioning rule of law; education;

From *Compass*, vol. 1, no. 1, 2003, pp. 28–32. Copyright © 2003 by Robert I. Rotberg. Reprinted by permission of the author.

health; and a framework conducive to economic growth. They ensure effective arteries of commerce and enshrine personal and human freedoms. They empower civil society and protect the environmental commons. Crucially, good leaders also provide their citizens with a sense of belonging to a national enterprise of which everyone can be proud. They knit rather than unravel their nations and seek to be remembered for how they have bettered the real lives of the ruled rather than the fortunes of the few.

Less benevolent, even malevolent, leaders deliver far less by way of performance. Under their stewardship, roads fall into disrepair, currencies depreciate and real prices inflate, health services weaken, life expectancies slump, people go hungry, schooling standards fall, civil society becomes more beleaguered, the quest for personal and national prosperity slows, crime rates accelerate, and overall security becomes more tenuous. Corruption grows. Funds flow out of the country into hidden bank accounts. Discrimination against minorities (and occasionally majorities) becomes prevalent. Civil wars begin.

It is easy in theory and in practice to distinguish among good, less-good, bad, and despicable leaders everywhere, especially in sub-Saharan Africa. Good leaders improve the lives of their followers and make those followers proud of being a part of a new Camelot. Good leaders produce results, whether in terms of enhanced standards of living, basic development indicators, abundant new sources of personal opportunity, enriched schooling, skilled medical care, freedom from crime, or strengthened infrastructures. Bad and despicable leaders tear down the social and economic fabric of the lands; they immiserate their increasingly downtrodden citizens. Despicable rulers, particularly, oppress their own fellow nationals, depriving them of liberty, prosperity, and happiness.

Poverty within the context of resource abundance, as in oil-rich Nigeria from 1975 to 1999, indicates inadequate leadership. Despicable leadership is exemplified by Mugabe's Zimbabwe, a rich country reduced to the edge of starvation, penury, and fear. Economic growth from a low base in the aftermath of civil war and in a context of human resource scarcity, as in contemporary Mozambique, signals effective leadership. The opening of a long-repressed society, with attention to education and a removal of barriers to economic entrepreneurship, as in post-dictatorship Kenya, is another sign of progressive leadership.

Botswana is the paragon of leadership excellence in Africa. Long before diamonds were discovered, the dirt-poor, long-neglected desert protectorate demonstrated an affinity for participation, integrity, tolerance of difference and dissent, entrepreneurial initiative, and the rule of law. The relative linguistic homogeneity of Botswana may have helped (but compare Somalia, where everyone speaks Somali, is Muslim, and there are clans but no separate ethnic groups). So would the tradition of chieftainship and the chiefly search for consensus after discussion among a *kgotla*, or assembly of elders. The century-old, deeply ingrained teachings of the congregational London Missionary Society mattered, too, and infected the country's dominant political culture. Botswana stands out in sub-Saharan Africa as the foremost country (along with Mauritius and South Africa) to have remained democratic in form and spirit continuously since its independence (in 1966). Throughout the intervening years it has conspicuously adhered strictly to the rule of law, punctiliously observed human rights and civil liberties, and vigorously attempted to enable its citizens to better their social and economic standings. A numerically small population (1.6 million) doubtless contributes to Botswana's relative success, and exploiting the world's richest gem diamond lodes—since 1975—has hardly made achieving strong results more difficult. But Angola, Gabon, and Nigeria all have abundant petroleum, without the same striking returns for their peoples.

Any examination of Botswana, especially before 1975, shows the value of well-intentioned, clear-eyed visionary leadership. Seretse Khama, heir to the paramount chieftaincy of the country's most important and largest ethnic polity, completed his bachelor's degree at Fort Hare College in South Africa in 1944, spent a year reading law at Balliol College, University of Oxford, and then studied for the bar at the Inner Temple in London. In 1948, he married Ruth Williams, a Briton, and returned home to take up his chieftainship. But the British colonial authorities prevented him from exercising the rights of paramountcy, and he and Ruth were exiled to Britain in 1951. Five years later, they were allowed to return, officially as commoners.

Khama came from a family of Bamangwato chiefs who were well regarded for their benevolence and integrity. His studies and his marriage may conceivably have reinforced those family traits. Being exiled might have embittered him, but Khama seems instead to have viewed exile as a mere bump along the road to leadership within the evolving context of Botswana's maturity from protectorate to nation. Whatever the combination of nature and nurture, when Khama (later Sir Seretse Khama) founded the Botswana Democratic Party (BDP) in 1961 and led his country to independence, he already held dear those values of deliberative democracy and market economic performance that proved a recipe for his young country's political, social, and economic success. Modest, without obvious narcissism, non-ostentatious as a chief and leader (unlike so many of his African contemporaries), and conscious of achieving a national, enduring legacy, Sir Seretse was able to forge a political culture for the emergent Botswana—a system of values governing the conduct of political affairs—that has endured during the peaceful and increasingly prosperous presidencies of Sir Ketumile Masire and Festus Mogae, his successors.

Sir Seretse had a largely implicit, understated, but nevertheless substantial program for his people and his country. He put that program into place gradually, never succumbing to external political whims (such as the affinity for Afro-socialism or crypto-Marxism that infected his peers elsewhere in Africa), instant panaceas (such as nationalizing his productive mineral industries in the disastrous manner of neighboring Zambia), or posturing ineffectively against the hideous crimes of apartheid in nearby South Africa. Indeed, Sir Seretse and Sir Ketumile were deft and decisive in their disapproving but non-antagonistic approach to South Africa. Sir Seretse engineered his control of Botswana's diamond resources without frightening off or limiting investment from South Africa. He trained his own Botswanan successors and empowered them, but gradually, and without overstretching indigenous human resource capacities. Sir Seretse took no shortcuts. He and his successors abided no abridgements of citizen rights.

Sir Seretse could have done otherwise. As a paramount chief beloved by his people and respected for his learning, he could have behaved as so many of his African peers behaved during the 1960s and 1970s. If he had arrogated more and more power to himself, "for the good of his people," there would have been few critics. The rest of Africa had largely followed President Kwame Nkrumah, Ghana's first president from 1957–1966, in renouncing colonial traditions of representative government and becoming autocrats. Even gentle Julius Nyerere of Tanzania and equally gentle, modest Kenneth Kaunda of Zambia by the 1970s were abandoning inherited democratic forms and substituting single-party, single-man rule in place of broad participation. They were depriving judiciaries of independence and legislatures of autonomy. Objectors were jailed. Newspapers were banned or bought out. State radio broadcast only the words of the rulers.

Not Sir Seretse. He adhered to the nostrums that were no longer in current use in nearly all of Africa. For him, there was an ethic of performance and good governance to which he adhered. Sir Seretse was conscious every day that he could do better than the leaders of next-door South Africa, where whites oppressed the majority and deprived most inhabitants of their human rights and civil liberties. For whatever set of personal and pragmatic reasons, Sir Seretse epitomized world-class qualities of leadership.

In very different circumstances, Sir Seewoosagur Ramgoolam, the first leader of Mauritius (an offshore member of the African Union), operated under the same internalized leadership rules as Sir Seretse. Ramgoolam was more explicit in charting his vision, however—more in the manner of Lee Kuan Yew of Singapore. When Sir Seewoosagur took the Mauritian prime ministerial reins immediately after independence in 1968 (remaining prime minister until 1976), he understood that the island nation's mélange of colors and peoples—a plurality of Tamil-speaking Hindu Indians, Urdu- and Hindi-speaking Muslims from India and Pakistan, Chinese, and indigenous Creole-speaking Franco-Mauritians, most of whom were descended from slaves—could not long survive in peace if he or others were anything but transparently democratic. He stressed open politics, nurtured social capital, welcomed a free press, and strengthened the rule of law inherited from Britain, and earlier from France. Sir Seewoosagur also sensed that Mauritius' economy, hitherto based entirely on exporting raw sugar, would have to be diversified and grow. He attracted new investors from Asia. Soon Mauritius was a major world textile manufacturer; an island without sheep became a dominant supplier of wool garments.

Once again, leadership was central to Mauritius' post-independence transition from a potentially explosive racial hothouse and a primary producer subject to the fluctuations of world markets into a bustling, prosperous, politically hectic sustainable democracy. Sir Seewoosagur could have attempted to follow the other possible road to peace and growth on a crowded island of 1.2 million—strong, single-man rule of the Lee Kuan Yew variety.

But the tactics that worked so well among overseas Chinese in Singapore might have been incendiary in Mauritius's multiethnic mix.

Likewise, without Nelson Mandela's inclusive leadership, black-run South Africa after 1994 would have been much more fractured and less successful in governing its apartheid-damaged peoples. Mandela's vision insisted on full rights for the majority, but without too abrupt a removal of minority economic privileges. It strengthened the rule of law, greatly broadened the delivery of essential services, largely maintained existing pillars of the economy such as transportation and communications networks, and slowly shifted away from the dominant command economy toward one that was more market driven.

Mandela, Khama, and Ramgoolam all led their nations democratically when they could have aggregated personal power. Their leadership model might have been more Asian—top-down, less open, less constitutional, and less multi-ethnic and multi-tribal—and still benevolent and thus accomplished. Instead, they demonstrated what few of their fellow African leaders then or since have demonstrated: that Africans are perfectly capable of building nations, developing sustainable democratic political cultures, and modernizing and growing their economies effectively. Given these particular individuals' disparate human and ethnic origins, and given their respective nations' very diverse colonial legacies, it makes no sense to assert that African traditional culture somehow inhibits the exercise of democratic leadership.

There must be other reasons for leadership gone wrong in Africa, especially for those men who begin as promising democrats and then emerge a term or two later as corrupt autocrats. Take Bakili Muluzi, president of Malawi, for example. After the thirty-year dictatorship of Dr. Hastings K. Banda ended, Muluzi led the new United Democratic Front (UDF) against Banda and his associates, promising a return to full-fledged democracy. Overwhelmingly victorious, the UDP and Muluzi took power in 1994 and governed reasonably effectively during their first five-year term. Educational and health services were expanded, civil society and an open press were embraced, the judges were released from their fetters, and serious steps were taken to improve the very poor country's economic performance. Muluzi presided genially over this peaceful and welcomed transition from autocracy in his country of twelve million people.

Muluzi's second term, from 1999, began well enough, although the sticky stain of corruption soon began to spread through the upper echelons of government and around the state house. Economic growth stagnated, not least because of decisions not made by the president and the arrangement of special deals for presidential associates. In 2002, he decided that Malawi would be better off if he broke the constitutional provision against a third presidential term, beginning in 2004. (President Frederick Chiluba, in Zambia, another post-dictatorial reformer, tried the same argument to keep his presidency, but was rebuffed by Parliament and the citizenry. In Namibia, however, President Sam Nujoma successfully breached his country's constitution and now serves a third term.)

Malawi's Parliament denied Muluzi the needed constitutional amendments twice, and Muluzi reluctantly backed away from a third-term attempt in 2003. Instead, he bulldozed the ruling party's executive committee into letting him hand-pick a questionable successor (illegal by party rules), and then compelled a suddenly called meeting of the UDF to amend the party's by-laws to give him that authority, and also to make him permanent chair of the party, in control of its finances. Key cabinet ministers resigned, and Malawi's politics were soon thrown into turmoil—all local matters which concern us little here. The main question is what causes a democrat to turn autocratic? Is it simply that absolute power corrupts absolutely, as Lord Acton said long ago?

Mugabe is another of many African leaders who began by governing plausibly (in 1980, in Mugabe's case), only to turn venal later. Admittedly, within a few years of assuming power in Zimbabwe, he had used a special military brigade to kill 20,000 to 30,000 followers of a key opponent. But the first eighteen years of Mugabe's prime ministerial and presidential leadership (the title changed in 1987) also brought enlarged educational and medical opportunities, economic growth, relatively modest levels of official corruption, and comparatively calm relations between the tiny white commercial farming community and black Africans. Throughout the period, Mugabe astutely gained more and more personal power. He used official terror to remove challengers, and state-supplied patronage to keep senior supporters in line. From about 1991, his onetime key backers say, Mugabe began behaving with more and more omnipotence and arrogance. He was reelected frequently as president, and his largely obedient party dominated Parliament. Every now and again one or two outspoken dissidents were tolerated.

By 1998, Mugabe was seventy-four years old, married for the second time, and increasingly cranky that Mandela's release from prison and assumption of the South African presidency had dimmed Mugabe's own attempt to be a major player in all-African politics. His second wife was known, too, to be avaricious, and by 1998 corruption at the highest levels of Zimbabwe had grown in scale and audacity. Mugabe sought to salt away wealth for his extended family. He also gave license to the corrupt activities of others so that he could control them, in the manner of Mobutu Sese Seko of the Congo. Then, in 1998, Mugabe unilaterally decided to send 13,000 soldiers to the Congo, ostensibly to assist Laurent Kabila, the rebel successor to Mobutu, to defend against a Rwandan-organized invasion. Mugabe also wanted to grab the diamonds, cobalt, cadmium, and gold of the Congo for himself.

With Zimbabwean troops in the Congo until 2003, and corruptly acquired funds fleeing to safe havens offshore, Mugabe and his cronies bled Zimbabwe until, by 2000, the foreign exchange coffers were largely empty and food and fuel shortages began to recur regularly. By then he had also unleashed thugs against commercial farmers, using an old tactic to mobilize indigenous support. This time, however, it failed to do so, especially in the cities. Mugabe lost a critical constitutional referendum in early 2000. By mid-year he had also come within a few seats of losing his party's parliamentary majority in a national election. The Movement for Democratic Change (MDC), led by Morgan Tsvangirai, had posed a formidable challenge and, indeed, claimed that Mugabe's party had falsified the votes in several key constituencies.

Mugabe lashed out furiously against the MDC, and attacked whites and blacks suspected of supporting the opposition. The country's once formidable rule of law became the law of the jungle, with Mugabe packing the Supreme Court and threatening High Court judges until they retired or resigned. Legions of hired thugs attacked white farmers and forcibly occupied the farms (despite High Court injunctions), thus depressing agricultural productivity. When Tsvangirai stood against Mugabe in the 2002 presidential election, he was defeated in a poll widely believed in Europe, the U.S., and among Zimbabweans to have been rigged. Even after such a disputed triumph, Mugabe persisted in victimizing MDC members and their presumed supporters. Having driven Zimbabwe to the brink of starvation in 2002 and 2003, he and his lackeys sought to deprive areas that had voted for the MDC of relief shipments of food.

By late-2003, Zimbabweans faced constant shortages of food and fuel. Unemployment had reached 80 percent and inflation 500 percent. The U.S. dollar, worth 38 Zimbabwean dollars in 2000, was being traded on the street for 5,000 local dollars. Hospitals operated without basic medicines. Schools were closed. President Bush and Secretary of State Colin Powell called for Mugabe's ouster, an unusual step, and so did Prime Minister Tony Blair. Zimbabwe's Council of Churches also railed at Mugabe, a Jesuit-trained Catholic. Tsvangirai, meanwhile, was indicted for treason and served some time in jail, as did many of his senior MDC colleagues. Mugabe, throughout, resisted entreaties to retire, as the once proud, wealthy country spiraled into decay.

These appalling details are less relevant, here, than seeking to explain why Mugabe and Muluzi, Chiluba and Nujoma, and also former President Daniel arap Moi of Kenya and many other African leaders perform adequately during their early elected terms and then, in their second terms or beyond, become despots. Is it the inevitability of Acton's aphorism, or some law of diminished accountability? Most African leaders, the Botswanan and South African presidents and the Mauritian prime ministers aside, travel in pompous motorcades, put their faces on the local currencies, and expect to see photographs of themselves in every shop and office. Almost invariably, the less legitimate the office and the less robust the country, the more ostentatious their displays and the more stilted their bearing and manner.

Is it the African reverence for "big men," a hangover from pre-colonial reverence for chieftainship, that turns democrats into despots and persuades obedient electorates to support the pretensions of their peers turned potentates? Would it help if the new nations of sub-Saharan Africa abandoned executive presidencies on the American model and reverted to pure parliamentary governing systems with ceremonial heads of state?

African leaders are driven by instincts no baser than those of their colleagues in Asia, Europe, or the U.S. But African electorates tend to acquiesce for long periods to the autocractic actions of their leaders. That acquiescence may stem from the sheer rawness of democracy in Africa, and from the absence of a long period of preparation for democracy, unlike in colonial India or the West Indies. The African press's lack of sophistication and independence is a contributing factor. Civil society is also weaker. Extensive public-sector patronage in most African countries also allows leaders to escape criticism until their leadership excesses are obvious.

Africa for the most part lacks a hegemonic bourgeoisie—a business class that is independent of government and capable of thriving without patronage and contracts; such independence lessens the zero-sum quality of a rule. In those few countries where there is that independence, as in South Africa but not yet Nigeria, a leader approaching the end of his term in office does not have to worry about taking the perquisites with him and looting the country before he and his colleagues go. In many other countries in Africa, especially the poorer ones, the incentives to grab it all are great.

Throughout most of Africa there is little expectation, thus far, that successors will be fair, that an incoming political movement will not necessarily victimize its predecessors, and that there is an acceptable role for former presidents and prime ministers—except, notably, in Mauritius and Botswana. In many places, too, there is as yet no sustainable democratic political culture. That is, whereas American and European politicians might want to behave as autocrats, they are restrained by the norms of their dominant political cultures and the likelihood of being found out. In Africa, shame is less apparent than a kind of entitlement. Once elected, or once chosen by a military junta to rule, the president confuses himself with the state—in some way thinking of himself as embodying and being the state. It is only the exceptional individuals like Khama or Mandela who can escape the deep psychological trap of constant sycophancy. Like Louis XIV, others come to accept their own importance as the suns around which their little countries revolve. Except in a few places like Botswana, where an early leader knew better and emulated President Washington's refusal to be royal, African leaders and their followers largely believe that the people are there to serve their rulers, rather than the other way around.

Fortunately, there are a handful of very new leaders in Africa who espouse an ethic of good governance. They are distinguished from their less democratic peers by a willingness to govern transparently, to consult with interest groups within their populations that are not their own, to create an atmosphere of tolerance and fairness in their official operations, and to strengthen the institutions of their societies. These promising new leaders include Presidents Abdoulaye Wade of Senegal, who had opposed previous methods of rule for decades and now rules consensually; Mwai Kibaki of Kenya, in opposition from 1992 to his election in late 2002; John Kuffour of Ghana, who has begun reducing the corrupt climate of his autocratic predecessor; and Prime Minister Pakalitha Mosisili of tiny Lesotho, who has modernized his country's methods of governance through a process of laborious national consultation.

Africa is not yet ready to parse distinctions between transactional and transformational leadership. It needs leaders in the first instance who serve whole nations, not just their tribes or ethnic groups or extended families. It needs leaders who embrace responsibility for the commonweal, and not for a group of associates who live off and puff up a country's all-commanding autocrat. It desperately needs new leaders who take Khama and Mandela as their models and embrace the Washingtonian-like restraints that they embodied.

Once there are a cadre of leaders who espouse and embody in their actions the democratic values that emboldened Khama, Mandela, and Ramgoolam, and now drive their successors and men like Wade, Kibaki, Koffour, and Mosisili, Africa will begin to move from despotism and denials of human rights to the era of democratic leadership. Given the timbre of Africa's younger leaders, and the spread throughout Africa of global bourgeois democratic values, that era may soon be at hand.

Arthur A. Goldsmith

Risk, Rule, and Reason: Leadership in Africa

Introduction

Sub-Saharan Africa is poorly led. The region has far too many tyrants and 'tropical gangsters,' far too few statesmen, let alone merely competent office-holders. Too often, these leaders reject sound policy advice and refuse to take a long and broad view of their job. They persecute suspected political rivals and bleed their economies for personal benefit. With a handful of exceptions, notably South Africa under Nobel laureate Nelson Mandela, countries in the sub-Saharan area are set back by a personalist, neopatrimonial style of national leadership (Aka, 1997).

Better leadership is not the cure-all for Africa's lack of development, but it would be an important step in the right direction. A few years back some observers saw hope in a new generation of supposedly benevolent dictators, such as Isaias Afwerki in Eritrea, Meles Zenawi in Ethiopia, or Yoweri Museveni in Uganda (Madavo and Sarbib, 1997; Connell and Smyth, 1998). Subsequent events (war between Eritrea and Ethiopia, invasion of the Congo Republic by Uganda) chilled the optimism (McPherson and Goldsmith, 1998; Barkan and Gordon, 1998; Ottaway, 1998). In most countries, it seems progressive leadership soon reverts to the more familiar form of autocratic one-man rule.

There is no shortage of macro-level explanations for this pattern. Author-itarian political traditions, lack of national identity, underdeveloped middle classes and widespread economic distress are among the sweeping, impersonal forces cited as factors that produce poor leader after poor leader. Foreign aid may have enabled some of these leaders to hang on longer than they would have otherwise, especially during the Cold War. This article instead takes a micro-level view of leadership. Without denying that macro-level social and economic factors bear on leaders' behaviour, I find it also useful to look at these people as individuals and to speculate about the incentives created by their environment.

In the tradition of political economy, we can begin with the assumption that African leaders are usually trying to do what they think is best for them-selves. We can posit that they choose actions that appear to them to produce the greatest benefit at least cost, after making allowances for the degree of risk involved. Such a leader also is capable of learning, and takes cues from what is happening to other leaders in neighbouring countries. He can improve his behaviour if he has to.

While no African leader fully exemplifies this rational actor model, all these individuals' behaviour can be illuminated by it. After all, even the best leaders have mixed, sometimes egoistic motives. To the extent that it represents reality, the rational actor model also may suggest how changing the political incentive system might induce African leaders to behave less autocratically.

I start this article by speculating about how these leaders might react to perceived levels of risk in their political environment. Next, I investigate the actual level of risk, guided by a new inventory that covers every major leadership transition in Africa since 1960. Then, I assess how risk appears to have distorted the way African leaders act in office. Finally, I consider the ways in which democratization may be changing political incentives for the better.

Leadership and Individual Motivation

Perhaps the most troubling thing about African leaders is their tendency to reject (or simply not follow through on) conventional economic advice (Scott, 1998). Africa is the graveyard of many well-intended reforms. The vacillating public attitude of Kenya's President Daniel Moi is emblematic. In March 1993, he rejected an International Monetary Fund (IMF) plan for being cruel and unrealistic. One month later, he reversed himself, and agreed to the plan. In June 1997, the IMF cut off lending to Kenya after Moi refused to take aggressive steps to combat corruption. Again, his initial reaction was defiance, swiftly followed by a more accommodating line.

Why are African leaders apt to resist advice to carry out market-friendly reforms that could boost national rates of economic growth? If one accepts the premise that, with sufficient time, open market policies will work in Africa, such a choice can look senseless. Certainly, no African leader would prefer to perpetuate mass poverty and economic stagnation in his country, which can only make governing more difficult. More to the point, perhaps, cooperating with the international financial institutions is the best way to assure continued diplomatic support and financial credit. Yet, many African leaders apparently see political rationality in choosing policies that are economically damaging or irrational. Miles Kahler (1990) refers to this as the 'orthodox paradox.'

Political economy offers a theory of micro-level behaviour that may explain the paradox. Mancur Olson (1993) argues that time is the key. According to this theory, the predicament facing any individual national leader is that the pay-offs to most economic reforms lie in the future, but he also has to hold on to power now. An insecure power base is likely to encourage either reckless gambling for immediate returns or highly cautious strategies to preserve political capital; it is unlikely to promote measured actions to obtain long-range returns. Whether a leader acts for the short or the long term, therefore, is influenced by his sense of the level of threat to his career.

A more technical way to understand a leader's intertemporal choices is to think of a 'political discount rate.' One of political economy's core ideas is that future events have a present value, which one can calculate by using a rate of discount. That rate of discount rises with risk and uncertainty. When an outcome is doubtful over time, it makes sense to mark down its present value. The more doubtful the outcome, the more valuable are alternative activities that yield immediate dividends, even if the expected return of those activities is low. Thus, under conditions of political uncertainty, the narrowly 'rational' leader will systematically forgo promising political 'investments'—ventures whose benefits he may not survive to reap. Whenever he is given a choice, according to this argument, such a leader will usually prefer current political 'consumption.' It follows that free-market reforms look like a poor bargain, requiring immediate political pain in exchange for distant (and therefore questionable) gain.

High political discount rates are also a possible explanation for the extensive and destructive political corruption seen in Africa. The Democratic Republic of Congo's Mobutu Sese Sekou is the archetype. The late dictator erased the line between public and private property, accumulating a vast personal fortune and bankrupting his country. His is an extreme case, yet every national leader has opportunities to profit individually from his office. According to the premises of political economy, it is the leader with the least certainty about his fate who has the strongest incentives to take his rewards now—and to take as much as possible. A more self-assured leader may calculate that it is safe to defer most personal financial gain until after he has left office. Some of the misuse of public office also may be due to the need to buy support from friends and extended family members. Olson (1993), for example, postulates that leaders with an insecure grip on power have an incentive to take steps to patronize favoured ethnic groups, often at the expense of national economic health. This sort of pork-barrelling is well known in Africa.

Political economy thus presents a cogent theory for why African rulers act the way they do. Short-term policy making and political corruption are 'rational' ways of trying to manage the risks associated with governing in an unsettled political system, as we typically find in Africa. According to this thesis, overly cautious or corrupt leaders may simply be attempting to maximize utility under conditions of personal and political uncertainty. Their assessment of risk is affected by their personal experiences and by their perceptions about larger trends in their country and region. Unfortunately for the social welfare, their effort to protect their individual interests has spillovers that hurt everyone else.

The issue for this article is whether the facts support this theory. First, is it true that African leaders face a high degree of risk? We can reasonably assume these people are tolerant of risk, or they would not have chosen political careers. Thus, we need to look for evidence of extraordinary occupational hazards for leaders. The second question is whether political risk in Africa is associated with 'bad' (anti-market) economic policy choices or with corruption. As we will see below, the answers to both questions seem to be affirmative: there is significant physical risk for leaders, and that risk correlates with anti-market policies and with corruption. Those two findings, in turn, suggest scope for enhancing the area's national leadership by reducing the risks of governing, a goal that may be abetted by democratization. . . .

Risk and Leaders' Behaviour

There is little doubt . . . that holding high office in Africa poses acute risks. To what extent do those risks affect leaders' behaviour, specifically their behaviour in the areas of economic reform and corruption? . . . That question is difficult to answer fully without detailed case studies of the individuals involved. In the absence of such information, however, we can look for approximate answers in national indicators of economic policy and corruption. To the extent we believe that country leaders control public policy or set the tone for public honesty, aggregate data may give us clues about how these leaders conduct themselves.

To represent a country's commitment to free market economics, I use the Heritage Foundation's Index of Economic Freedom (Johnson et al., 1999). The index is calculated by aggregating country scores on 10 policy indicators and measures of the business climate. Depending on their scores, countries are categorized as free (none in Africa), mostly free, mostly unfree, or repressed. While I do not see eye to eye with the Heritage Foundation on many subjects, I suspect that these categories offer a good approximation for how fully countries comply with IMF-style structural adjustment programs. My grouping of countries is based on the average economic freedom rating for 1995–1999.

I hypothesized earlier in this article that low-risk environments would tend to produce more reform-minded leaders, or at least leaders who would be more willing to go along with economic reform in exchange for financial credit. . . . [A] correlation exists between the hazards of leadership and the degree of 'economic freedom.' Leaders in the so-called mostly free countries were the least likely to be overthrown, killed, arrested or exiled. Leaders in the mostly unfree and repressed countries, by contrast, experienced a greater number of negative outcomes. Low political risk and liberal economic programmes seem to go together in Africa.

Correlation does not prove causation, especially in making inferences about micro-level behaviour based on macro-level data. We cannot say whether a safer political environment encourages leaders to opt for the market, or conversely, whether leaders who opt for the market make their political environment safer (though the latter possibility seems less likely, at least in the short run). In either case, however, the results are consistent with political economy theory.

What is the relationship between political risk and corruption? For a measure of the latter, I use Transparency International's Corruption Perception Index for 1999. Transparency International is a watchdog organization formed to help raise ethical standards of government around the world. It compiles an annual index that assesses the degree to which public officials and politicians are believed to accept bribes, take illicit payment in public procurement, embezzle public funds, and otherwise use public positions for private gains. The index is based on several international business surveys, using different sampling frames and varying methodologies (Transparency International, 1999). While Transparency International is careful to point out that the rankings only reflect perceptions about corruption, I find it reasonable to assume that they correspond roughly to reality.

I have conjectured that leaders in the riskier African countries would have the greatest propensity to use their public offices for personal ends. Once more, the data lend support to my hypothesis. . . . The pattern is striking. There has never been a successful coup in the less corrupt group of countries. None of their ex-leaders has been arrested or exiled, and only one was killed while in office (South Africa's Verwoerd). The more corrupt countries, by contrast, have many coups and many leaders who suffered personally upon losing power.

As with the economic freedom index, these correlations do not prove that a hazardous political environment encourages leaders to become corrupt. The opposite is also plausible: corrupt rulers seem likely to invite coups and to bring personal suffering on themselves. To the extent that risk and corruption are related, the relationship between the two probably is mutually reinforcing. The important point for this article is that the observed association of risk and corruption conforms to what you would expect, based on the assumption of 'rational' behaviour among national rulers. Without overstating the case, the correlation lends support to a political economy account of poor leadership in Africa.

Democratization and Improved Leadership

Political economy also suggests that one solution to poor leadership is to make the political environment less hazardous. A safer environment would reduce the incentives to engage in political misbehaviour and, in principle, encourage more responsible and forward-looking activity. In this context, Africa's recent moves toward more pluralistic national political systems, where people can express their political opinions and take part in public decisions, are reasons for hope. It is fashionable—and correct—to observe that democracy has shallow roots in most African countries (Joseph, 1997; van de Walle, 1999). Much of the impetus for reform comes from abroad, from the region's creditors. Yet, when we observe the patterns of leadership transitions, it is hard to deny that genuine changes are taking place.

No sitting African leader ever lost an election until 1982, when Sir Seewoosagur Ramgoolam of Mauritius was voted out. Since then, 12 more incumbents have been turned out of office by voters—accounting for about one-sixth of the leadership transitions in the 1990s. The threat of losing an election also may account for the increasing rate of leader retirements—nine in the 1990s versus only eight in the previous three decades.

Democratization appears to be altering the outcomes of the many coups that still occur. In the past, the new heads of military juntas often declared themselves permanent leaders (sometimes after doffing their uniforms and becoming 'civilians'). Now, it is becoming the norm for coup leaders quickly to organize internationally acceptable elections—and, more importantly, to honour the results afterwards (Anene, 1995). Recent examples include Niger and Guinea-Bissau. . . . The fact we see more transitions of this type in the 1990s is an indirect reflection of the region's growing democratization.

[There are] additional reasons to think that contemporary presidential elections are not simply façades in many countries. The entire sub-Saharan region had only 126 elections for top national office in the 30 years through 1989. Most of those were show elections, with an average winner's share of close to 90%. Conditions have changed significantly in the 1990s. There were 73 leadership elections during that decade, or more than half as many as in the three prior decades. All but five of sub-Saharan countries were involved. Equally important, the winner's share dropped to an average of about two-thirds of the votes cast. Such results would be considered landslide victories in the developed world. No president in the history of the United States has ever reached two-thirds of the popular vote. Still, in African terms, the tendency clearly is toward greater competitiveness at the ballot box.

The classic liberal defences of free and fair elections are that they give voice to majority demands and that they are a means for recruiting new leadership talent. Political economy and African experience suggest three additional benefits, all associated with reducing the hazards leaders face.

First, elections have the virtue of softening the penalties of losing political office. The defeated candidate in an election campaign, as opposed to the victim of a coup plot, is far less likely to be executed, jailed or exiled by his successor. By providing a low-risk avenue of exit, elections thus reduce the stakes in political competition. If the arguments in this article are correct, that would free African leaders to take a more purposeful, pragmatic view of their jobs.

A second benefit occurs if elections become institutionalized, and take place according to a schedule. Countries that hold regular elections reduce speculation about when (and how) the next political transition is likely. Again, the probable impact in Africa would be to change the political calculations made by the region's chief power holders, to allow them to worry less about how to hold onto power and to think more about the long term. Predictable political transitions might also reduce anxiety among private investors, and thus mitigate the harmful political business cycle that exists in some countries.

The third benefit stems from the more rapid turnover among national rulers that results when elections become a regular part of political experience. As leaders come to see their jobs less as an entitlement and more as a phase in their careers, that actually may liberate them to 'do the right thing,' and not always feel forced to do what is politically expedient. Merilee Grindle and Francisco Thoumi (1993) have remarked on this phenomenon among lame-duck presidents in Latin America. Knowledge that their positions are transitory can, somewhat ironically, concentrate the incumbents' attention on how best to leave a lasting legacy. Similar results are possible in Africa.

Concluding Observations

Before multi-party competition and elections can have these positive effects on leaders, Africa's competitive political systems must become institutionalized. This has yet to happen in most countries, according to the results of Samuel Huntington's (1968) 'two-turnover test.' Huntington notes that institutionalized democracies prove themselves by repeatedly carrying out peaceful transfers of power through the ballot box. The first time an opposition leader replaces an incumbent power holder does not necessarily establish a tradition of peaceful political change. It is only after the new incumbent is defeated and leaves office that one can begin to be confident that constitutional procedures have taken root.

Second turnovers are almost unheard of in Africa. Botswana has not had one. The same party has ruled that country since independence. Mauritius has had two election-based leadership turnovers, but many observers question whether that island nation properly deserves classification in the region. Bénin is the only other African country where incumbent power holders have twice lost elections. The dictator Mathieu Kérékou fell to Nicéphore Soglo in 1991, but he regained the presidency by defeating Soglo in the election 5 years later. Kérékou's continued role raises some doubt whether Bénin's second transition indicates much other than the persistence of narrow, personalistic politics in that country.

Nonetheless, the last decade does offer hope that some African societies will be able to establish more orderly systems of political competition. That could change the incentives for African leaders, and encourage them to act more responsibly and even-handedly. As a means of redressing decades of oppression and economic stagnation, that cannot happen soon enough.

POSTSCRIPT

Is Corruption the Result of Poor African Leadership?

In this issue, Robert Rotberg attributed Africa's troubles to the fact that "African leaders and their followers largely believe that the people are there to serve their rulers, rather than the other way around." In contrast, Arthur Goldsmith viewed the mismanagement and corruption of African leaders as a structural problem in which individual leaders are responding to the incentive structures around them. These two interpretations of the situation appear to be quite different. Yet, while Rotberg eschewed the political economy or structural perspective, he concluded his argument by both calling for new leaders who will serve whole nations *and* pointing out the conditions that allow the base instincts of many African leaders to prevail. In pointing to the adage that "absolute power corrupts absolutely," Rotberg evokes a universal rather than particularistic (i.e., conditions unique to Africa) explanation. Furthermore, by asserting that this tendency is unleashed by a weak civil society and the lack of democratic traditions, he acknowledges that the environment in which African leaders operate contributes to the problem. This is quite different than other internalist arguments that often seek to explain corruption in Africa in terms of cultural factors. These conditions are also considered at a scale that is consistent with the micro-level sphere that Goldsmith investigates. In sum, the differences between the two arguments may be more about style and emphasis than substance.

A less-discussed element of corruption is the extent to which international corporations may contribute to the problem in their dealings with African governments. In some instances, it is standard business practice for companies to provide financial incentives (or bribes) to government officials in order to win contracts, obtain permits, or gain certain rights. Interestingly, it is the government officials who are often accused of corruption, while little is heard of the companies' role in this process. An example of an exception to this involves a Canadian company that was convicted by the Lesotho high court for bribing a senior Lesotho government official in order to win contracts on that country's $8 billion Lesotho Highlands Water Project (a joint project of the governments of Lesotho and South Africa). Commenting on the case, a South African official said that it is often assumed that corruption is "a peculiarly African problem. This case shows that such a perception is wrong. It takes two to tango." ("Government Cracks Down on Western Corruption," *New African*, December 2002.)

For those interested in further reading on this topic, good examples of the structural perspective on corruption include an article by M. M. Munyae and M. M. Mulinge in a 1999 issue of the *Journal of Social Development in Africa* entitled "The Centrality of a Historical Perspective to the Analysis of Modern Social Problems in Sub-Saharan Africa: A Tale from Two Case Studies," and an article by N. I. Nwosu in a 1997 issue of the *Scandinavian Journal of Development Alternatives* entitled "Multinational Corporations and the Economy of Third World States." There are roughly two types of particularlistic or internalist explanations regarding corruption in Africa—one is negative and the other is positive. The first, essentially negative perspective views internal African characteristics as flaws that contribute to a universally accepted problem known as corruption. An example of this perspective is an article by M. Szeftel in a 2000 issue of the *Review of African Political Economy* entitled "Clientelism, Corruption and Catastrophe." The second, more positive or postmodern interpretation often views "corruption" as a relative concept that is culturally defined. In other words, what is viewed as corruption in one culture is not necessarily corruption in another. Examples of this perspective include a volume by Chabel and Daloz entitled *Africa Works: Disorder as Political Instrument* (Indiana University Press, 1999) and a 2001 article by Jeff Popke in the *African Geographical Review* under the title "The 'Politics of the Mirror': On Geography and Afro-Pessimism."

ISSUE 19

Are African-Led Peacekeeping Missions More Effective Than International Peacekeeping Efforts in Africa?

YES: David C. Gompert, from "For a Capability to Protect: Mass Killing, the African Union and NATO," *Survival* (Spring 2006)

NO: Nsonurua J. Udombana, from "Still Playing Dice with Lives: Darfur and Security Council Resolution 1706," *Third World Quarterly* (2007)

ISSUE SUMMARY

YES: David C. Gompert, an adviser to Refugees International and a senior fellow at the RAND Corporation, believes that the African Union could be effective in Sudan if adequately supported. He believes that Africans are willing to commit combat forces to stop the killing in Darfur because they are more deeply affected by such abuses than Europe or North America. In fact, the unwillingness of the great powers to create a standing UN peacekeeping force is illustrative of a weak commitment of the West to intervene in Africa.

NO: Nsonurua J. Udombana of Central European University is more critical of the African Union peacekeeping mission in Sudan. He believes this mission has failed in Darfur because it suffers from several weaknesses, including problems of command and control, logistical support, operational practice, and lack of funds.

Certain major powers (most notably the United States, the United Kingdom, and France) and the United Nations (UN) have supported peacekeeping missions in war-torn areas of Africa over the past several decades. The UN has supported a number of such missions in Africa, with the effort in Mozambique often considered to have been more successful, and those in Somalia and Rwanda receiving poor reviews. In light of mixed or bad peacekeeping experiences, the UN Security Council has found it increasingly difficult to authorize new peacekeeping missions. Due to the West's growing reluctance to intervene in Africa, African states have increasingly taken matters into their own hands. Most of these African efforts have been undertaken under the auspices of the African Union (AU).

Any discussion of the AU must be preceded by a brief discussion of its predecessor, the Organization of African Unity (OAU). The OAU was formed in 1963—a time when many African states had recently achieved independence. Its major purpose was to foster unity and provide a collective voice for African nations. The OAU also aimed to eradicate all remaining forms of colonialism on the continent. Some (most notably Kwame Nkrumah of Ghana) even saw the OAU as a vehicle for a federation of African states. The OAU suffered from the fact that it had no real power to enforce its decisions. Its policy of noninterference in the affairs of nonmember states was also problematic. Given its inability to protect basic human rights, and its uncritical acceptance of several dictators on the continent, the OAU was dissolved in 2002. It was replaced by the AU, a supranational organization of 53 African states established in 2001. The AU aims to support democracy, basic human rights, and sustainable economies and to bring an end to conflicts on the continent. It eventually aims to have a single currency (the Afro) and an integrated defense force.

The AU's first peacekeeping mission was to Burundi in 2003. They currently have peacekeepers in the Darfur region of Sudan and in Somalia. The AU began sending peacekeeping troops to Darfur in 2005, and had 7,000 peacekeepers in place in mid-2007. Their mission has been extremely challenging because they must patrol an area roughly the size of France and are poorly funded and badly equipped. In June 2006, the U.S. Congress appropriated $173 million to support the AU peacekeeping force. Many Western advocacy organizations have called for UN peacekeepers to supplement or replace the AU force. In July 2007, the UN Security Council approved a resolution to create a joint African Union and UN peacekeeping mission known as the the United Nations African Union Mission in Darfur (UNAMID). It has a mandate of 12 months and is to begin deploying in October 2007. This 26,000 member force is to be merged with the 7,000 member AU force at the end of 2007.

In this issue, David Gompert, an adviser to Refugees International and a senior fellow at the RAND Corporation, argues that the AU could be effective in Sudan if adequately supported. He notes that the idea of creating a standing UN peacekeeping force has been in discussion for years and gone nowhere. He sees this as illustrative of a weak commitment of West to intervene in Africa. He believes that Africans (and the AU) are willing to commit combat forces to stop the killing in Darfur because they are more deeply affected by such abuses than Europe or North America. He acknowledges, however, that the AU needs the logistical and military support of an organization like NATO.

Nsonurua Udombana of Central European University is more critical of the African Union peacekeeping mission in Sudan, calling it "patently ineffective." He believes this mission has failed in Darfur because it suffers from several weaknesses, including problems of command and control, logistical support, operational practice, and lack of funds. He would rather urge the UN Security Council to unite and compel the government of Sudan to accept a UN deployment of peacekeepers.

YES

David C. Gompert

For a Capability to Protect: Mass Killing, the African Union and NATO

While condemning the Sudanese government and its Janjaweed henchmen for the horrors they have visited on the people of Darfur Province, the United States and its NATO allies have opted to back an African Union (AU) peacekeeping force to stop the atrocities. But the AU force lacks the firepower and authority to succeed. Darfur makes clearer than ever that it takes combat forces, not ordinary peacekeepers, to stop large-scale killing. If the West will not intervene militarily in Africa to prevent future Rwandas and Darfurs, it should help Africans build their own capability to do so.

Meanwhile, the US government's Global Peace Operations Initiative (GPOI) aims to add 75,000 peacekeepers worldwide by 2010, most of them African. While the GPOI is welcome—more peacekeeping capacity is obviously needed—it does not, as conceived, promise to create combat capabilities sufficient to stop mass violence when there is no peace to keep. Now that we know that the scourge of genocide has jumped from the twentieth century into the twenty-first, the world needs multilateral military capabilities for forcible intervention against large-scale atrocities, displacement and death. A partnership between NATO and the AU, with UN support, to develop AU forces able to stop genocide would be a step towards this end. Because the AU is far from having such capabilities today, such a partnership could take years to bear fruit, but by the same token should be launched as soon as possible.

Since the UN's birth after the Second World War, a precondition of peacekeeping has been the consent of the warring parties to accept peacekeepers and honour cease-fire or peace agreements, thus creating permissive conditions and obviating the need for peacekeepers to fight and take casualties. But the forces, legal authority and operating rules of engagement that are adequate for such consensual situations are inadequate for violent ones, such as civil war and state failure, in which mass atrocities usually occur, especially when the local government is complicit.

Repeated crises have shown that peacekeeping is either precluded or, if in place, incapable of stopping mass killing. Peacekeepers did not prevent Serbian ethnic cleansing of Bosnian Muslims or the massacre of Muslim men and boys at Srebrenica. Peacekeepers could not halt the rampaging of the 'West Side Boys' of Sierra Leone prior to the arrival of British paratroopers, nor prevent the slaughter of Tutsis and moderate Hutus in Rwanda. Peacekeepers of the AU Mission in Sudan (AMIS) may discourage but cannot stop Janjaweed burning, raping and killing in Darfur; indeed, AMIS personnel have been abducted and murdered.

This is not to disparage peacekeeping, which can help prevent conflict and provide security for post-war reconstruction. However, there is an additional requirement for military intervention forces that are strong enough that they do not need the consent of the parties, the cooperation of the offending government, or the self-restraint of killers—strong enough even to deliver violence to killing forces if there is no other way to stop genocide. To limit the international community to the ability to end mass atrocities only under *permissive* conditions is to ignore that mass atrocities occur under *non-permissive* conditions. Peacekeeping might be enough to keep peace between two obliging sovereigns but is not enough to keep a vicious sovereign from warring on its own people.

The Responsibility, Right and Capability to Protect

The persistence of mass killing and the incapacity of traditional peacekeeping to stop it have fuelled political support for important new international legal principles: *the responsibility to protect* and *the right to protect* human beings that are undefended or even slaughtered by their government. The former principle means that states, acting together, *should* intervene to stop atrocities; the latter means that they *may* do so. The concept underlying these principles is that sovereignty cannot be a shield behind which to commit or permit crimes against humanity, especially when illegitimate regimes rely on such crimes to stay in control.

These ideas challenge centuries-old tenets of unqualified sovereignty and prohibition of intervention. Hard-nosed statesmen and pundits may fear that such a compromise of sovereignty will produce international chaos, and authoritarian regimes—China, for instance—sneer at the *responsibility and right to protect* for plainly self-serving reasons. Consequently, the responsibility and right to protect face political, legal and practical hurdles and will likely be debated for years before being broadly accepted.

Yet these disagreements are no reason to delay the development of an international *capability to protect*. Multilateral military forces adequate to stop mass killing would, at the very least, provide the option to act in cases when there is international consensus to do so. Moreover, having such capabilities would increase pressure on states and groups to desist from atrocities and allow international humanitarian intervention or else face forcible action anyway. A government that is condoning or abetting mass killing, like the Sudanese government is today, would likely be more amenable to cooperation if it knew that the international community could act effectively to stop the killing whether it cooperates or not.

Attempts by the UN to extend peacekeeping into non-permissive situations, such as Bosnia and Somalia, have run into institutional and material barriers. Neither the UN Charter nor the peacekeeping capabilities and mandates that are predicated on it enable forcible international action to stop mass atrocities. Consequently, countries (e.g., the United States, the UK, Australia and France) and coalitions (e.g., NATO and ad hoc groupings) with superior military forces are authorised to undertake especially demanding and dangerous interventions, ranging from the liberation of Kuwait to air strikes in Bosnia to British intervention in Sierra Leone to current operations in Afghanistan. In Kosovo, the UN not only lacked the means to stop Serbian ethnic cleansing of Albanian Kosovars, but also could not reach agreement in the Security Council in support of NATO's view that the Belgrade regime's savagery had compromised its sovereignty and justified intervention—forcing NATO to take vigilante action.

Thus, the international community has had to outsource forcible interventions precisely when they are most needed—when innocents are being slain or uprooted in large numbers. The problem, of course, is that nations, including the United States, and coalitions, including NATO, may opt not to intervene in the very situations that involve mass human-rights violations because they are taking place in Africa or other places that are not deemed important enough to warrant forcible intervention on grounds of national interest. As we know, genocide can happen where the powerful Western democracies have no vital interests.

So we have cases like Rwanda and Darfur in which peacekeepers are available but militarily inadequate while militarily adequate forces are unavailable. This suggests a need for multilateral intervention forces, readily available to the UN or the AU—Africa being the focus here—that can stop mass killing and cleansing when an international consensus exists to do so. Such forces would reduce both the burden and reliance on the United States and its allies to use deadly force when they have no vital stakes. This is not to say that the Western democracies should never intervene solely to save lives—they cannot plausibly, or morally, claim that genocide is Africa's problem, not theirs—but rather to point out that counting on them is unwise and, with the right effort, could be unnecessary.

Military Superiority

The scale of this challenge should not be underestimated, because the combat capabilities of intervening forces will now become more critical. Forces dispatched to protect human beings under non-permissive, non-consensual conditions must be militarily superior to those of any of the hostile parties, including a recalcitrant sovereign state. Operational superiority is not a standard that has applied, or need apply, to traditional peacekeepers; because they are not intended to fight, they are generally not equipped, trained or organised to fight. But military superiority is crucial for forcible multilateral humanitarian intervention because of the presumption of armed opposition, the need to deter such opposition, and even the intention of initiating hostilities against killing forces if need be.

The superiority of humanitarian intervention forces over opposing forces is achievable. More often than not, the perpetrators of mass atrocities are killers but not fighters, in the sense of disciplined troops prepared to stand up to combat forces. Rag-tag Bosnian Serbs, machete-wielding Hutus and militia of the old regime in East Timor all folded when confronted by real soldiers. Janjaweed marauders signed up to rape and kill, not to fight and die in Darfur. The notion that intervening forces will face fierce and formidable enemies at great risk of casualties and failure is not borne out by experience or analysis. While there is plenty of evidence that killing forces flee instead of fight Western forces, there are anecdotal indications that they are not keen to shoot it out with good African troops either. With Western help, capable African combat troops—South Africans, Rwandans, Nigerians, Senegalese, Kenyans and others—can be improved and melded into effective multilateral humanitarian-intervention forces. While this would have the effect of making superior African militaries stronger still, it would do so in a multilateral—'denationalised'—framework and under Western influence.

Broadly speaking, military superiority has two components: operationally superior deployed forces, under expected conditions; and escalation dominance, through some combination of deployable reinforcements and deterrent forces. The requirement for superior deployed capabilities implies a need to strengthen the military qualities of forces available to operate under the UN or AU. More specifically, in the context of sizeable non-permissive humanitarian crises, the standard of superiority of intervening forces demands some measure of:

- accurate and persistent strategic awareness;
- fast deployment and employment, e.g., air and high-speed land mobility;
- precision fire-power from land, gunship and other strike systems;
- information dominance: tactical intelligence collection, fusion and dissemination;
- command and control, including deployable communications with adequate bandwidth and connectivity to global grids;
- flexible and ample logistics; and
- able combat leadership, decision-making and cognition at all levels.

Of course, the size and weight (armour) of humanitarian intervention forces must be adequate to ensure their superiority and survivability. However, the better they are in the qualities just listed, the smaller and lighter they can be and still have an operational advantage and escalation dominance. Indeed, small, light, mobile, aware forces can have appreciable operational advantages over large, heavy, ponderous ones in negotiating difficult terrain, covering large territory, responding to warnings of attack, manoeuvring and visiting effective but discriminating fire on opposing forces. All else being equal, speed and knowledge are more important than mass and lethality in operations aimed at stopping dispersed violence in remote areas. The essential quality of humanitarian intervention forces is that they be networked—able to exploit the operating concepts and technologies of the information age.

Over the years, the idea of creating a standing UN peacekeeping force has surfaced but gone nowhere. If there is no consensus to create a UN permanent peacekeeping force for permissive missions, there will surely be none to create a standing humanitarian combat intervention force under supranational control. However, this does not prevent taking action to ensure the availability of such a capability when needed. With the military application of information technology, especially data connectivity, it is becoming increasingly easy to generate forces at relatively short notice that can join up and operate effectively together by virtue of their ability to 'plug and play.' This is especially so if they are light, air-mobile forces at high readiness in training and materiel, and provided arrangements and preparations to generate them have been made and rehearsed.

NATO as Model and Partner for the AU

Generating and using superior military forces cannot be done via the established way of assembling peacekeeping forces, which depends mainly on ad hoc recruitment from sundry countries that offer spare non-combat forces of uneven quality. A better model for force generation can be found by looking at NATO, which maintains few standing military capabilities but has obligations, standards and agreed procedures, frequently exercised, to generate forces quickly for joint action upon warning and a political decision to do so. Because of such preparations, NATO is able to join up earmarked national forces into a coherent multilateral force with reasonable predictability and speed, which adds to deterrence.

In essence, whereas UN members are, fundamentally, obligated only to *abide* in peace and security, NATO members have committed themselves by treaty, under certain circumstances, to *provide* peace and security, even if it means going to war. True, NATO includes a commitment of forces that may go beyond what could be expected for countries that are not formal allies bound to defend one another but instead members of an inclusive international organisation like the AU. However, what is so interesting about NATO is not the legal obligation of its members to furnish forces for common defence of allied territory (under Article 5 of the NATO Treaty) but instead the emerging understandings and procedures for planning and generating forces for expeditionary operations (under Article 4) when there is not a binding commitment. Even as NATO relies increasingly on its members to make good on non-obligatory plans to prepare and provide forces for Article-4 contingencies, Africans and others could learn and borrow from these practices.

Under such an approach, the relevant international organisation—the AU, for our purpose here—would need permanent control of only modest command and control cells and information systems in order to permit quick force generation. It would need to issue quality, readiness and interoperability standards and common communications protocols. Of course, prior political understandings would be needed with at least some members to the effect that they accept the responsibility to protect and, in principle, can be counted upon to provide forces as planned. Such understandings are implicit in the current effort to form regional peacekeeping brigades of the Africa Standby Force under AU auspices, though these forces lack fire-power, deployability, interoperability and efficient force-generation processes. For African humanitarian intervention forces to be able to deploy and fight decisively to stop mass killing, force-generation principles and practices like NATO's would need to be adopted.

Even such arrangements for planning, preparing and generating superior military forces for non-consensual humanitarian intervention might not suffice if unexpectedly strong opposition is encountered or if fighting drags on or expands. This suggests a need for escalation options sufficient to leave no doubt that the international intervention will succeed militarily. Reinforcements may come in the form of additional ground forces, air strikes, intelligence, command and control, and logistical capabilities. It is less crucial that these capabilities be under the control of the AU (or UN) than that their availability be a certainty to all concerned, including hostile forces. Such escalation dominance can and probably must be provided by a powerful partner.

There are, of course, ample military capabilities among the United States and certain European NATO members to establish and if need be exercise escalation dominance in most situations of international intervention to stop mass atrocities. The distance, scale, logistical demands and firepower requirements will determine whether the United States must participate; under many circumstances, the combination of NATO command and non-US forces would lend sufficiently credible escalation dominance. Designation of a lead country would be natural, though a direct relationship between NATO and the AU would broaden participation and enhance (while admittedly also complicating) command and control.

As for escalation strategies, NATO air strikes in Bosnia demonstrated the decisive effect of the use of precision tactical air-power against forces that are interfering with peacekeepers on the ground. If air-power is not enough, the new net-centric capabilities and operating concepts being adopted by NATO militaries can provide joint (air–ground–naval) forces to reinforce African humanitarian intervention forces that find themselves in trouble. This would be a natural mission for the NATO Response Force.

Political understandings would have to be reached prior to an intervention such that adequate forces of NATO or individual members will be ready to escalate. Such understandings, made known, would help ensure the success of intervening forces, deter opposition to them, and even reduce the need for escalation. Mass-killing forces and their leaders will not be eager to face Western military power. The arrival of US Marines in Liberian waters clearly helped convince Charles Taylor to exit the country and permit the UN to insert a peacekeeping force. The concept of a humanitarian-intervention partnership between NATO and the AU offered here would enable the Western powers, including the United States, to become a steady wholesaler instead of an inconsistent retailer of security—a more acceptable and sustainable role politically both in the West and in the view of Africans.

In time, the EU might be in a position to help the AU maintain and use humanitarian intervention forces. For now, it is lacking in experience in forcible military action, in force planning and force generation, in combat command and control, and in escalation capabilities, apart from those of member states that are also members of NATO. More important than these shortcomings is that the EU does not include the United States, the participation of which is important for technical—e.g., its global information grid—and political reasons. Just as it would be a mistake for the United States to pursue a partnership with the AU to the exclusion of its European allies, it would be a mistake for the Europeans to pursue such a partnership to the exclusion of the United States. The aim of supporting the AU is best served by the Atlantic allies working together. . . .

Although NATO and several of its members already have the military capacity to intervene to stop genocide in Africa, neither they nor, for that matter, the Africans favour that option. NATO could help the AU make the capability to protect a reality. As noted earlier, the United States and several European countries now have underway programmes to expand peacekeeping capacity, mainly but not only in Africa. Beyond this, NATO can provide:

- expanded and focused training and other assistance to those AU members prepared to earmark quality *combat* forces for humanitarian military intervention, as many African countries have already agreed to do in connection with the several African regional brigades of the AU's Africa Standby Force (at a minimum, NATO can provide coordination and some military standards among the national assistance programmes of its members);
- sharing of expertise, models and procedures for planning, preparing, generating and integrating stand-by national forces for forcible multilateral action when needed; and
- capabilities and plans for operational support, strengthening of command and control, reinforcement and escalation dominance.

The principal instruments that NATO would use to carry out such a programme are its military commands (under Allied Command Operations), its international staff, and those special operations forces (SOF) of members with competence in developing indigenous forces. In this regard, the US government is considering the idea of proposing a NATO SOF capability. While intended for direct counter-terrorist operations, NATO SOF could also be highly useful for improving and working with African forces, as some are doing on a small scale already. Overall, NATO provides an ideal venue for the Atlantic democracies to pool military efforts to help Africa and other regions where peace often fails and mass suffering occurs. And nothing could be more consistent with NATO's new aim of extending security beyond Europe.

Readers might wonder whether Africans themselves are willing to commit combat forces to stop genocide when the West is not. The answer is that African countries pay a high price for mass human-rights abuses and killing on their continent—a price that European and North American countries do not pay, at least not directly. Flows of hundreds of thousands of refugees, cross-border incursions of militia groups, social and political upheaval, and damage to already struggling economies can hurt entire regions, as has happened in

West Africa, Central Africa and the Horn of Africa since the end of the Cold War. As long as wars and mass atrocities continue to destabilise parts of Africa, it makes sense for Africans to play a leading role in intervening to stop them. While they have far to go in building consensus and capabilities to protect their region from conflict, Africans are putting much greater effort into and reliance upon the AU than they ever did its feeble predecessor, the Organisation of African Unity.

Although Africans are indeed taking greater responsibility to stop the killing that has plagued their people and harmed development, they need help to create multilateral military capabilities to stop genocide, especially if the West is disinclined to intervene. To get this right, Africans should not hesitate to look to NATO for assistance. The time is ripe for NATO to respond with a major initiative to help the AU create and stand behind a capability to protect.

Nsonurua J. Udombana  **NO**

Still Playing Dice with Lives: Darfur and Security Council Resolution 1706

The Darfur genocide is the first in the 21st century and may eventually be the best known in history. 'Never in human history has a genocide in progress been so visible,' although it remains invisible to the obdurate and politically motivated doubters. By the time the genocide claims its last victim, the UN Security Council, which has the primary mandate of maintaining international peace and security, will be suffering from resolution fatigue. The Council has already adopted about 10, mostly lame, resolutions bearing on the Darfur crisis. Its latest release was Resolution 1706 of 31 August 2006, adopted by a vote of 12 in favour, none against; China, Russia, and Qatar predictably abstained. These 10 resolutions do not include statements by the Security Council presidents, numerous reports by the UN Secretary-General and the infamous Report of the UN Commission of Inquiry on Darfur. They also exclude resolutions, declarations, and decisions of the African Union (AU) and its Peace and Security Council (PSC).

The avalanche of global and regional resolutions, statements, communiqués and mission reports evidences, in principle, a desire to see an end to the pogrom in Darfur. In practice, these exercises, not to say the several ceasefire and peace agreements signed by belligerents, have done little to stem what the PSC calls 'the ever-deteriorating security situation on the ground.' In the past three years the Security Council has repeatedly expressed its alarm that the violence in Darfur has continued unabated. It expresses 'its utmost concern over the dire consequences of the prolonged conflict for the civilian population in the Darfur region as well as throughout Sudan, in particular the increase in the number of refugees and internally displaced persons.' It condemns 'all violations of human rights and international humanitarian law . . . in particular the continuation of violence against civilians and sexual violence against women and girls.' The Council has expressed every sentiment except a desire to take a strong action to save victims of the Darfur genocide.

Resolution 1706 is expected to change the status quo ante. It authorises the expansion of the UN Mission in Sudan (UNMIS) to Darfur—without prejudice to its existing mandate and operations—and its deployment to replace the patently ineffective African Union (AU) Mission in Sudan (AMIS) and support

From *Third World Quarterly*, vol. 28, no. 1, 2007, pp. 97–106, 110–111, omit notes. Copyright © 2007 by Taylor & Francis Journals. Reprinted by permission via Rightslink.

the effective implementation of the Darfur Peace Agreement (DPA). It requests the Secretary-General to arrange rapid deployment of additional capabilities to enable UNMIS deploy in Darfur and calls on UN member states to ensure such an expeditious deployment—subject to consent by the Government of National Unity (GNU), a euphemism for the National Islamic Front/National Congress Party in Khartoum! The resolution requests the Secretary-General to consult jointly with the AU and with parties to the DPA—including the SPU—on a plan and timetable for a transition from AMIS to UNMIS.

The AU and its PSC welcomes the proposed transition, but the government of Sudan views the move as an encroachment on its sovereignty, an attempt at recolonisation and the climax of efforts to undermine the DPA. Notwithstanding the promise that UNMIS shall have a strong African participation and character, Omar al-Bashir, the strongman of Sudan, threatens an all-out war against the UN should it attempt forcible deployment of peacekeepers in Darfur. Sudan's tough talk 'is a simple matter of bristling at the perceived loss of sovereignty,' but it also is an act of resentment against the West for not rewarding Sudan's so-called attempt at modernisation. Khartoum, says *The Economist*, would like to see its co-operation with the Central Intelligence Agency (CIA) in the war against terrorism recognised; its name removed from the US State Department's list of state sponsors of terrorism; and Congress lift its trade embargo. These wishes are not being granted, hence the hard-line posturing over Resolution 1706.

This article interrogates Resolution 1706 in the context of the continuing mayhem in Darfur and the AU's first major experiment at peacekeeping through AMIS. How well has AMIS performed? Could it have done better with a more robust mandate and enhanced military strength? How correct is it to say that AU member states are incapable of pulling resources together for effective regional peacekeeping? What is the legal basis for the proposed transition from AMIS to UNMIS and what will be the role of the AU and its PSC in this new enterprise? Does Resolution 1706 translate into clear rules of engagement? Given the nature of the Sudanese state, is governmental consent an absolute necessity for troop deployment, or merely desirable? This article attempts to provide answers to these and related questions.

The next part assesses AMIS's experiment at peacekeeping in the context of its mandate. The following section analyses Resolution 1706 and UNMIS. The final part draws some conclusions from the narrative and analysis.

Darfur and the AU Mission in Sudan

Darfur, a large, arid region in western Sudan, is the symbol of a nation's retreat from humanism, the ugly story of a failed state and, by extension, of a failed continent, where so much is, and rightly should be, expected, but which offers so little peace, security, stability and prosperity to its citizens. Darfur is a painful reminder that a peaceful order is still a far cry in Africa, where rulers, intoxicated by the narcotics of power, rule their countries in anger with unrelenting persecution and smite their peoples in wrath with unceasing blows. Death, devastation and displacement have defined Darfur since early 2003, when two rebel movements—the Justice and Equality Movement (JEM) and the Sudan Liberation Movement/Army (SML/A)—attacked government positions in Sudan. In response, the Sudanese government and its Janjaweed criminal alliance unleashed a terrifying violence on Darfur and ethnic Darfurians, aimed clearly at changing the demography of the region.

Darfur is characterised by a criminality of the most unimaginable invention. The narrative 'has become numbingly, terrifyingly familiar:' the Janjaweed monsters—the vicious instrument of ethnic destruction—are killing farmers wholesale and cutting down fruit trees; slaughtering innocent civilians and poisoning wells with their corpses; burning villages; raping women; 'cutting babies out of mother's wombs and playing with the foetuses in front of them.' Sudan's aircraft destroy whole villages with impunity. Between 80% and 90% of all villages of the Fur, Massalit, Zaghawa and Birgid tribes have been destroyed, in a conflict that has poisoned inter-ethnic relations and threatened the stability of neighbouring countries, especially Chad. Humanitarian workers remain 'targets of brutal violence, physical harassment and rhetorical vilification.' Over 250 000 civilians have died since the onset of the genocide. Millions have had their livelihoods destroyed and have become refugees and internationally displaced persons (IDPS)—a meaningless classification that ought to have gone with the 20th century. Up to 2000 refugees a day flood into Darfur's camps of despair, many of them 'seriously malnourished, dehydrated or in critical need of medical assistance.' Unfed and thirsty, many sleep without shelter in the dust . . . but why go on?

Suffice to state that Darfur is a byword for human misery. It also exposes the façade of collective regional peacekeeping in Africa, given AMIS's failure to make a real difference to the security situation on ground. This section evaluates the AU's 'pioneering experiment' in peacekeeping, after briefly examining the meaning of the concept in international law.

The Concept of Peacekeeping

Peacekeeping is a mechanism for addressing a wide variety of threats to international peace and security. Boutros-Ghali defines it as 'the deployment of a [UN] presence in the field, hitherto with the consent of all parties concerned, normally involving [UN] military and/or police personnel and frequently civilians as well. Peacekeeping is a technique that expands the possibilities for both the prevention of conflict and the making of peace.' The practice of peacekeeping evolved in response to the cold war deadlock that paralysed the Security Council and prevented it from functioning as designed. The UN Charter made no direct reference to it, thereby creating some difficulties in defining where peacekeeping lies on the spectrum between consent and coercion, between passivity and force. However, Dag Hammarskjøld, the former Secretary-General who proposed the first UN peacekeeping force, described the practice as a 'Chapter Six and a Half' operation, falling in between Chapter VI measures for peaceful resolution of conflicts and Chapter VII enforcement action. The International Court of Justice (ICJ) also confirmed the legality of peacekeeping actions in 1962, in both Suez and the Congo.

Peacekeeping functions have traditionally ranged from observance of ceasefire, to demarcation lines or withdrawal of force agreements, but the practice has acquired a certain elasticity in recent years, as new circumstances have made new demands on the military capacity of the institutions of security governance. Increasingly modern peacekeeping has become 'a catch-all term,' extending to election monitoring, delivery of humanitarian supplies, assisting the national reconciliation process, and rebuilding a state's social, economic and administrative infrastructure. Both before and after the Cold War the UN has consistently employed peacekeeping operations as its primary mechanism for restoring peace and security and the Security Council increasingly mandates such operations to protect civilians. Cox argues that the future credibility of the UN depends on successful peacekeeping operations, but their efficacy as a tool for restoration of peace remains debatable.

Many regional organisations have also undertaken peacekeeping operations, both collaboratively with the UN and independently of it. The proliferation of intra-state conflicts after the Cold War began to weigh down the international security framework, in terms of both finance and logistics, necessitating the need for burden sharing with the UN. Besides, these intra-state conflicts have regional specificities and being grassroots mechanisms, regional bodies are better placed to conduct short, robust stabilisation or peacekeeping and peace-enforcement operations before the UN has the chance to undertake a multidimensional mission. The North Atlantic Treaty Organisation (NATO), for example, has deployed peacekeeping missions in recent years, including the Implementation Force (IFOR) in 1995, Stabilisation Force (SFOR) in 1996 and Kosovo Force (KFOR) in 1999. There are NATO-led missions elsewhere, including—presently—Afghanistan.

One of the key tasks of the AU is to promote peace, security and stability in Africa, which its Constitutive Act acknowledges as a prerequisite for the implementation of Africa's development and integration agenda. One of the AU's organising principles is 'peaceful resolution of conflicts among Member States of the Union through such appropriate means as may be decided upon by the Assembly.' The PSC is the current 'appropriate means' by which the AU attempts to resolve conflicts. The PSC was inaugurated in May 2004, after the entry into force of its Protocol. The PSC Protocol establishes an African Standby Force (ASF) with a mandate to intervene in a member state in respect of grave circumstances or at the request of a member state in order to restore peace and security.

The ASF shall also provide humanitarian assistance to alleviate the suffering of civilian populations in conflict areas and support efforts to address major natural disasters. The ASF shall perform other functions as the PSC or AI Assembly might direct, including observation and monitoring missions and other types of peace support missions. It shall also engage in deployment missions in order to prevent 1) a dispute or a conflict from escalating; 2) an ongoing violent conflict from spreading to neighbouring areas or states; and 3) the resurgence of violence after parties to a conflict have reached agreement. Finally, it shall engage in peace building, including post-conflict disarmament and demobilisation.

AMIS and Its Mandate

AMIS was established April 2004; it was then called the Ceasefire Commission (CFC). Its mandate consisted of planning, verifying and ensuring the implementation of the rules and provisions of the N'djamena Agreement; defining the routes for the movement of forces in order to reduce the risks of incidents; requesting appropriate assistance with de-mining operations; receiving, verifying, analysing and judging complaints related to possible violations of the ceasefire; developing adequate measures to guard against such incidents in the future; and determining clearly the sites occupied by the combatants of the armed opposition and verifying the neutralisation of the armed militias. The badly drafted CFC Agreement also provided for an AU Monitoring Mission as the operational arm of the CFC. It provided for Military Observers (MILOBS), which 'may be lightly armed.' The AU Monitoring Mission was to be deployed on the basis of the commitment of the Sudanese government, JEM and SML/A—the principal antagonists in the Darfur crisis—to ensure the protection and safety of the MILOBS.

AMIS initially comprised 80 military observers and a protection force of 600 troops. The PSC enlarged this number to 3320 in July 2004 and, on 20 October 2004, the force level was beefed up to around 7000 personnel, made up of 686 military observers, 4890 troops, and 1176 civilian police. The PSC further increased the force level on 28 April 2005 to its current total authorised strength of 6171 military personnel and 1560 civilian police, because of the 'precarious security situation and persistent attacks against civilians.'

As modified, AMIS's mandate is to monitor and observe compliance with the N'djamena Agreement and all such agreements in the future; to assist in the process of confidence building; and to contribute to a secure environment for the delivery of humanitarian relief, and beyond that, to the return of IDPS and refugees to their homes, in order to assist in increasing the level of compliance of all parties with the N'djamena Agreement and to contribute to the improvement of the security situation throughout Darfur. It is also authorised to 'protect civilians whom it encounters under imminent threat and in the immediate vicinity, within resources and capability, it being understood that the protection of the civilian population is the responsibility of the GoS [Government of Sudan].' Even this new mandate is less robust than expected, probably because of a lack of capacity and of prudential calculations about the overall prospects for peace in Sudan. Nevertheless, the Security Council expressed its support for AMIS and urged UN member states 'to provide the required equipment, logistical, financial, material, and other necessary resources.'

The first phase of AMIS's deployment ended in June 2005, with total personnel of 2635, comprising 452 military observers, 1732 troops, 40 CFC members/international support staff, 413 civilian police personnel and 12 members of the Darfur Integrated Task Force (DITF). The second phase was scheduled to end in August 2005, with deployment of 6171 military personnel, with an appropriate civilian component including up to 1560 civilian police. The third phase was a follow-up mission, with projected personnel of 12 300. Instructively, Rwanda—a country that had experienced the horrors of genocide—was the first

African state to contribute troops towards the AMIS. Similarly, as of 26 July 2005, 'three (3) days before the deadline established by the AU plan of deployment' of enhanced AMIS, Rwanda had completed the deployment of its first battalion of 680 troops in Darfur and 'made itself ready to start the deployment of its second Battalion.'

AMIS was scheduled to withdraw from Sudan by the end of September 2006 after the expiration of its current mandate. Its term has now been extended until the end of 2006, a temporary concession that the PSC managed to obtain from the government during the September 2006 UN General Assembly meeting in New York. The Arab League reportedly pledged to fund the interlude, while the UN would provide logistics and other material support. Al-Bashir insisted that he would not accept a UN force designed to place 'Sudan under mandate, a sort of trusteeship.' Thus, a patient that submits to global medication in Southern Sudan insists on a dangerous self-medication on its Western frontier, notwithstanding that the disease has deepened into a septic abscess. Khartoum's rhetoric is obviously short on logic and long on mass psychology, but it has so far cowed the Security Council and prevented it from taking robust action to end the genocide.

A Preliminary Evaluation

AMIS is the first continental experiment at peacekeeping under the AU legal framework. What has been its achievement for the short period of its operation in Darfur? Certainly, AMIS serves as a moral presence and encourages a modicum of restraint in Sudan. However bad things presently look in Darfur, they could have been far worse without AMIS. The Security Council itself acknowledges AMIS's role in reducing large-scale organised violence in Darfur. AMIS has also been able to offer limited protection to aid workers, which has enabled the delivery of essential humanitarian assistance, without which the genocide would have been worse. Ultimately, success at peacekeeping is judged by the probability of ending a crisis and achieving a lasting peace. Thus, even when all allowances have been made, AMIS is not the firewall that will stop unremitting atrocities in Darfur.

AMIS has failed in Darfur because it has been incapacitated by numerous weaknesses, including problems of command and control, logistical support, operational practice and a lack of funds. Its patently inadequate budget has largely been supported from the outside—by the EU, the UK and Germany, with the USA providing some logistics. The UN also provides some strategic support, through its Assistance Cell in Addis Ababa. Nevertheless, AMIS remains grossly incapacitated and is 'watching helplessly while civilians are slaughtered.' The terms and conditions under which it operates have largely been negotiated with, if not defined by, the Sudanese government, thereby compromising one of the cardinal rules of peacekeeping—objectivity and impartiality. According to Krizner, 'The African Union [Mission in Sudan] is not motivated. Culturally they don't identify with the Muslim population in the region. They have little sympathy for Darfurians. And, ultimately, they are controlled by the government in Khartoum. When they receive reports of Janjaweed attacks they wait for days before investigating.'

AMIS does not have the ability even to resist attacks against its members, let alone the refugees it is supposed to protect. A group of AMIS contingents was taken hostage by one of the rebel factions in October 2005; four were killed, others were released a day later. A few days later the UN withdrew its non-essential aid workers from west Darfur, acknowledging that it could not protect even its own staff. The government is still conducting offensive military over-flights in north Darfur and troop deployments to north, south and west Darfur, in direct violation of the DPA. Darfur has been subjected to renewed aerial bombing that brings greater misery to a population that has already endured much pain and anguish.

Although the AU is reportedly expanding its peacekeeping operation by 4000 to bring the force level of AMIS to 11 000 by the end of 2006, the failure to make a serious impact on the security situation in Darfur has created a 'categorical imperative' for the deployment of UNMIS. A promise of protection becomes destructive where real force is not available to back up the offer. To save face, the AU insists on playing a lead role in the overall Darfur peace process, 'including the conduct of the Abuja Peace Talks and the Darfur—Darfur dialogue and consultation provided for by the Declaration of Principles (DOP) signed in Abuja on 5 July 2005, as well as in the implementation of existing and future agreements between the parties.' Africa is not bereft of a sense of irony.

In retrospect, is there really a need for a new peacekeeping force in Darfur? Is it not possible for the global community to reinforce AMIS by further enhancing its mandate, strengthening its mobility, providing more funding and better equipment for it to deal with the crisis? The present writer finds fallacious the claim that the AU and its member states cannot assemble and finance an effective African force to keep the peace in Darfur. If a few West African states were able to prevent anarchy in Liberia and Sierra Leone, under the auspices of the Economic Community of West African States Monitoring Group (ECOMOG), common sense show that a continental organisation of 53 member states should be able to mobilise and deploy a credible force in Darfur, with necessary logistical support. There is simply no political will among states that make up the AU.

UNMIS and Resolution 1706

The spiralling violence and displacement in Darfur led senior UN officials and human rights NGOS to warn of an imminent humanitarian catastrophe, compelling the Security Council to adopt Resolution 1706. An earlier attempt at bolstering the peacekeeping efforts in Darfur was Resolution 1590, which established UNMIS to support implementation of the Comprehensive Peace Agreement (CPA) between the Sudanese government and the Sudan People's Liberation Movement/Army (SPLM/A) in Nairobi, Kenya on 9 January 2005. The CPA brought to an end the 30-year-long civil war in Southern Sudan that led to two million deaths and four million displacements. It allows the south to determine, in a referendum to be conducted in six years from 2005, whether to remain in a united Sudan or to set up an independent state, based, obviously, on the inherent right of peoples to self-determination. 'When a people,' says Soyinka, 'have been subjected to a degree of inhuman violation for which there is no other word but genocide, they have the right to seek an identity apart from their aggressors.'

Resolution 1590 authorised UNMIS to have up to 10 000 military personnel and an appropriate civilian component, including up to 715 civilian police personnel. It requested UNMIS to 'closely and continuously liaise and coordinate at all levels with the . . . [AMIS] with a view towards expeditiously reinforcing the efforts to foster peace in Darfur, especially with regards to the Abuja peace process and the [AMIS].' On 16 May 2006 the Council adopted Resolution 1679, which called on the AU and UN to agree upon requirements necessary to strengthen the capacity of AMIS to enforce the security arrangements of the Darfur Peace Agreement, 'with a view to a follow-on [UN] operation in Darfur.'

In July 2006 the UN Secretary-General presented a report on Darfur to the Security Council. The report described the main elements of the DPA and identified the implementation priorities. It also identified some risks and challenges involved in deploying a UN peacekeeping force in Darfur; elaborated on support the UN could offer to AMIS to enhance its ability to protect civilians and implement the Agreement; proposed a mandate and mission structure for a UN operation in Darfur; and provided details on various components of the proposed mission and their specific functions. Finally, the report recommended 'a large, agile and robust military force' that entails 'high troop density to provide wide area coverage; high mobility to move forces rapidly in response to developing crises; and robust military capability to deter and defeat spoilers.'

The DPA was intended to instil some new dynamics into resolving the Darfur conflict, but it faces formidable challenges. To start with, the DPA has not received support from IDPS and many others affected by the conflict. Two of the parties involved in the Abuja talks—the UNMIS and the Abdelwahid faction of the SLM/A—remain outside the DPA, despite extension of the signing deadline to 31 May. Only the government delegation and the SLM/A faction led by Minni Minawi signed the Agreement. The lack of unanimity prevents the Cease-fire Commission and the Joint Commission—two mechanisms established by the DPA—from making real progress on implementation. Elements opposed to the DPA have reportedly mobilised sections of Darfurians against the Agreement. Rebels have further split into factions, attacking government locations, all of which provides the government with a ready excuse for continuous attacks and air raids, 'under the pretext that the civilian population had to be protected.' Current attacks follow long-standing patterns of destruction, forced displacement and violence against civilians.

The remainder of this section examines UNMIS's mandate for Darfur and the principles underpinning Resolution 1706.

UNMIS's Mandate

Resolution 1706 defines the mandate of UNMIS in Darfur. It authorises the expansion of the existing UNMIS by up to 17 300 military personnel and an appropriate civilian component including up to 3300 civilian police personnel and up to 16 Formed Police Units. UNMIS's mandate is to support implementation of the DPA and the N'djamena Agreement. Consequently, it 'shall take over from AMIS responsibility for supporting the implementation of the [Agreements] upon the expiration of AMIS' mandate but in any event not later than 31 December 2006.' Specifically, UNMIS shall monitor and verify implementation of the Agreements by the parties; observe and monitor the movement of armed groups and redeployment of forces in areas of UNMIS deployment; investigate violations of Agreements and report them to the Ceasefire Commission; co-operate and co-ordinate, together with relevant actors, with the Ceasefire Commission through provision of technical assistance and logistical support; and maintain a presence in key areas, including buffer zones, camps and demilitarised zones, in order to promote the re-establishment of confidence and deter violence.

UNMIS shall also monitor trans-border activities of armed groups along the Sudanese borders through regular ground and aerial reconnaissance activities; assist with development and implementation of a comprehensive and sustainable programme for disarmament; assist parties, in co-operation with other actors, in preparations for, and conduct of, referendums provided for in the DPA; and assist parties, in co-ordination with the AU, in promoting understanding of the DPA and the role of UNMIS, through an effective public information campaign.

UNMIS is to co-operate closely with the Chair of the Darfur–Darfur Dialogue and Consultation (DDDC), through technical support and co-ordination of other UN agencies to this effect, and assist parties to the DDDC to adopt an all-inclusive approach to reconciliation and peace building; assist parties to the DPA in promoting the rule of law (including an independent judiciary) and the protection of human rights; and ensure an adequate human rights and gender presence, capacity and expertise within UNMIS to carry out human rights promotion, civilian protection and monitoring activities that pay attention to the needs of women and children. Finally, UNMIS is to facilitate and co-ordinate the voluntary return of refugees and IDPS and provide humanitarian assistance through the establishment of necessary security conditions in Darfur; contribute towards international efforts to protect, promote and monitor human rights in Darfur and co-ordinate efforts towards protecting civilians, especially vulnerable groups; assist parties to the Agreements with humanitarian de-mining assistance, technical advice, co-ordination and mining awareness programmes; and assist in addressing regional security issues, including through the establishment of a multidimensional presence in key locations.

Significantly the Security Council, acting under Chapter VII of the UN Charter, authorises UNMIS to use 'all necessary means' it deems within its capabilities to: protect UN personnel, facilities, installations and equipment; ensure the security and freedom of movement of UN personnel, humanitarian workers, assessment and evaluation commission personnel; prevent disruption of the implementation of the DPA by armed groups, without prejudice to the responsibility of the GoS; protect civilians under threat of physical violence;

and seize or collect arms or related materials whose presence in Darfur is in violation of the Agreements and the measures imposed by resolution 1556, and to dispose them as appropriate.

The Security Council is now known to be evasive when invoking Chapter VII, and Resolution 1706 is reflective of this ambivalence. The reluctance to identify the precise legal basis for its resolutions has led to arguments regarding the legal basis of Security Council operations, since different articles of Chapter VII give different legal outcomes. However, the particular article envisaged could be discovered from the wordings of each resolution. Resolution 1706 authorises UNMIS 'to use all necessary means,' a phrase that appears in Article 42 of the UN Charter. This provides that, where non-forceful measures prove inadequate in restoring international peace and security, the Council can 'take such action by air, sea, or land forces as may be necessary to maintain or restore international peace and security,' including 'demonstrations, blockade, and other operations by air, sea, or land forces of Members of the [UN].' It may be concluded that Resolution 1706 represents a robust mandate for UNMIS, albeit in the context of peacekeeping. . . .

Conclusion: Playing Dice with Lives

Almost everyone agrees that Darfur is a humanitarian catastrophe—the worst in recent memory, according to the UN—and that its magnitude requires sustained assistance and engagement by the global community. It is also not in doubt that the Sudanese government cannot be trusted to guarantee the security of lives in Darfur, being a participant in the mayhem. As Polgreen has noted, 'one of the assumptions of the [DPA] was that the government had control over the Arab militias and the power to disarm them. This is based on a deeper assumption that the interests of Darfur's Arabs would be tended to by Khartoum. Neither is turning out to be true.'

Although Darfur has pushed the moral sensibilities of the global community beyond limits, it still hibernates between sympathy and apathy. During the recent 34-day carnage in Lebanon—between the Hezbollah guerrillas and Israel—Darfur literally drifted to the margins of public discourse. As in Iraq, death has become too commonplace to matter. As in Rwanda, the world watches while the catastrophe worsens. The Security Council is busy playing dice with lives, holding endless closed-door meetings and passing countless, often self-serving, but utterly meaningless resolutions. These failings in stopping mass atrocities across the globe cast a shadow over the UN's policing mechanism. The Security Council's hesitancy to take robust actions to suppress Africa's often brutal conflicts has led to missed opportunities to stem bloodshed and damage to the economic and social fabric of many countries, as happened with the UN's ambivalence in the Democratic Republic of Congo in 1997. The Council, says the UN High-level Report, 'has not always been equitable in its actions, nor has it acted consistently or effectively in the face of genocide or other atrocities. This has gravely damaged its credibility.' Wedgwood likens the Council to 'the US Senate in the late 18th century—an often opaque body delivering often opaque results.'

The signals emitting from the AU are even more depressing—and sometimes laughable. Pursuant to a decision taken at the AU Assembly's January 2006 Khartoum Summit, al-Bashir is preparing to assume leadership of the AU in January 2007. That exceptionally absurd decision reads:

> The leaders expressed their appreciation for the initiative taken by HE President Omar Hassan Al-Bashir to accept the postponement of his term of Chairmanship for the [AU] until 2007. The leaders consider this gesture to be a true reflection of the great sense of responsibility and leadership demonstrated by President Al-Bashir. The leaders agreed after extensive consultations that the Sudan will assume the Chairmanship of the Union in the year 2007.

When that happens, when a tyrant who speaks mainly the language of violence, hatred and division takes over the mantle of leadership of a continental organisation whose goals include the promotion of peace and security, then al-Bashir will become a vendor of peace and the protagonist of an African renaissance! It will be interesting to see 'how best' the AU under Chairman al-Bashir will 'engage the Sudanese authorities and other stakeholders . . . to expedite the peace process in Darfur.' Such political undercurrents show that group solidarity matters more in Africa than the protection of human and peoples' rights. Caught up in this crossfire, this pincer movement, are millions of refugees and IDPS who face the torture of hunger and misery daily and whose future remains bleak. These victims of ethnic hatred are crying out for compassion and rescue, whatever the helmet of rescuers, green or blue.

The global community should stop playing dice with Darfur lives and should end its fruitless megaphone diplomacy in Sudan. 'The situation in Darfur,' says the *Minnesota Daily*, 'is a tragedy of epic proportions, and the world's conscience is on trial.' Organising endless peace talks in Abuja and adopting countless resolutions in New York—with a tsunami of publicity—are no longer sufficient; they never were. Darfur requires a muscular multilateralism, because it is a contest between terror and resistance. Western states, including the EU, have the capacity to put real pressure on the Sudanese government to end the killing in Darfur, but they have chosen to play Pontius Pilate. These states should recall their troubled history—including the two brutal world wars of the 20th century—and realise that tyranny knows neither colour nor race.

POSTSCRIPT

Are African-Led Peacekeeping Missions More Effective Than International Peacekeeping Efforts in Africa?

The debate about African peacekeeping initiatives is not simply about military effectiveness. Some see it as an abdication of Western responsibility in Africa, or an unwillingness to intervene due to a combination of past failures and stereotypes about conflict in Africa.

In the case of the United States, two watershed events for American peacekeeping efforts in Africa were the failed military intervention in Somalia in 1993 and the lack of intervention during the Rwandan genocide in 1994. In the twilight of the first Bush administration in 1992, the United States intervened militarily in Somalia to dispense food aid that was not being effectively distributed due to the presence of armed militias. The American force was then reduced under the new Clinton administration and made part of a UN operation with a mandate to protect the delivery of humanitarian assistance and disarm the warring factions. Following the deaths of 26 Pakistani soldiers, the United States sent in army rangers and delta force commandos to try to capture a particularly problematic warlord named Mohammed Aideed. Then in October 1993, 18 U.S. soldiers were killed and 50 wounded in a street war that also cost the U.S. military two Black Hawk helicopters. With the sight of a dead U.S. army ranger being dragged through the streets of Mogadishu on international television, then president Bill Clinton ordered the withdrawal of all U.S. troops. This debacle is now commonly referred to as "crossing the Mogadishu line." It resulted in a new U.S. Presidential Directorate (#25) stating that the United States should not become involved in a war unless there was a clear national interest and the conflict could be won.

The events in Somalia had a chilling effect on U.S. interest in addressing the humanitarian crisis in Rwanda in 1994. Following the death of Rwandan President Juvenal Habyarimana in a mysterious plane crash in April 1994, the systematic elimination of Tutsis and moderate Hutus was begun by hard-line elements in the military and Habyarimana government. With most expatriates having left the country, and the meager UN force ordered to protect itself, it is estimated that over a million Tutsis and moderate Hutus were murdered while hundreds of thousands fled the country as refugees. Following these atrocities, the minority Tutsis amassed their forces and began a process of retribution. While the situation in Rwanda eventually was quelled, the inaction of the U.S. and other foreign powers has been roundly condemned. The specter of

Somalia and Rwanda continue to haunt U.S. peacekeeping efforts in Africa today, with a fear of failure lingering from the first case, and the regret of inaction from the latter.

One response of the United States has been to support the development of an African peacekeeping force so that it no longer needs to intervene on the African continent. Some feel that this is an abdication of responsibility, or worse yet, an unwillingness to intervene because of stereotypes about Africans and the nature of ethnic conflict. In his essay "The Politics of Naming: Genocide, Civil War, Insurgency" (*London Review of Books*, March 2007), the prominent African intellectual and native of Uganda, Mahmood Mamdani, is critical of the language used to describe the conflict in Darfur as compared to similar atrocities in Europe. Yet he is also leery of a humanitarian intervention by the major powers in Darfur, Sudan. He argues that a lasting peace cannot be built on such interventions and is rather more supportive of the African Union's efforts to broker power-sharing arrangements.

Contributors to This Volume

EDITOR

WILLIAM G. MOSELEY is a professor of geography, and former director of the African studies program, at Macalester College in Saint Paul, Minnesota, where he teaches courses on Africa, environment, and development. He received a B.A. in history from Carleton College, an M.S. in environmental policy and an M.P.P. in international public policy from the University of Michigan, and a Ph.D. in geography from the University of Georgia. He has worked for the U.S. Peace Corps, the Save the Children Fund (UK), the U.S. Agency for International Development, the World Bank Environment Department, and the U.S. Department of State. His research and work experiences have led to extended stays in Mali, Zimbabwe, Malawi, Niger, Lesotho, and South Africa. He is the author of over 60 peer-reviewed articles and book chapters that have appeared in such outlets as *Ecological Economics*, *Proceedings of the National Academy of Sciences of the United States of America*, the *Geographical Review*, *The Geographical Journal,* and *Geoforum*. He also has written pieces for the popular press that have been published in the *International Herald Tribune*, *The Christian Science Monitor*, the *Philadelphia Inquirer*, the *San Francisco Chronicle*, the *Minneapolis Star Tribune*, and the *Chronicle of Higher Education*. He is the lead author of several edited collections, including *African Environment and Development: Rhetoric, Programs, Realities* (Ashgate, 2004), *The Introductory Reader in Human Geography: Contemporary Debates and Classic Writings* (Blackwell, 2007), and *Hanging by a Thread: Cotton, Globalization and Poverty in Africa* (Ohio University Press, 2008). He previously served as editor of the *African Geographical Review*.

AUTHORS

FUAMBAI AHMADU, originally from Sierra Leone, is an anthropologist and postdoctoral fellow in the Department of Comparative Human Development at the University of Chicago. Dr. Ahmadu studies male and female initiation rites in Africa and has published numerous book chapters on the subject.

ANITA ALBAN is an external lecturer of health economics at the Institute of Public Health at the University of Copenhagen. Her research interests include cost-effectiveness of HIV/AIDS and TB interventions in Africa and Asia, access to drugs for poor communities, and mathematical models of health care in developing countries. She has been widely published in books and scientific journals on topics of strategic health plans, HIV/AIDS, and access to health services. She works for the DANIDA, DFID, the World Bank, UNDP, UNAIDS, and WHO.

KOFI A. ANNAN, originally from Ghana, was the seventh secretary-general of the United Nations. He received a Nobel Peace Prize in 2001, and he has advocated for peace, human rights, and HIV/AIDS research within the international community. He has worked extensively in peacekeeping negotiations and more recently in fighting terrorism.

SIMON ANSTEY was a researcher and Ph.D. student at the Centre for Applied Social Science in Zimbabwe, where he studied environmental and political processes in Mozambique. He has worked for the World Conservation Union in Jordan and now acts as a consultant on natural resource projects in southern Africa. He is the head of Terrestrial Programme, WWF Coastal East Africa Network at WWF, and the director of ResourceAfrica UK.

PETER J. BALINT is an associate professor of environmental policy in the Department of Public and International Affairs and the Department of Environmental Science and Policy at George Mason University. His primary research focus is community-based natural resource management, which he studies in Central America, Central Asia, and Eastern and Southern Africa.

JOEL D. BARKAN is a professor of political science at the University of Iowa and senior consultant on governance at the World Bank. He is also a resident fellow at the Woodrow Wilson International Center for Scholars in Washington, D.C. From 1992 to 1994, he served as the regional democracy and governance advisor for East and Southern Africa to the United States Agency for International Development. His research interests include democratization, macroeconomic reform in developing countries, and electoral processes. Since 1966, Barkan has worked in Ethiopia, Ghana, Kenya, Lesotho, Namibia, Nigeria, South Africa, Tanzania, Uganda, and Zambia.

MICHAEL BRATTON is a distinguished professor of political science and African studies at Michigan State University. His research interests include comparative politics, public administration, and African politics. He is the author or coauthor of *The Local Politics of Rural Development: Peasant and Party-State in Zambia* (University Press of New England, 1980), *Governance and Politics in Africa* (Lynne Rienner Press, 1992), *Democratic Experiments in Africa: Regime Transitions in Comparative Perspective* (Cambridge University Press, 1997), and *Public Opinion, Democracy and Markets in Africa* (Cambridge University Press, 2005). He has also published over 60 articles and chapters. Bratton is a founder, former executive director, and now senior advisor to the Afrobarometer, a cross-national survey research project on public opinion in Africa.

PADRAIG R. CARMODY is a lecturer in geography at Trinity College Dublin and a research associate at the Institute for International Integration Studies. His research focuses on the effects of globalization and economic liberalization on African countries. He has published numerous articles about poverty alleviation, development strategies, and economic liberalization and is the author of *Tearing the Social Fabric: Neoliberalism, Deindustrialization, and the Crisis of Governance in Zimbabwe* (Heinemann, 2001).

JUDITH CARNEY is a professor of geography at the University of California in Los Angeles. She is the author of numerous peer-reviewed articles that have appeared in such outlets as the *Journal of Ethnobiology*, *Progress in Human Geography*, and *Human Ecology* and has won many awards for her work. Her book, *Black Rice: The African Origins of Rice Cultivation in the Americas* (Harvard University Press, 2001), won the African Studies Association's Herskovits Award in 2002.

CHAKA CHIROZVA is a lecturer and project facilitator at the University of Zimbabwe. His work revolves around community-based natural resource management, with special focus on active community participation in governance.

MARCUS COLCHESTER is the director of the Forest Peoples Programme of the World Rainforest Movement. His primary work has involved securing the rights to land and livelihood of indigenous peoples. He has been a fellow in the Pew Fellows Program in Conservation and the Environment, an associate editor for *The Ecologist* magazine, and honorary advisor on development in the Amazon to the Venezuelan government.

LIZ CREEL is a population specialist and senior policy analyst at the Population Reference Bureau.

ANDREW CREESE is a senior health economist at the World Health Organization's Department of Essential Drugs and Medicines Policy. His interests include health systems financing, economic evaluation, and pharmaceutical policy issues. He has worked in Ghana and Malawi, as well as in other African nations.

JARED DIAMOND is professor of physiology and geography at University of California at Los Angeles (UCLA). His research interests include biogeography, and geography and human society. He is best known for his Pulitzer Prize–winning book, *Guns, Germs and Steel: The Fates of Human Societies* (W. W. Norton, 2005) and its follow-up, *Collapse: How Societies Choose to*

Fail or Succeed (Penguin Books, 2006). In 1999, Diamond was awarded the National Medal of Science.

AWA DIOUF is a senior gender advisor to the United Nations Development Programme (UNDP) in Burundi.

KATHERINE FLOYD is a health economist in the Stop Tuberculosis Department at the World Health Organization.

ARTHUR A. GOLDSMITH is professor of management at the University of Massachusetts, Boston. His research interests are in governance, institutional development, and comparative public administration. He is the author of *Building Agricultural Institutions: Transferring the Land-Grant Model to India and Nigeria* (Westview, 1990). Goldsmith has served as a senior fellow at John W. McCormack Institute of Public Affairs since 1998, and he served as a senior research fellow at Harvard University in 2004 and 2005.

DAVID C. GOMPERT is currently the principal deputy director of National Intelligence. He has served as an advisor for Refugees International and a senior fellow at the RAND Corporation, a nonprofit research and policy institution. He served as senior advisor for national security and defense for the Coalition Provisional Authority in Iraq in 2003 and 2004, and worked extensively for the State Department and National Security Council. As a researcher, Gompert published books on international affairs, national security policy, and information technology.

ROBIN M. GRIER is an associate professor of economics and area coordinator for Latin American studies at the University of Oklahoma. She has published numerous articles in journals such as *Economic Inquiry* and *Public Choice*. Her areas of specialization include international finance, development, and Latin American economics.

LORNA GUINNESS is a lecturer in the health policy unit at the University of London's School of Hygiene and Tropical Medicine. She is also a visiting research fellow at the Australian Centre for Economics Research in Health at Australian National University. She previously worked as an economist for UNAIDS in Geneva. Her research interests include economic evaluation, health systems, HIV/AIDS, sexual health, and sexually transmitted diseases.

YAN HAIRONG is an assistant professor of anthropology and East Asian languages and cultures at the University of Illinois at Urbana-Champaign. She is the author of *New Servants, New Masters: Migration, Development, and Women Workers in China* (2008).

DUNCAN CLINCH HEYWARD was a Carolina rice planter during the early twentieth century and also served as the governor of South Carolina for four years from 1903 to 1906.

PHILIP J. HILTS is a prize-winning health and science reporter for *The New York Times* and *The Washington Post*. He teaches science journalism in the graduate school at Boston University and has also taught undergraduates at the University of Botswana. Among other books, Hilts is the author of *Rx for Survival: Why We Must Rise to the Global Health Challenge* (2005).

CAREY LEIGH HOGG is a program officer for Vital Voices, a nonprofit organization, which identifies, trains, and empowers emerging women leaders and social entrepreneurs around the globe. She has a master of science from the London School of Economics and Political Science. She previously worked for a number of different international development organizations, including the All Party Parliamentary Group on the Great Lakes Region of Africa and Operation Smile International. She has written extensively on her research interests, which include human rights, conflict studies, and international development, specifically in the Great Lakes region of Africa and Southeast Asia.

HUMAN RIGHTS WATCH is a nonprofit organization supported by contributions from private individuals and foundations worldwide. The organization is the largest of its kind based in the United States. Human Rights Watch researchers conduct fact-finding investigations into human rights abuses in all regions of the world. They then publish these findings in dozens of books and reports every year. The aim is to generate extensive coverage in local and international media that will help to embarrass abusive governments in the eyes of their citizens and the world.

LUCY JAROSZ is an associate professor of geography at the University of Washington. Her research interests include food and agriculture, rural poverty and inequality, rural development, and environmental change. She has conducted fieldwork in Madagascar and South Africa. Her articles have appeared in such outlets as the *Journal of Rural Studies* and *Antipode*.

BABOUCARR KOMA works for the Private Sector Development, Investment and Resource Mobilization division of the African Union Commission.

PRADEEP KURUKULASURIYA is a senior technical advisor for the United Nations Development Programme (UNDP). He is from Sri Lanka. Since 2005, he has worked with UNDP to develop and manage climate change adaptation projects in Africa funded by the Global Environment Facility (GEF). Kurukulasuriya also works as the regional technical advisor for climate change adaptation at UNDP-GEF Regional Coordinating Unit in Pretoria, South Africa.

PAUL E. LOVEJOY is a fellow of the Royal Society of Canada, the Canada Research Chair in African Diaspora History, and a distinguished research professor of history at York University in Toronto, Ontario, Canada. His research interests include the trans-Atlantic slave trade, diaspora studies, and slavery in Africa. He has published more than 20 books on these topics, including *Transformations in Slavery: A History of Slavery in Africa* (Cambridge University Press, 2000).

JUDITH MASHINYA is a senior program officer in the World Wide Fund for Nature (WWF) Conservation Leadership Programs. She manages the African conservation capacity-building activities and also provides technical support to WWF Southern Africa's Miombo Ecoregion Program. Before beginning her work with the WWF, Mashinya worked as an assistant secretary and undersecretary for the Parliament of Zimbabwe and then as an international policy and marketing advisor for Africa Resources Trust. She is a native of Zimbabwe.

ROBERT MATTES is an associate professor of political studies and director of the Democracy in Africa Research Unit in the Centre for Social Science Research at the University of Cape Town. He is also cofounder and codirector of Afrobarometer (a survey of Africans' attitudes towards issues including democracy and markets) and an associate with the Institute for Democracy in South Africa. His research interests include the development of democracy/democratic political culture in Africa and the impact of race and identity on politics in South Africa. He is the author of *The Election Book: Judgment and Choice in the 1994 South African Election* (Idasa, 1996).

CHEIKH MBOW is an associate professor and research fellow at the University of Dakar in Dakar, Senegal, where he teachers about land use, remote sensing, and GIS. Most recently, he worked as the coordinator of the steering committee on the flooding issue in Dakar for the National Water Partnership of Senegal.

OLE MERTZ works in the Department of Geography and Geology at the University of Copenhagen in Denmark. He is also the coordinator of the Research Network for Environment and Development there.

THANDIKA MKANDAWIRE, originally from Malawi, is the director of the UN Research Institute for Social Development (UNRISD). From 1986 through 1996, he was executive secretary of the Council for the Development of Social Science Research in Africa (CODESRIA), headquartered in Dakar. He is an economist who has published extensively on structural adjustment, democratization, and social sciences in Africa.

MICHAEL MORTIMORE is a geographer who taught at Nigerian universities between 1962 and 1986, and subsequently was a research associate at Cambridge University and the Overseas Development Institute. He is currently with Drylands Research. He has performed research and published numerous books on the topic of environmental management by smallholders in the dry lands of Africa. He is the author of *Roots in the African Dust: Sustaining the Dry Lands* (1998) and the coauthor of *Working the Sahel: Environment and Society in Northern Nigeria* (1999) and *More People, Less Erosion: Environmental Recovery in Kenya* (1994). In 2008, Mortimore won the Robert McC. Netting Award from the Association of American Geographers.

DAMBISA MOYO is a Zambian economist who has worked at both Goldman Sachs and the World Bank. She is the author of *Dead Aid: Why Aid Is Not Working and How There Is a Better Way for Africa* and the forthcoming book *How the West Was Lost: Fifty Years of Economic Folly—And the Stark Choices Ahead.* Moyo has been widely published in *Financial Times*, *The Economist* magazine, and *The Wall Street Journal*. *Time Magazine* named her one of "100 most influential people in the world."

JOHN MURTON is a British diplomat currently serving as high commissioner to Mauritius and nonresident ambassador to Madagascar and Comoros.

His Ph.D. thesis on the social and economic impacts of population growth in the Machakos and Kitui districts in Kenya won the Audrey Richards Prize for best British African studies thesis in 1997.

JOSEPH O. OKPAKU, Sr. is the president and CEO of the Telecom Africa International Corporation. He has served as an advisor to the United Nations ICT Task Force and a member of the Digital Divide Task Force of the World Economic Forum. Okpaku is the author of *Information and Communications Technologies for African Development* (2003).

FRANCIS Y. OWUSU, originally from Ghana, is an associate professor of community and regional planning at Iowa State University. His research interests include sustainable development, urban and community development, environmental planning, and policy. He is the director of the Africa Specialty Group of the Association of American Geographers, and a member of the Global Development Network.

FUDZAI PAMACHECHE is an economist and the acting director of the Southern African Development Community Directorate on Trade, Industry, Finance, and Investment. He resides in Gaborone, Botswana.

ELIZABETH POWLEY is an independent consultant in gender and postconflict reconstruction. She served for two years as the director of Institute for Inclusive Security's Project Rwanda. She is an adjunct professor at a number of universities and the founder and executive director of the nonprofit organization Every Child is My Child.

ANETTE REENBERG teaches landscape and agricultural geography in the Department of Geography and Geology at the University of Copenhagen.

LIZ RIHOY is the director of the Zeitz Foundation Programme.

ROBERT I. ROTBERG is president of the The World Peace Foundation. He has previously served as director of the Program on Intrastate Conflict and Conflict Resolution at Harvard University's John F. Kennedy School of Government. He has also taught political science and history at both Harvard and the Massachusetts Institute of Technology. His research focuses on political and economic issues of developing countries, especially Africa and Southeast Asia. He has authored *The Founder: Cecil Rhodes and the Pursuit of Power* (Oxford University Press, 1988) and *The Rise of Nationalism in Central Africa* (Harvard University Press, 1965).

BARRY SAUTMAN is an associate professor of social science at the Hong Kong University of Science and Technology. His research interests include communist and postcommunist systems, Chinese politics, China–Africa relations, China–U.S. relations, and international law.

RICHARD A. SCHROEDER is an associate professor of geography and former chair of African Studies at Rutgers University. He is the author of *Shady Practices: Agroforestry and Gender Politics in The Gambia* (1999) and *Producing Nature and Poverty in Africa* (coeditor) (2000). He has published numerous articles in outlets such as the *Annals of the Association of American Geographers*, *Economic Geography*, and *Africa*.

APOORVA SHAH is a research fellow at the American Enterprise Institute for Public Policy Research. His work focuses on policy and development in India and Pakistan.

CAROL B. THOMPSON is a professor of political economy at Northern Arizona University. Her research interests include international political economy, environment, and Southern Africa. She is on the executive board of the Association of Concerned Africa Scholars and has testified in U.S. congressional hearings about policy toward Africa. Her books include *Biopiracy of Biodiversity—Global Exchange as Enclosure* (Africa World Press, 2007).

JOHN THORNTON is a professor of history and African American studies at Boston University. His research and teaching have focused on Africa and the Middle East. His books include *The Kingdom of Kongo: Civil War and Transition, 1641–1718* (University of Wisconsin Press, 1983), *Africa and Africans in the Formation of the Atlantic World, 1400–1680* (Cambridge University Press, 1992, second expanded edition, 1998), and *The Kongolese Saint Anthon: Dona Beatriz Kimpa Vita and the Antonian Movement, 1684–1706* (Cambridge University Press, 1998).

MARY TIFFEN is a historian and socioeconomist at Drylands Research. She is interested in long-term change and development, interdisciplinary research, and social and economic interactions with technology. She authored *The Environmental Impact of the 1991–92 Drought on Zambia: Report* (1994) and *The Enterprising Peasant: Economic Development in Gombe Emirate, North East Nigeria* (1976). She also coauthored *More People, Less Erosion: Environmental Recovery in Kenya* (1994).

NSONURUA J. UDOMBANA is a professor of international law at the University of Uyo in Nigeria. He is also visiting professor at the Central European University, the International University College of Turin, and the University of Pretoria. His research interests include international law, human rights, and the African judicial process. He has published many articles. He is the author of *Human Rights and Contemporary Issues in Africa* (Malthouse Press, 2003).